Integrating New Technologies for *Striga* Control

Towards Ending the Witch-hunt

Integrating New Technologies for *Striga* Control

Towards Ending the Witch-hunt

edited by

Gebisa Ejeta
Purdue University, USA

Jonathan Gressel
Weizmann Institute of Science, Israel

World Scientific

NEW JERSEY • LONDON • SINGAPORE • BEIJING • SHANGHAI • HONG KONG • TAIPEI • CHENNAI

Published by

World Scientific Publishing Co. Pte. Ltd.
5 Toh Tuck Link, Singapore 596224
USA office: 27 Warren Street, Suite 401-402, Hackensack, NJ 07601
UK office: 57 Shelton Street, Covent Garden, London WC2H 9HE

Library of Congress Cataloging-in-Publication Data
Integrating new technologies for striga control : towards ending the witch-hunt /
edited by Gebisa Ejeta & Jonathan Gressel.
p. cm.
Includes bibliographical references and index.
ISBN-13: 978-981-270-708-6 (hardcover : alk. paper)
ISBN-10: 981-270-708-5 (hardcover : alk. paper)
1. Witchweeds--Integrated control. I. Ejeta, Gebisa. II. Gressel, Jonathan.
SB615.W58I58 2007
632'.52--dc22

2007016770

British Library Cataloguing-in-Publication Data
A catalogue record for this book is available from the British Library.

Printed by Mainland Press Pte Ltd

PREFACE

The parasitic witchweeds (*Striga* species) are the scourge of agriculture in much of Africa and parts of Asia and with even one small part of the USA. *Striga* attacks the major cereal grains and legumes in sub-Saharan Africa, on average halving the already very low yields of subsistence farmers. The *Striga* problem has been a major reason for keeping crop productivity at or below subsistence, leaving poor farmers with no way out of a situation that is only getting worse. For many decades, research approaches on *Striga* targeted eradication, suppression, or breeding for host crops that support fewer emerged *Striga* plants. Decades of such effort have led to few successes.

More recently, basic research efforts that focused on the more fundamental biology of the parasite and its association with its hosts have led to a far better understanding of the enemy. That in turn led to series of successes in the field that are being expanded slowly throughout Africa. Highly successful weeds such as *Striga* have a tendency to evolve resistance to all types of control. Ways to circumvent these evolutionary pitfalls need to be crafted so these technologies remain sustainable and not fail. As no single method is likely to offer a lasting solution, it was clear that proven methods must be integrated with each other. However, integration is often an anathema to basic scientists who are taught to alter single variables in their experiments.

We thus brought together key leaders for a symposium in Addis Ababa, Ethiopia in early November, 2006 to deal with the development of the new, integrated, knowledge–based control strategies, including those new successes deployed in the field, as well as those with promising strategies currently under development. These experts discussed how these strategies can be integrated with each other to develop more durable and sustainable methods that will be useful for decades to come. They were met by an audience of practitioners with expertise in the field, who will assist in integrating these solutions.

The chapter authors, leaders in the field, who have been supplying the basic biology, genetics, biochemistry, and the molecular information offer insights on the technologies they generated in how to deal with *Striga*. These chapters were the basis for lectures that formed the core of the symposium. The chapters were all peer-reviewed prior to publication of this book. Other scientists (molecular biologists, breeders, agronomists, and social scientists) who are key in the fight against *Striga* participated in structured panel discussions that were useful to provide continuity and integration between the ideas of the various chapters. The messages from these discussions addressing important issues of technology development and transfer, roles of biotechnology and conventional science as well as technology and national policy have been summarized in the epilogue.

The editors sincerely thank the authors of the chapters for timely submission of their manuscripts and for their excellent cooperation in going through the accelerated pace of the review procedure with diligence and patience. Special thanks go to those who anonymously and selflessly served as peer reviewers.

This book could not have been written nor the symposium held without the financial support of the United States Agency for International Development (USAID) through a grant to the International Sorghum and Millet (INTSORMIL) collaborative research support program. Grateful acknowledgement is extended to the Ethiopian Institute of Agricultural Research (EIAR), Purdue University, and INTSORMIL for providing the organizational logistics. Appreciation goes to BASF, EARI, and Alemaya University for the generous hospitality during the symposium. Especial acknowledgements go to Mrs. Katy Ibrahim for her impeccable organization and service in the transcontinental and local arrangements for the symposium. The tireless efforts of the local host committee ably and competently chaired by Dr. Tesfaye Tesso were crucial for the success of the symposium.

The editors dedicate this book to their late colleague, Dr Larry Butler, a pioneer in promoting basic sciences for the fight against Striga. Larry was motivated to advance science for the sake of the rural poor.

Gebisa Ejeta and Jonathan Gressel, editors
Nairobi, February 2007

CONTENTS

Preface v

Section I Introduction — The Witches' Curse **1**

Chapter 1 The *Striga* Scourge in Africa: A Growing Pandemic 3
Gebisa Ejeta

Section II Biology and Chemistry — The Needed Basics **17**

Chapter 2 Biology of Host-Parasite Interactions in *Striga* Species 19
Patrick J. Rich and Gebisa Ejeta

Chapter 3 Host Detection by Root Parasites: Insights from Transciptome Profiles 33
John I Yoder, Russell Reagan, Alexey Tomilov, Natalya Tomilova and Manuel Torres

Chapter 4 Germination of *Striga* and Chemical Signaling Involved: A Target for Control Methods 47
Zhongkui Sun, Radoslava Matusova and Harro Bouwmeester

Chapter 5 Chemicals Involved in Post-Germination Inhibition of *Striga* by *Desmodium*: Opportunities for Utilizing the Associated Allelopathic Traits 61
John A. Pickett, Zeyaur R. Khan, Ahmed Hassanali and Antony M. Hooper

Chapter 6 Genetic Diversity of *Striga* and Implications for Control and Modeling Future Distributions 71
Kamal I. Mohamed, Jay F. Bolin, Lytton J. Musselman and A. Townsend Peterson

Section III Knowledge-based Breeding — Translating Information to Products **85**

Chapter 7 Dissecting a Complex Trait to Simpler Components for Effective Breeding of Sorghum with a High Level of *Striga* Resistance 87
Gebisa Ejeta, Patrick J. Rich and Abdallah Mohamed

Chapter 8 Breeding Maize for Broad-Based Resistance to *Striga Hermonthica* 99
Abebe Menkir, Baffour Badu-Apraku, Chabi G. Yallou, Alpha Y. Kamara and Gebisa Ejeta

Chapter 9 Molecular Markers for Analysis of Resistance to *Striga Gesnerioides* in Cowpea 115
Michael P. Timko, Bhavani S. Gowda, Jeremy Ouedraogo and Boukar Ousmane

Chapter 10 Introgression of Genes for *Striga* Resistance Into African Landraces of Sorghum 129
Issoufou Kapran, Cecile Grenier and Gebisa Ejeta

Section IV Biotechnology: Opening New Frontiers **143**

Chapter 11 Success with the Low Biotech of Seed-Coated Imidazolinone-Resistant Maize 145
Fred Kanampiu, Alpha Diallo, Michael Burnet, Haron Karaya and Jonathan Gressel

Chapter 12 Marker-Assisted Selection for *Striga* Resistance in Sorghum 159
Cécile Grenier, Yahia Ibrahim, Bettina I. G. Haussmann, Daniel Kiambi and Gebisa Ejeta

Chapter 13 The Molecular Genetic Basis of Host Resistance to *Striga* Species: A Way Forward 173
Julie D. Scholes, Philip J. Swarbrick, Jon Slate and Malcolm C. Press

Chapter 14 Effects on *Striga* Parasitism of Transgenic Maize Armed with RNAi Constructs Targeting Essential *S. Asiatica* Genes 185
Anic de Framond, Patrick J. Rich, John McMillan and Gebisa Ejeta

Section V Agronomic Options: The First and Essential Line of Control and Policy Considerations 197

Chapter 15 An Integrated *Striga* Management Option Offers Effective Control of *Striga* in Ethiopia 199
Tesfaye Tesso, Zenbaba Gutema, Aberra Deressa and Gebisa Ejeta

Chapter 16 Integrating Crop Management Practices for *Striga* Control 213
Joel K. Ransom, Abdel Gabar Babiker and George D. Odhiambo

Chapter 17 Cultural and Cropping Systems Approach for *Striga* Management — A Low Cost Alternative Option in Subsistence Farming 229
Fasil Reda and J. A. C. Verkleij

Chapter 18 Field Developments on *Striga* Control by *Desmodium* Intercrops in a "Push-Pull" Strategy 241
Zeyaur R. Khan, Charles A. O. Midega, Ahmed Hassanali and John A. Pickett

Chapter 19 Integrated *Striga* Management to Meet Sorghum Demand in Tanzania 253
Ambonesigwe M. Mbwaga, Charles Riches and Gebisa Ejeta

Chapter 20 *Striga* Economics 265
Hugo De Groote

Section VI Biocontrol: Untapped Potential? 281

Chapter 21 Biocontrol Using *Fusarium Oxysporum*; A Critical Component of Integrated *Striga* Management 283
Fen D. Beed, Steven G. Hallett, Julien Venne and Alan K. Watson

Chapter 22 Genetically Enhancing the Virulence and Efficacy of Plant Pathogens for Biological Control of Parasitic Plants 301
David C. Sands, Alice L. Pilgeram, Brian M. Thompson and Matthew M. Kirkpatrick

Chapter 23 Transgenic Biocontrol Agents to Overcome Evolutionary Barriers 313
Jonathan Gressel, Sagit Meir, Yoav Herschkovitz, Hani Al-Ahmad, Olubukola Babalola and Ziva Amsellem

Section VII Epilogue 325

Chapter 24 Epilogue — Will There be Integrated *Striga* Control? 327
Jonathan Gressel and Brhane Gebrekidan

Index 339

Section I

Introduction — The Witches' Curse

CHAPTER 1

THE *STRIGA* SCOURGE IN AFRICA: A GROWING PANDEMIC

Gebisa Ejeta

Department of Agronomy, Purdue University, West Lafayette IN, USA
E-mail: gejeta@purdue.edu

Witchweeds (*Striga* spp.), endemic parasitic weeds of sub-Saharan Africa are steadily increasing their geographic domain and level of infestation, and bewitching plants they invade, thereby greatly reducing crop yield. They have become a widely acknowledged scourge. The *Striga* problem undermines the struggle to attain food security and economic growth in the continent. Countries with nascent infestation of *Striga* only 25 years ago are now showing heavy annual losses of crop yield. Rough estimates are nearly 300 million people in sub-Saharan Africa are adversely affected by Striga, and up to 50 million hectares of crop lands in the continent show varying degrees of Striga infestation. Areas of otherwise productive agriculture have been abandoned because of this scourge. Crops previously unaffected by *Striga* are now showing serious infestation. *Striga* is, therefore, fast becoming a pandemic of serious proportions in Africa because of its vast geographic spread and its economic impact on millions.

1. Introduction

1.1. *The Problem*

Striga has long been recognized as the greatest biological constraint to food production in Africa as nearly 100 million hectares of the African savannah are infested annually with *Striga*. Although these parasites attack several crops, the brunt of the ravage has fallen on the staple crops of the poor in the African savanna, namely maize, sorghum, pearl millet

(*Pennisetum glaucum*), upland rice, and cowpeas (*Vigna unguiculata*). *Striga* damage to crops is often severe because of its remarkable bewitching effect on crops it invades. The *Striga* problem in Africa is exasperated by its exquisite adaptation to the climatic conditions of the semi-arid tropics, its high fecundity, and longevity of its seed reserves in tropical soils. The growing conditions in sub-Saharan Africa permit timely breakdown of seed dormancy and conditioning of *Striga* seeds and exposure to exudates from host seeds planted around them. Its many flower stalks each produce and deposit a new supply of tens of thousands of tiny seeds to an already enormous seed bank (Chapter 2). The large number of parasitic seeds produced increases the chance that some *Striga* seeds will find a suitable host. Every year some seeds germinate, some revert to dormancy, some remain in the soil unconditioned, while others are added from the new growth, continually enriching the seed reserve in the soil. The type of crop cultivars grown has a direct influence on *Striga* infestation. The best practice is long term rotational cultivation of cereal crops with legumes or other crops unaffected by the parasite. Continuous use of susceptible crop cultivars without protective amendments leads to disastrous levels of heavy infestation, crop failures, and build up of the *Striga* seed reserve in the soil.

Local landraces are often described as having tolerance to *Striga*. However, the moderate level of tolerance exhibited by local landraces in the past may no longer provide protection at higher levels of infestation. Local landraces behave similarly to susceptible cultivars at high infestation levels, supporting more *Striga* and bringing more parasitic plants to set seed, further enriching the soil seed bank. Where the *Striga* infestation is high, only cultivars with high levels of *Striga* resistance would provide protection and help diminish the seed bank.

1.2. *The* Striga *Scourge*

The production of crops under African soils and climatic conditions is wrought by a number of agronomic and management challenges. *Striga* is only one of several biotic, climatic, and edaphic problems reducing crop yield, directly marginalizing capacity for food production in the continent. However, severe *Striga* infestation appears to render African farmers helpless and often bewildered, even though they are otherwise very resilient and adaptive. Because of the seemingly sudden build up of *Striga*, its dramatic bewitching effects on crop plants, and widespread dreadful affliction and devastation, African farmers recognize *Striga* less as a biological constraint to crop production and more as a scourge handed from above.

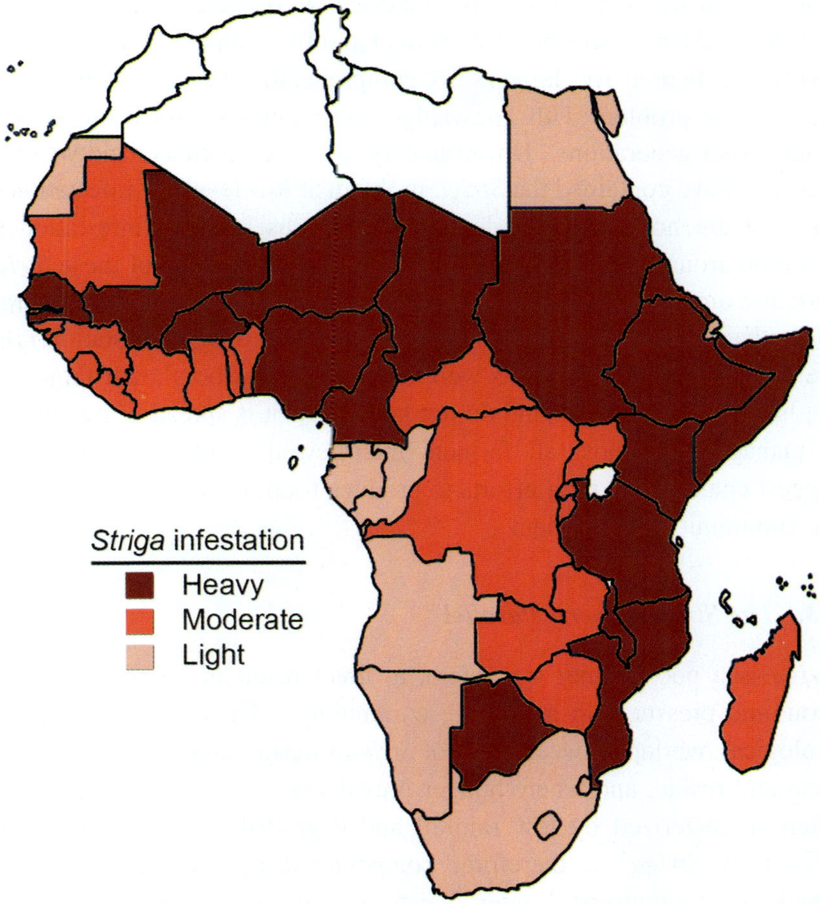

Figure 1. Striga *infestation in Africa is most severe in the most food insecure areas. One or more species of the parasite are found in crop lands and/or grasslands of nearly all African countries below the Sahara. Adapted from a report by Gressel and colleagues.*[8]

Across the continent, farmers ascribe local names to *Striga* that so aptly translate to effects on humans of evil spirits or serious illnesses. While the parasite is still invisible underground, it causes the crop to suddenly turn sickly, apparently bewitched. Its annual occurrence and wide geographic spread affecting large proportions of farmlands and populations also give *Striga* the appearance and feel of a natural pandemic. What is most baffling to African farmers is that *Striga* is not a new problem, as it has long been a common occurrence in crop fields. Yet, the sudden expansion of its spread and the rampant high infestation levels are harder for farmers to comprehend. Farmers continue to manage the problem with knowledge and practices passed on to them from earlier generations. Unfortunately, practices such as hand weeding that may have contained the *Striga* problem at low levels of infestation in the past are not making a dent in the serious rampant infestation so common around them today. Surveys conducted in some of these *Striga* endemic areas (e.g. Chapter 20) reveal the seriousness of this scourge. More than half of African farmers recognize that *Striga* infestation is on the increase on their farms as well as on their neighbors', that they have not had it so bad before, and do not know how it is spread or how best it is managed. Almost all farmers interviewed identify *Striga* as the biggest challenge in their efforts to produce food for their families and in the community around them.

1.3. *The Striga-Poverty Parallel*

Striga is a poor farmer's problem, a direct result of demographic and economic pressures in a farming community. There is a near perfect ecological overlap between areas of *Striga* infestation and where the poor farm and reside, and where hunger prevails (Fig. 1). These regions are often characterized by low rainfall and degraded, infertile soils. The impact of *Striga* is, therefore, compounded by its predilection for attacking crops already under moisture and nutrient stress, conditions that are very common in much of the semi-arid tropics. There is growing evidence that the *Striga* problem is worsening across sub-Saharan Africa.[1] The world's population is projected to grow to 8 billion in 2025, stabilizing at about 8 billion people by the middle of the 21st century.[2] During the same period Africa's population is projected to rise to 1.5 billion in 2025 and 2 billion by 2050, although the projected population growth may vary from region to region. These are ominous trends for African agriculture unless corrected by the introduction of modern technologies to accelerate a concomitant growth in food production. Poor African farmers have limited access to formal education or vital information. They are generally recalcitrant to adopting new technologies, and are risk averse. As a result, rapid population growth in rural Africa has created pressure on availability of farm lands, forcing crop agriculture onto marginal lands with poor soil fertility, poorly drained soils, and soils with acute moisture stress. The *Striga* problem has been worsened and crop yields reduced by farming practices with shortened or no fallow periods, little or no use of inorganic fertilizers because of cost, and low values paid for the crop, increased use of mono-cropping, and continuous cultivation without traditional practices of crop rotation and intercropping systems. Farmers with crop fields severely infested with *Striga* often resort to abandoning their fields, contributing to an already severe pressure on availability of farm lands. *Striga* has expanded to cover a wider ecological range encroaching into previously un-infested crop lands and invading new crops.[3] These realities have worsened the *Striga* problem raising it into a growing pandemic and an agricultural scourge of significant proportions to subsistence farmers in very many African countries (Fig. 1).

2. Distribution and Impact of Striga

2.1. *Geographic and Species Distribution*

The genus *Striga* includes over 40 species, of which 11 species are considered parasitic on agricultural crops (Chapter 6). Africa is thought to be the center of origin for *Striga*. The prevalence and extent of genetic diversity of a species in a particular geographic area, where more forms appear and specialized associations are observed, are often indicators of origin for plant species. The vast tropical savannah between the Semien mountains of Ethiopia and the Nubian hills of Sudan has the greatest biodiversity of sorghum and pearl millet, the two crops that *Striga* readily infests, as well as that of the parasite population itself. This area is recognized as the center of origin for sorghum and pearl millet[4] and may also be the home of the two species of *Striga* affecting cereals, namely *S. hermonthica*, and *S. asiatica*. The species that is specially adapted as a pest of legume crops, *S. gesnerioides*, may have originated in western Africa. Today, *Striga* is found in almost all regions of sub-Saharan Africa, except in areas where rainfall is too high or in high altitude areas where temperatures may be too low for development of the parasite,[5] but is most severe in infertile, nutrient depleted soils with low organic matter content.

S. hermonthica has the largest geographical distribution. With its obligate out-crossing behavior and its large plant stature, it is the species that causes the greatest crop damage. *S. hermonthica* is found in much of sub-Saharan African with particular prevalence in western, central, and eastern Africa, as well as parts of the south-western part of the Arabian Peninsula across the Red Sea.

S. asiatica has its widest distribution in the eastern and southern Africa. It is also found in Asia, particularly in southern India, as well as the United States and Australia. *S. gesnerioides* occurs in Africa, the Arabian Peninsula, the Indian subcontinent, and has also been introduced to the United States.[6] This species causes its greatest economic damage on legume crops widely grown in western Africa, particularly cowpeas. Both *S. gesnerioides* and *S. asiatica* are self-fertile resulting in apparent genetic variability observed as distinct morphotypes as well as parasitic specialization. They appear to be distinctly less variable than the obligately out-crossing *S. hermonthica*.

2.2. *Dispersal and Expansion of Infestation*

There have been limited studies on the modes of spread and dispersal of *Striga* seeds. Farm practices as well as human and animal movements across geographic areas have been implicated as factors responsible for spreading parasitic weed seeds. Crop seeds are a major vehicle for *Striga* seed dispersal, with 20-40% of seed lots in the market contaminated by

Striga seeds.[7] Most African farmers grow their own "seed", saving grain from a previous crop. Yet, there is always a significant activity of seed exchange among farmers within and among distant neighborhoods. Grain consignments distributed as "food aid" and "seed aid" often result in widespread serious *Striga* infestations.[8] Even when improved crop cultivars are deployed, there are no functional seed production and dissemination programs in most African countries. Where public seed agencies are in place, they are often non-functional, producing under-par quality seeds, or are not producing enough to meet needs. The private seed industry is in its infancy in Africa, but well-functioning seed enterprises that supply certified, brand quality seed to farmers are badly needed. While true-to-type, quality-controlled seed is essential for increasing productivity, it will have the additional benefit of limiting the spread of parasitic weed seeds. Better farm sanitation, handling of farm equipment, and management and movement of crop residues on farm also are important in curtailing dispersal and spread of parasitic weed seeds. Educational programs are vital that promote the value of quality seed, proper sanitation, and handling of equipment and crops as a way to effectively address this important problem in *Striga* endemic regions.

There seems to be physiological specialization in *Striga*, as some strains cross-infect host plant species and others do not. In some places *Striga hermonthica* "strains" that infect sorghum are different from those that infect millet or maize. Where there has been distinct geographic or ecological separation of regional crop cultivation in Nigeria, Niger, Burkina Faso, and most of West Africa, it appears that different specialization of host strains has developed. In such situations, a crop introduced to the new region will initially be unaffected by strains of *Striga* that are there, but will gradually succumb to infestation. Is this specialized physiological adaptation or the result of introduction into the new area of strains with a capacity to attack the new crop? In the early 1980s pearl millet was introduced to the eastern Sudan from the western part of the country where it is mostly grown. It was unaffected by the strains of *Striga* in the region. Several years later, pearl millet was equally attacked by *Striga* east or west of Khartoum. In other areas, there are no separate geographic bands for strains specific to a crop species. Very often when sorghum and millets are grown in the same region, both crops show similar degrees of *Striga* infestation, although it is not always easy to determine if these strains are different, related, or the same. Occasionally, an unusual pattern of *Striga* infestation is observed, such as the situation in western Eritrea. Although the area is long known as a hot spot for *Striga* infestation, and sorghum and pearl millet are major crops of the country. Sorghum is always highly infested, yet pearl millet in Eritrea has been cultivated totally free of *Striga*. Crops that were never known to be affected, such as barley, wheat, and tef have been seriously infested.[20] The bases for some of these observations and events are not well known, and merit serious investigation. We also have information accumulating on the inheritance of host plant resistance in several crops, but little is known of the genetics of virulence of parasite populations. The diversity and structure of parasitic populations in the economically important *Striga* species is hardly understood.

2.3. *Economic Importance and Impact*

That parasitic weeds are significant constraints of crop production in much of Africa is widely recognized. However, hard data on the extent of spread, yield losses, impact on the economy and welfare of nations or households have not been available, except for the few reports[3,7,9-12] that continue to be repeatedly cited. These estimates, rough as they are, have examined national and regional impact of parasitic weeds and have been useful, but they pose some limitations in extrapolating to continental impact. In general, average yield losses due to *Striga* are estimated at or above 50%. The total area under severe to moderate *Striga* infestation had been estimated to range from 30 to 50 million hectares.[3] Nevertheless, estimates on the extent of crop damage in a country or region in the African continent vary depending on prevalent cultural practices, the crop cultivar, and degree of infestation.[13] Much spread has occurred since these early and rough estimates were made, but no new figures have emerged. The degree of *Striga* infestation is most severe in eastern Africa where invasion by the parasite is expanding at an alarming rate, often resulting in total loss of crops in any given crop season. Expansion of *Striga* infestation has also increased in western Africa.

The annual crop production in the savanna regions of Africa alone was estimated to have a significant negative impact on the food supply of over 100 million people two decades ago[14], and the situation is getting worse.[3] The Food and Agricultural Organization (FAO) of the United Nations estimates that, across the continent, *Striga* causes annual losses in excess of US$7 billion, adversely affecting over 300 million people.[1]

3. *Striga* Management Options

It is generally believed that with abundant resource commitment, parasitic weeds can be managed in agriculture. However, the sufficiency of currently available technologies for effective *Striga* management is debatable. Strategies may be directed to the alternative management options of *Striga* control, containment, or eradication. Control of *Striga* is slow but feasible. *Striga* damage and infestation can be somewhat alleviated with well-managed practices and measures that fit the local knowledge, economy, as well as labor capacities, and practiced for several seasons. Four independent *Striga* control approaches, namely cultural, chemical, genetic, and biological options have been widely investigated and developed, and are described in other chapters. In most cases, these control measures have had limited success. Effective control of *Striga* has been difficult to achieve through conventional hand or mechanical weeding as the parasite exerts its greatest damage bewitching the crop before its emergence above ground, and providing evidence for host plant infection. Suggested cultural practices involving crop rotation, trap cropping, intercropping, or multi-year fallow, are not adopted.[15] These practices are perceived by poor farmers as unaffordable or uneconomical, labor intensive, impractical, or not congruent with their other farm operations. An intriguing new concept dubbed "push-pull", which employs co-planting of different crop species between rows in a maize field (repellent, push), and another crop around the field (attractant, pull) to control two major biotic problems of maize (stem, insects and *Striga*) is discussed in Chapter 18. Its wide adoption will hinge on finding suitable companion and trap crops that fit into the farming systems of target communities. Many of these management options are effective practices that not only offer *Striga* control, but also build up soil fertility, organic matter, as well as enhance overall soil health. These practices also require several years of repeated and continued application before their effects are realized through a significant rise in annual grain yield, or as an apparent reduction in level of infestation. The use of resistant crop cultivars is the most economically feasible and environmentally friendly means of *Striga* control. *Striga* resistant cultivars have been bred in a number of crops, as discussed in later chapters. However, cultivars with immunity to *Striga* have not been found in any host crops. Multiple genes for *Striga* resistance, found so far only in sorghum, have been pyramided in cultivars that also possess desirable genes for agronomic and grain quality traits (Chapter 7).

Biocontrol of *Striga* has also recently emerged as a potential control measure (Chapter 21). Natural enemies of *Striga* have been found in insects, fungi, and bacteria. However, current biocontrol agents are probably not effective enough to be deployed *per se*, but are applicable as part of an integrated approach. Seed coating of non-transgenic maize with a low dose of herbicide was recently developed and released in Kenya, but will need a slow release mechanism to last the whole season in long season maize (Chapter 11).

In general, only a few of these control measures have been widely adopted or commercialized. *Striga* continues to inflict significant yield losses on staple crops of Africa impacting the state of food security in Africa. The reasons for limited adoption of these control practices include limited knowledge of the problem, its biology, the lack of labor or resources to make the needed investment, an uncertainty of potential control, and a return to investment, and an unwillingness to make the long-term investments.[16]

A new infestation of *Striga* can also be contained into a small geographic area, again with sufficient resource commitment. The most successful experiment of containment took place in the United States, where a seemingly nascent level of *S. asiatica* infestation was discovered in the state of South Carolina.[17] *Striga* infestation was beginning to spread when a highly organized campaign started including effective quarantine of seed movement, sanitation, and application of costly chemical applications to destroy emerged parasitic weeds above ground as well as seeds in the soil. Nevertheless, it took more than 40 years and over US$250 million to contain *Striga* infestation into two counties of the Carolinas.[18]

Eradication of *Striga* or other invasive species is difficult to achieve especially after a major infestation. Examples of successful eradication of invasive species are hard to find.[19] Eradication is unattainable because several small unnoticed centers of nascent infestation may spread into larger areas. Attacking nascent foci becomes valuable if long-term eradication is to be attained. This is particularly true in species such as *S. hermonthica* with its large plant size and immense seed production capacity.

4. Investments in *Striga* Control

Significant advances have been made in understanding the biology of parasitic weeds and in devising technologies that can be used for their control. However, this progress has been slow and inadequate to significantly impact lives of rural farmers. A major reason for this insufficient progress is the lack of significant investment into the research and control of *Striga*. Past investments have not been commensurate to the magnitude of the problem. The *Striga* problem has become too big for any resource commitment by national programs of many developing nations. There is also an insufficient scientific base to address the problem in a meaningful way in many of these countries. National or regional efforts directed to *Striga* management will need to place a mix of scientific expertise with resource support. Intractable biological problems such as *Striga* infestation require new knowledge. Such resources may not exist in the emerging scientific and development organizations of most developing countries, and may require external input or regional cooperation.

There have been several, albeit intermittent, international investments into *Striga* research. The International Development Research Center (IDRC) of Canada provided much of the early support in *Striga* research in the 1970s and 1980s, particularly with their uncommon model of providing national talent in developing countries with direct resource support and encouragement. Early advances in breeding of *Striga* resistance in sorghum were catalyzed by this IDRC investment. The British Overseas Development Agency (ODA) as well as the German technical support program (GTZ) have also provided significant support to *Striga* research. Much of this research was devoted to testing and retesting promising cultural practices. Agronomic research tends to be location specific. Basic research was badly needed to understand how *Striga* parasitizes and where its weaknesses lie.

The most sustained and significant resource support for *Striga* research has been provided by the United States Agency for International Development (USAID) and the Rockefeller Foundation. These two agencies should be credited for the development and deployment of the only two commercially launched technologies for *Striga* control, namely *Striga*-resistant sorghum cultivars deployed as an integrated *Striga* management technology in Ethiopia, and the seed-coating of imazapyr-resistant maize in Kenya.

5. The Current State of Knowledge

Although successes in on-farm *Striga* control have been limited, the global research community has laid a good foundation of knowledge about the nature of the parasite pointing to potential avenues for intervention. There has been an enhanced understanding of the biology of the parasite and its interaction with crop plants and other hosts. This knowledge-base continues to be built as can be gathered from the chapters in this book. Knowledge is also emerging on the basis of specificity of *Striga* adaptation to different hosts and to different environmental conditions. New molecular biological tools with potentially promising applications for *Striga* control, coupled with the emergence of genomic sequences of major agronomic crops. These serve as hopeful signs that effective management and control of *Striga* may be nearer than the horizon.

There is also much that is not yet known about the parasite. The behavior of the parasite under natural conditions is not well understood. We do not fully understand why *Striga* behaves erratically in nature. The interaction with the environment in which *Striga* so readily thrives and the extent to which these variables determine annual crop infestation are also not well understood. The germination of *Striga* in the field or even in laboratory conditions is not predictable, nor do we always know what parameters to alter to obtain predictable results. We have insufficient and often conflicting information on how the soil pH, soil microbial activity, soil organic matter, and the degree of soil mineralization impact *Striga* infestation.

6. The Challenge

There is no doubt that, if left unchecked, the *Striga* problem in Africa will continue to lead to disaster. *Striga* will ruin farm communities and destroy fragile ecosystems that are managed by poor subsistence farmers, turning an already precarious state of food production into an even greater continental crisis.

There is sufficient knowledge to develop interim technologies for the control of *Striga*, but more needs to be learned for even greater impact. This book is written with the premise that none of the currently available technologies can provide sustainable solution to the *Striga* problem. Conquering this scourge requires a good mix of disciplinary approaches towards the development of an integrative approach that will bring together multiple control factors. The theme and purpose of the conference that led to this book was to promote integrated *Striga* control as a sure way to synergize scientific approaches and generate greater impact. Successful experiences were shared and the challenge for the next generation of integrated technologies was extended. Hopefully, this challenge will be met and the growing scientific talent pool and the prowess of the emerging scientific capabilities and tools will be able to conquer this scourge and avert disaster.

References

1. http://www.fao.org/wairdocs/tac/x5749e/5749eOc.htm#1.2.%20cereals
2. W. Lutz, W. Sanderson, and S. S. Scherbov. *Nature* **412**, 543 (2001).
3. A. M. Emechebe , J. Ellis-Jones, S. Schulz, D. Chikoye, B. Douthwaite, I. Kureh, G. Tarawali, M. A. Hussaini, P. Kormawa, and A. Sanni. *Expl. Agric.* **40**, 215 (2004).
4. H. Doggett, Sorghum, 2nd Ed: (Longman, Harlow UK, 1988).

5. J. Sauerborn In: Proc. 5th Intl. Symp. on Parasitic Weeds, Eds. J. Ransom, L. J. Musselman, A. D. Worsham, and C. Parker (Nairobi, CIMMYT. 1991) p. 137.
6. L. J. Musselman and C. Parker. *Adansonia* **20**, 431 (1981).
7. D. K. Berner, K. F. Cardwell, B. O. Faturoti, F. O. Ikee, and O. Williams. *Plant Dis.* **78**, 402 (1994).
8. J. Gressel, A. Hanafi, G. Head, W. Marasas, A. Obilana, T. Souissi, and G. Tzotzos. *Crop Protect.* **29**, 661 (2004).
9. S. S. M'Boob. In *Striga* Improved management in Africa. (eds. T.O. Robson and H. R. Broad). (FAO, Rome, 1989).
10. S. T. O. Lagoke, V. Parkinson, and R. M. Agunbiane., In Proc. of Intl. Workshop on Combating Striga in Africa. Ed. S. K. Kim, (IITA, Ibadan, 1991).
11. J. Kroschel. In: Advances in Parasitic Weed Control at On-farm Level. Vol I. Joint Action to Control Striga in Africa,. eds J. Kroschel, H. Mercher-Quarshie, and J. Sauerborn. (Margraf Verlag, Weikersheim, 1999).
12. G. L. Hartman. *Plant Dis.* **75**, 494 (1991).
13. C. Parker and C.R. Riches, Parasitic Weeds of the World: Biology and Control (CAB International, Wallingford, 1993).
14. S. S. M'Boob, In: Improved *Striga* management for Africa. Ed. T. Robson and H. Broab. (FAO, Rome,1986) p. 190.
15. A. Oswald. *Crop Protect.* **24**, 333 (2004).
16. D. M. Joel, Y. Hershenhorn, H. Eizenberg, R. Aly, G. Ejeta, P. J. Rich, J. K. Ransom, J. Sauerborn and D. Rubiales. *Hort. Rev.* **33**, 267 (2007).
17. R. E. Eplee. *Plant Dis.* **65**, 951 (1981).
18. R. G. Westbrooks and R. E. Eplee. In: Invasive Species and Biodiversity Management. Eds. O. T. Sandulun, P. J. Schei, and A. Viken (Kluwer, Dordrecht, 1999) p. 283.
19. R. N. Mack and W. M. Lonsdale. In: Turning the tide: the eradication of invasive species. Eds. C. R. Veitch and M. N. Clout (IUCN, Gland, 2005).
20. F. Reda, G. Wolde-Wahid, B. Zemichael, and W. Bayou. *Pest Manage. J. Ethiopia* 1, 43 (1997).

Section II

Biology and Chemistry — The Needed Basics

CHAPTER 2

BIOLOGY OF HOST-PARASITE INTERACTIONS IN *STRIGA* SPECIES

Patrick J. Rich* and Gebisa Ejeta

Department of Agronomy, Purdue University, West Lafayette, IN 47907, USA
**E-mail: pjrich@exchange.purdue.edu*

In this chapter we present background information on the biology of *Striga* species from seed after-ripening through germination, parasitism, and seed production. As an obligate root parasite, *Striga* is dependent on its host and therefore modulates its development to correspond with its host's life cycle. *Striga* seeds have specific dormancy and environmental conditioning requirements that must be met before they germinate. Germination of *Striga* proceeds in response to various chemicals exuded by host plants. Differentiation of radicle cells into the haustorium is also cued by host rhizosphere chemistry. Both germination and haustorial initiation need to occur very near host roots for parasitic attachment. Post-attachment haustorial development allows the parasite to establish vital vascular connections as well as metabolic and osmotic linkage with the host plant. Finally, the *Striga* matures and produces numerous seeds completing the life cycle. Many details of *Striga* biology have yet to be discovered. As this occurs, our ability to formulate integrated control strategies should improve.

1. Overview of the *Striga* Life Cycle

The life cycle of *Striga* is intimately linked to that of its hosts. A generalized overview of the *Striga* life cycle is presented in Figure 1. *Striga* seeds pass through a period of dormancy and generally can not germinate in the season in which they are produced. They are released from dormancy through a process called conditioning or preconditioning during which species specific temperature and moisture requirements must be met. Once conditioned, germination of *Striga* proceeds in response to signals derived from host plants. Radicle cells differentiate into distinctly parasitic organs called haustoria in response to different host derived signals. The haustorium begins its function in attachment and after attachment develops into an organ of acquisition and metabolism of host-derived nutrients and water. Shoot development follows and *Striga* eventually emerges above ground, matures and sets seed. Proposed influences of the host that mediate the developmental transitions are indicated in Figure 1 by bold arrows radiating from the sorghum plant in the center of the schematic representation. Details of each developmental stage are presented in the following sections.

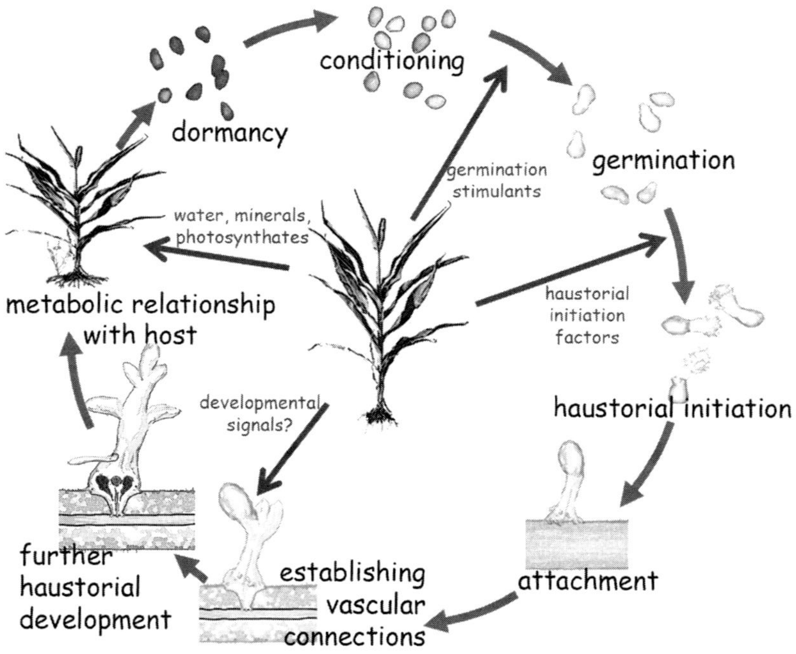

Figure 1. The Striga *life cycle.*

2. Seed Dormancy and After-Ripening

The seeds of obligate parasitic weeds are tiny relative to facultative parasitic weeds and those of free-living angiosperms. Seeds of *Striga asiatica* are typically 0.33mm long and weigh 3.7μg, those of *S. hermonthica* are 0.38mm and 7.1μg.[1] Energy reserves in such small seeds are limited and only sufficient for a short period of autonomous growth. It is therefore a matter of survival that parasitic weed seeds germinate at the proper time corresponding to the growing season of their potential hosts. *Striga* seeds can reportedly retain their viability for decades,[2] survival rates up to 14 years have been demonstrated for *S. asiatica*,[3,4] although intervals of only 2 years are reported for *S. hermonthica* seeds.[5] These estimates, based on concentrated batches of buried seed exhumed and assayed for viability after intervals, probably under estimate longevity as the crowding promotes degradation by pathogens that thrive on the concentrated seed source.[6] The large numbers of seeds produced by most *Striga* spp., even surviving only a few years, would elevate the soil seed bank to damaging proportions, particularly in a continuous cropping situation.

Striga seeds have an after-ripening requirement and cannot germinate in the season in which they were produced.[7] This requirement prevents newly matured *Striga* seed from germinating too late in a growing season, when host plants capable of supporting a parasitic plant to maturity are scarce. Additionally, *Striga* seeds can enter a state of secondary dormancy (see Chapter 4) when environmental conditioning has prepared them to germinate but no host has stimulated them to do so. The ability to revert to the dormant state, which can be reversible after a period of desiccation,[8] ensures longevity in the soil by avoiding elevated states of respiration and saving limited seed reserves for the committing step of germination.

3. Conditioning

After-ripened *Striga* seeds will not germinate until they have passed through a preconditioning period. Peak germination of *S. asiatica* seed occurs *in vitro* after 10-15 days of soaking in water at a temperature of 28°C.[9] The duration and temperature optima for the conditioning period vary with species and generally reflect the environmental conditions of their geographical origins at the beginning of their natural growing season. With *Striga*, a native of the semi-arid tropics, seeds are conditioned in the wet soils of the rainy season, when suitable host plants, including sown crops, are beginning to germinate. Preconditioning is discussed further in Chapter 4.

4. Germination

Preconditioned, after-ripened seeds germinate in response to chemical stimulants in the rhizosphere of potential hosts. A germinated *Striga* can only survive in the free-living state for a few days,[10] because it must rely solely upon its small seed reserves. Storage lipids of *S. asiatica* seeds are about 37% of the seed weight[11] and are conservatively used during the germination process and free-living radicle elongation period.[12] In the absence of a host, the *Striga* radicle will elongate up to 2-3mm over about 4 days and elongation may be sustained for as long as 10 days on the limited seed reserves.[13] Radicle growth is without branching, mainly by cell elongation, with longitudinal cell divisions. Chemotropism guides the growing radicle towards the potential host root,[14] which may result from the gradient of the germination stimulant.[15] Detailed studies of chemotropism are lacking.

Various natural stimulants of *Striga* seed germination have been identified over the recent decades. The most active of these are the strigolactones, able to induce germination of *Orobanche* and *Striga* seeds at nanomolar or even picomolar concentrations. The abundance of strigolactones in plant root exudates is low but they are widespread among plants, produced by both hosts and non-hosts of *Striga* spp. The ubiquitous nature of strigolactones is probably due to their recently discovered role as hyphal branching factors for arbuscular mycorrhizal (AM) fungi.[16] These organisms colonize plant roots forming symbiotic relationships. Eighty percent of terrestrial plants are estimated to form associations with AM fungi.[17] AM fungi benefit their hosts by improving nutrient and water uptake which often translates into increased plant biomass and crop yield. Strigolactones are discussed in detail in Chapter 4.

There is some debate in the literature over the chemistry of germination stimulation in sorghum-*Striga* associations.[18,19] An alternative germination stimulant in the sorghum-*Striga* association is the hydroquinone dihydrosorgoleone,[20] named **s**orghum **x**enognosin for ***S**triga* **g**ermination (SXSg).[12,19] This compound is not produced by other hosts of *Striga*.[21] Dihydrosorgoleone is considered to be the immediate precursor of the alleochemical sorgoleone, which is formed by autoxidation of the hydroquinone to the benzoquinone as it is exuded into aerated soil.[20,22] Only the hydroquinone form (dihydrosorgoleone) acts as a *Striga* germination stimulant.[20,23,24] Sorgoleone production is associated with fully elongated and living root hairs of primary, secondary and adventitious roots.[25] From the perspective of the parasite, germinating in response to transient signals present at the most proximal areas of the rhizosphere would mean that there is a living root nearby.

Arguments against the naming of dihydrosorgoleone as *the* SXSg center on observations that micromolar quantities (as opposed to the nanomolar quantities of strigolactones) are required for *Striga* germination stimulation[18] and variation for sorgoleone production among sorghum cultivars is not associated with the germination stimulant activity of sorghum cultivars to *S. asiatica*[26] (although in another survey of 25 other sorghum genotypes, a nearly 30-fold variation was reported[27]). The proponents of SXSg counter by noting that the amount of dihydrosorgoleone may be modulated by varying amounts of the biosynthetically related resorcinol in the root exudates of these sorghums which could enhance its persistence in the rhizospere.[22,23] Although strigolactones have been measured in sorghum root exudates,[28,29] and some genotypic variation for strigolactones exuded from sorghum roots observed,[30] a direct association between host genotype strigolactone exudation and *Striga* germination stimulant activity has yet to be demonstrated.

Root exudate composition is influenced by soil type and plant-microbe ecology[31]. The incredibly versatile *Striga* appears to be able to germinate with both a fungal stimulant, the strigolactones, and with dihydrosorgoleone, an immediate precursor to sorgoleone that possesses antifungal properties[32]. The plant parasite can sense its potential host before committing its limited resources amidst a variety of soil conditions and microflora.

In addition to the strigolactones and hydroquinones, several other natural products have been shown under laboratory conditions to induce germination in obligate root parasites. These include various plant growth regulators like cytokinins, ethylene,[33] and jasmonates.[34] Other compounds as diverse as methionine and inositol can also induce *Striga* germination.[35] Concentrations of these compounds required to induce germination are generally orders of magnitude higher than strigolactones.

5. Haustorial Initiation

In order to attach to their hosts, the obligate root parasites must form a special organ called the haustorium, from the Latin *haurire*, to drink. We use the term haustorium in the broadest sense to include all functions of this distinctly parasitic organ from attachment and penetration of the host root through acquisition and processing of host-derived vital substances throughout the life of the parasite. With haustorial formation the apical meristem of the *Striga* radicle switches from cell divisions in a longitudinal direction to radial divisions resulting in a swelling and proliferation of hair-like projections. Chemical stimulants in the host rhizosphere called haustorial initiation factors trigger this developmental transition. It is particularly important that this transition occurs very near the host root, as further radicle elongation stops with haustorial formation. Remaining seed reserves are rapidly consumed once newly germinated *Striga* are exposed to haustorial initiation factors.[11]

Haustorial initiation factors are different from compounds that stimulate *Striga* seed germination. Kinetin, simple phenolic compounds and quinones like 2,6-dimethoxy-1,4-benzoquinone (DMBQ) are quite active haustorial initiators,[36] but their presence in exudates is only detectable when host roots are mechanically damaged.[37] Parasitic preferences for particular haustorial initiation factors may be species and strain specific (see Chapter 3). Quinones and other reactive oxygen species are involved in defense mechanisms against pathogens.[38] Similar to the situation in which an immediate precursor of allelopathic and antifungal sorgoleone is used as a germination stimulant, *Striga* spp. might exploit these defense responses to determine the proximity and viability of host roots.

Thigmotropic responses (directed growth resulting from a tactile stimulus) may be involved in haustorial initiation of *Striga*. *Striga asiatica* root cultures growing in a liquid medium did not form haustoria when DMBQ was added. However, when an equal concentration of DMBQ was supplied to these cultures on solid media, haustoria readily formed.[39] We have observed haustorial formation of *S. asiatica* germinated with ethylene on pieces of felt.

6. Attachment

Striga asiatica and *Agalinis purpurea* are typical of many root parasites in that the haustorium that develops on the radicle is covered with hair-like projections. These haustorial hairs have a rough papillate surface and secrete a hemicellulose-based adhesive that fixes *Striga*[40] and *Orobanche*[41] to the host root. The binding that occurs is strong and durable. Haustorial hairs of mechanically detaching *Agalinis* resulted in a tearing away of host root cell wall fragments.[40] Attachment is apparently not specific, as *S. asiatica* will attach to host or non-host roots,[42] and *Agalinis purpurea* will attach to other plant parts, as well as wood, glass, and plastic if these are placed in contact with young haustoria.[40] *Striga asiatica* growing in agar will occasionally attach to each other. Newly induced haustoria can attach to a host root in as few as 6 hours after induction but their ability to attach is lost if they have not contacted a host root within 72 hours.[40] Attachment competency is associated with the sticky coating of the haustorial hairs which may require chemical or tactile signals from the host root to maintain.[36] *Striga* enter the penetration phase of development as cells of the haustorium divide and flatten to the surface of the host root, but this occurs on non-host roots as well.[42] Perhaps it is simply the anchorage or characteristics common to most roots that signal the penetration phase of haustorial development upon attachment.

7. Establishing Vascular Connections

Penetration of the root and tapping host nutrients must occur rapidly, as seed reserves are waning after germiation, radicle elongation and differentiation of the pre-attachment haustorium. During post-attachment haustorial development, the remaining seed storage lipids are mobilized in *S. asiatica*.[43] *Striga asiatica* typically requires 6 days after contacting a sorghum root to reach the vascular core.[42] Upon attachment, the haustorium undergoes a series of changes that characterize the penetration mode of development. The main center of differentiation shifts to the centrally located cells of the attached haustorium.[36] The distal-most protoplasmic cells of this region divide and form a wedge. Cells of this wedge then elongate and penetrate the epidermis of the root to which the haustorium is attached.[42] Penetration through the cortex is correlated with enzymatic activity that breaks down wall components of the host cortical cells.[1,44] Whether these enzymes are secreted by the invading haustorial cells or induced in the host is not clear.[45] Upon reaching the endodermal barrier, the invading cells proliferate in rows.[44] Penetration of the sorghum endodermis is typically delayed for 3-4 days in *S. asiatica*, during which further subcellular changes occur in the invading haustorium.[42] With breaching of the endodermis, intrusions into host vascular elements occur. In the hemiparasitic *Striga*, these intrusions occur mainly in larger xylem elements.[46] Upon penetration of the xylem vessels, the invading haustorial cells lose their protoplasts and undergo wall changes transforming them into water conducting elements continuous with host xylem.[46] The holoparasitic *Orobanche crenata* develop sieve pore connections with the phloem system of its host.[47] No direct connections with host phloem have been observed in *Striga*.[44,46,48]

Attachment appears to be a prerequisite of the transition to the penetration phase of haustorial development, which may involve additional chemical or tactile signals from the host root. No attempts to chemically induce cellular changes associated with the penetration phase succeeded in unattached haustoria.[36] *Striga asiatica* can penetrate the epidermis and at least part of the cortex of several non-host roots, suggesting that at its earliest stages, the penetration phase is triggered by factors not unique to suitable host plants.[42] However, sustained cellular development that allows intrusion to the point of vascular connection may depend on host-supplied factors.

8. Further Haustorial Development

The haustorium continues to mature upon successful establishment of vascular connections, forming distinct structures. The haustorium and surrounding parasitic tissue swells after successfully establishing vascular connections with its host. The surrounding tissue in *S. gesnerioides* becomes a structure called a tubercle similar to those post-attachment organs of *Orobanche* spp. In longitudinal sections, the haustorium of *S. hermonthica* on maize shows lobed structures called hyaline bodies which are composed of organelle-rich cells and extracellular deposits.[49,50] Development of xylem elements in the haustorium only occurs after penetration, implying that an additional host derived signal is necessary for triggering this transition.[51]

The haustorium that originated from the apical meristem of the radicle is called the primary haustorium. This primary haustorium functions throughout the life of the *Striga* spp. Adventitious roots form at the stem base of growing *Striga* plants and secondary haustoria may develop from lateral positions on these, providing additional connections with the host. Hundreds of these secondary haustoria may develop on an individual *Striga* plant by maturity.[1] These secondary haustoria are similar in form and function to primary haustoria.[50]

9. Metabolic Relationship with Host

The parasite can obtain the factors it needs for continued growth and development from its host after the establishment of vascular connections. *Striga* spp. are classified as hemiparasites because upon emergence, they are capable of photosynthesis. *Orobanche* spp. lack photosynthetic capacity and are therefore classified as holoparasites. Even with carbon-fixing capabilities, much of the *Striga* life cycle is spent underground where photosynthesis cannot occur. Albino *Striga* plants can reach maturity[1] as can *Striga* plants kept in darkness,[44] which shows that *Striga* can obtain all its carbon needs from its host. Photosynthetic activity of green *S. hermonthica* is low.[52] Growth and photosynthesis measurements collected from *S. hermonthica* on graminous hosts suggest that the parasite cannot sustain growth without host-supplied carbon.[53] As much as 85% of the carbon in *S. hermonthica* leaves is host derived.[53]

Established *Striga* have no direct phloem connections with their host[46] and so must obtain their carbon needs from the host xylem sap or through other apoplastic pathways.[54] Although root parasitic plants are in contact with the soil, they may use the organic forms of nitrogen already reduced by their hosts.[55] *Striga hermonthica*, however, is capable of reducing nitrogen sufficiently to support shoot growth without a host in tissue culture media when nitrate is the sole nitrogen source.[56,57] Much has yet to be learned about the manner in which carbon and nitrogen are sequestered by either xylem or phloem feeders.[58] Maintenance of higher transpiration rates than their hosts may contribute to solute flux into *Striga*.[54] Xylem sap contains several potential carbon sources in the form of organic acids, amino acids, soluble carbohydrates and plant growth regulators.[54] Availability of specific nitrogenous and organic solutes may contribute to the host preference of parasitic plants. There is some evidence for selective uptake of solutes from the host.[59] Parasitic plants have different soluble carbohydrate reserves, often in the form of polyols, that differ from the soluble carbohydrates of their hosts.[55] *Striga* and *Orobanche* contain mannitol which can account for as much as 75% of their soluble sugars.[60] In addition to their role as carbon storage units, the polyols accumulated by parasitic plants may act as osmoprotectants and osmoregulators that maintain water potentials below that of their hosts.[55]

Striga spp. are notorious for their tendency to adversely effect the growth and development of their hosts. The effects are manifested soon after attachment (see Chapter 13). Symptoms often mimicking drought stress are associated with *Striga* infestation.[1] This may be due to perturbations in plant growth regulator balance in the host, particularly elevated ABA concentration in host xylem sap.[61] ABA may be responsible for increased root:shoot ratios, reduced leaf expansion and reduced stem elongation reported in maize and sorghum infected with *Striga*.[62] *Striga* can also reduce photosynthesis in their graminous hosts.[62] How this benefits the parasite is unclear, but there is a general tendency of the root hemiparasites to negatively effect host photosynthesis while holoparasites tend to enhance or have a neutral effect on their hosts.[62]

10. Maturity and Seed Production

The cotyledon leaves emerge from the seed coat within a day after vascular connections are established between *Striga* and its host.[42] Scale leaf pairs continue to initiate along the growing stem and within 6 weeks the young shoots emerge above ground.[63] Morphological differences between *in vitro* cultures of *Striga* with and without exposure to sorghum roots suggests that parasitic plant architecture is influenced by host factors. *Striga asiatica* cultured on medium previously used to culture sorghum roots developed haustoria, branched shoots and multiple shoot-borne adventitious roots. When grown on the same medium but which had not been exposed to sorghum roots, a radicle-derived root system developed but shoot development was delayed for several weeks.[64] Flowering occurs within 6 weeks of emergence and is day neutral.[63] Some *Striga* spp. are primarily self-pollinating, for instance, *S. asiatica*, while others are primarily outcrossers like *S. hermonthica*.[65] *Striga* fruits (capsules) contain mature seeds in as little as 2 weeks after pollination. Thus *Striga* completes its life cycle within 10-16 weeks.[63] The number of seeds produced by a single *Striga* plant ranges from 25,000 in *S. forbesii* to 200,000 in *S. hermonthica*.[1] Large quantities of long-lived seeds assure the parasite genetic adaptability to changes in host availability, resistance and population dynamics.

11. Conclusions

Much has been investigated and learned about the fascinatingly interactive biology of Striga and its hosts, and there is much left to be studied. The emerging knowledge has been useful for learning about plant growth and development, and has provided avenues for attack in the control of the parasite in crop fields. As a weed *Striga* is a target for destruction. The more we know about its biology, the better our chances for killing it. Several control options are presented in the following chapters, including host plant resistance, herbicides, biological agents, improved agronomic practices and biotechnological methods. This chapter provides some background for those options. Although we have learned much about the biology of *Striga* species, much has yet to be discovered. Gene expression profiles of the parasite (Chapter 3) and hosts of *Striga* (Chapter 13) should reveal much about virulence and resistance to aid in breeding host crops that are free from *Striga*. The ability of *Striga* to produce numerous and long-lived seeds will quickly overwhelm any control measures based on a single mechanism. It is therefore vital that control is attempted through multiple strategies integrated into packages suitable and available to African agriculture where *Striga* has become a major production constraint.

References

1. C. Parker and C. R. Riches, *Parasitic Weeds of the World: Biology and Control* (CAB International, Walingford, 1993).
2. H. Doggett, in Advancing Agricultural Production in Africa, Proceedings, CAB's First Scientific Conference, Arusha, Tanzania, Ed. D.L. Hawksworth (CAB, Slough, 1984), p. 81.
3. A. R. Saunders, South Africa Department of Agriculture Science Bulletin **128**, 1 (1933).
4. F. F. Bebawi, R. E. Eplee, C. E. Harris, and R. S. Norris, *Weed Res.* **32**, 494 (1984).
5. G. Gbehounou, A. H. Pieterse, and J. A. C. Verkleij, *Weed Sci.* **51**, 940 (2003).
6. T. A. van Mourik, T. J. Stomph, and A. J. Murdoch, *J. Appl. Ecol.* **42**, 299 (2005).
7. C. A. Kust, *Weeds* **11**, 247 (1963).
8. A. Mohamed, G. Ejeta. L. G. Butler and T. L. Housley, *Weed Res.* **38**, 257 (1998).
9. G. Ejeta and L. G. Butler, *Afr. Crop Sci. J.* **1**, 75 (1993).
10. L. G. Butler, in *Allelopathy: Organisms, Processes and Applications,* Ed. Inderjit, K. M. M. Dakshini and F. A. Einhellig (American Chemical Society, Washington, 1995), p. 158.
11. M. Chang and D. G. Lynn, in *Allelochemicals: Role in Agriculture and Forestry*, Ed. G. R. Walker (American Chemical Society, Washington, 1987), p. 551.
12. L. S. Boone, G. Fate, M. Chang and D. G. Lynn, in *Parasitic Plants*, Ed. M. C. Press and J. D. Graves (Chapman and Hall, London, 1995) p. 14.
13. J. L. Riopel and W. V. Baird, in *Parasitic Weeds in Agriculture. Vol 1 (*Striga*)*, Ed. L. J. Musselman (CRC Press, Boca Raton, 1987) p. 107.
14. C. N. Williams, *Ann. Bot.* **25**, 407 (1961).
15. M.-P. Dubé and A. Olivier, *Can. J. Bot.* **79**, 1225 (2001).
16. K. Akiyama, K.-i. Matsuzaki, and H. Hayashi, *Nature* **435**, 824 (2005).
17. A. Besserer, V. Puech-Pagès, P. Kiefer, V. Gomez-Roldan, A. Jauneau, S. Roy, J.-C. Portais, C. Roux, G. Bécard, and N. Séjalon-Delmas, *PLoS Biology* **4,** 1239 (2006).
18. S. C. M. Wigchert and B. Zwanenburg, *J. Agric. Food Chem.* **47**, 1320 (1999).
19. A. G. Palmer, R. Gao, J. Maresh, W. K. Erbil and D. G. Lynn, *Annu. Rev. Phytopathol.* **42**, 439 (2004).

20. M. Chang, D. H. Netzly, L. G. Butler and D. G. Lynn, *J. Amer. Chem. Soc.* **108**, 7858 (1986).
21. X. Yang, B.E. Scheffler and L. A. Weston, *J. Exp. Bot.* **55**, 2251(2004).
22. G. D. Fate and D. G. Lynn, *J. Am. Chem. Soc.* **118**, 11369 (1996).
23. J. Erickson, D. Schott, T. Reverri, W. Muhsin, and T. Ruttledge, *J. Agric. Food Chem.* **49**, 5537 (2001).
24. D. H. Netzly, J. L. Riopel, G. Ejeta, and L. G. Butler, *Weed Sci.* **36**, 441 (1988).
25. C. Hauck, S. Muller, and H. Schilknecht, *J. Plant Physiol.* **139**, 474 (1992).
26. B. A Siame, Y. Weerasuriya, K. Wood, G. Ejeta and L. G. Butler, *J. Agric. Food Chem.* **41**, 1486 (1993).
27. A. Awad, D. Sato, D. Kusumoto, H. Kamioka, Y. Takeuchi, and K. Yoneyama, *Plant Growth Reg.* **48**, 221 (2006).
28. F. E. Dayan, *Planta* **224**, 339 (2006).
29. D. E. Hess, G. Ejeta, and L.G. Butler, *Phytochemistry* **31**, 493 (1992).
30. C. I. Nimbal, J. F. Pedersen, C.N. Yerkes, L. A. Weston, and S.C. Weller, *J. Agric. Food Chem.* **44**, 1343 (1996).
31. A.M. Hirsch, W. D. Bauer, D. M. Bird, J. Cullimore, B. Tyler and J. I. Yoder, *Ecology* **84**, 858 (2003).
32. Y. Suzuki, Y. Kono, T. Inoue, and A. Sakurai, *Phytochemistry* **47**, 997 (1998).
33. A. D. Worsham, in *Parasitic Weeds in Agriculture. Vol 1 (*Striga*)*, Ed. L. J. Musselman (CRC Press, Boca Raton, 1987) p. 45.
34. K. Yoneyama, M. Ogasawara, Y. Takeuchi, M. Konnai, Y. Sugimoto, H. Seto and S. Yoshida, *Biosci. Biotechnol. Biochem.* **62**, 1448 (1998).
35. J. C. G. Galindo, F. A. Macias, M. D. Garcia-Diaz and J. Jorrin, in *Allelopathy: chemistry and mode of action of allelochemicals*, Ed. F.A. Macias, J. C. G. Galindo, J. M. G. Molinillo and H. G. Cutler (CRC Press, Boca Raton, 2004) p. 125.
36. J. L. Riopel and M. P. Timko, in *Parasitic Plants,* Ed. M. C. Press and J. D. Graves (Chapman and Hall, London, 1995) p. 39.
37. J. L. Riopel, W. V. Baird, M. Chang, and D. G. Lynn, in *Witchweed Research and Control in the United States*, Ed. P. F. Sand, R. E. Eplee and R. G. Westbrooks (Weed Science Society of America, Lawrence, 1990) p. 27.
38. C. Lamb and R. A. Dixon, *Annu. Rev. Plant Physiol. Plant Mol. Biol.* **48**, 251 (1997).
39. S. J. Wolf and M. P. Timko, *Plant Sci.* **73**, 233 (1991).
40. W. V. Baird and J. L. Riopel, *Protoplasma* **118**, 206 (1983).
41. D. M. Joel and D. Losner-Goshen, *Can. J. Bot.* **72**, 564 (1994).
42. M. E. Hood, J. M. Condon, M. P. Timko and J. L. Riopel, *Phytopathology* **88**, 70 (1998).
43. M.L. Menetrez, R.C. Fites and R.F. Wilson, *J. Amer. Oil Chem. Soc.* **65**, 634 (1988).
44. W. E. Rogers and R .R. Nelson, *Phytopathology* **52**, 1064 (1962).
45. J. Visser and I. Dörr, in *Parasitic Weeds in Agriculture. Vol 1 (*Striga*)*, Ed. L.J. Musselman (CRC Press, Boca Raton, 1987) p. 91.
46. I. Dörr, *Ann. Bot.* **79**, 463 (1997).
47. I. Dörr, I. and R. Kollmann, *Bot. Acta* **108**, 47 (1995).

48. U. Neumann, B. Vian, H. C. Weber, and G. Sallé, *Protoplasma* **207**, 84 (1999).
49. P. S. Mallaburn and G. R. Stewart, in *Parasitic Flowering Plants*, Ed. H. C. Weber and W. Forstreuter (Proceedings of the 4th International Symposium on Parasitic Flowering Plants, Marburg, 1987) p. 523.
50. A. L. Gurney, D. Grimanelli, F. Kanampiu, D. Hoisington, J. D. Scholes and M. C. Press, *New Phytol.* **160**, 557 (2003).
51. J. I. Yoder, *Planta* **202**, 407 (1997).
52. I. Cechin and M. C. Press, *Plant Cell Environ.* **16**, 237 (1993).
53. J. D. Graves, A. Wylde, M. C. Press and G. R. Stewart, *Plant Cell Environ.* **13**, 367 (1990).
54. M. C. Press and J. B. Whittaker, *Phil. Trans. R. Soc. Lond. B* **341**, 101 (1993).
55. M. C. Press, in *Parasitic Plants* Ed. M. C. Press and J. D. Graves (Chapman and Hall, London, 1995) p. 103.
56. I. Igbinnosa, K. F. Cardwell and S. N. C. Okonkwo, *Eur. J. Plant Path.* **102**, 77 (1996).
57. A. Rousset, P. Simier and A. Fer, *Plant Biol.* **5**, 265 (2003).
58. J. M. Hibberd and W. D. Jeschke, *J. Exp. Bot.* **52**, 2043 (2001).
59. J. S. Pate, K. C. True and E. Rasins, *J. Exp. Bot.* **42**, 441 (1991).
60. M. C. Press, N. Shah and G. R. Stewart, in *Biology and control of* Orobanche, Ed. S. J. ter-Berg (LH/VPO, Wageningen, 1986) p. 96.
61. D. S. H. Drennen and S. O. El Hiweris, in *Proceedings of the Second Symposium on Parasitic Weeds* Ed. L.J. Musselman, A.D. Worsham and R.E. Eplee (North Carolina State University, Raleigh, 1979) p. 144.
62. J. R. Watling and M. C. Press, *Plant Biol.* **3**, 244 (2001).
63. G. R. Stewart, in *Parasitic Weeds in Agriculture. Vol 1 (*Striga*)*, Ed. L. J. Musselman (CRC Press, Boca Raton, 1987) p. 77.
64. T. Cai, A. G. Babiker, G. Ejeta and L. G. Butler, *J. Exp. Bot.* **44**, 1377 (1993).
65. L. J. Musselman, C. Parker and N. Dixon, *Beit. Biol. Pflanz.* **56**, 329 (1981).

CHAPTER 3

HOST DETECTION BY ROOT PARASITES: INSIGHTS FROM TRANSCRIPTION PROFILES

John I Yoder*, Russell Reagan, Alexey Tomilov, Natalya Tomilova and Manuel Torres

Department of Plant Sciences, University of California-Davis
Mail drop 3, Davis, California, USA
**E-mail: jiyoder@ucdavis.edu*

The implementation of biotechnology-based solutions for *Striga* control will be facilitated by a detailed knowledge of the genetic factors controlling successful parasitism. We are investigating the genetics of parasite – host plant interactions with the goal of identifying critical parasite genes that can be targets for engineered resistance in crop plants. Our approach is to identify parasitic plant genes that are transcriptionally regulated during different stages of parasitism and evaluate their functions in transgenic parasite roots. In collaboration with the Department of Energy Joint Genome Institute, we are sequencing cDNAs enriched for transcripts either up or down regulated in roots of the facultative parasite *Triphysaria* after contact with host roots, host root exudates, or purified haustorium-inducing factors. Over 9000 ESTs representing about 4000 unique genes are available at GenBank and the Pscroph database (http://pscroph.ucdavis.edu). This web site also provides BLAST functions for homology searches and keyword searches for functional annotations. We identified a set of genes that likely encode functions critical for successful parasitism using a combination of bioinformatics and cDNA array data.

1. Introduction

The successful parasitism of host roots by parasitic plants is a multistage process moderated by physical and chemical signals exchanged between the two plants.[1,2] Each of these developmental stages is a potential target for engineering host resistance against parasitic weeds. Unfortunately so little is known about the genes regulating host-parasite interactions that directed approaches to engineering host resistance are largely premature. One approach to identify genes essential for plant parasitism is to first identify genes that are regulated during host-parasite interactions and then determining their function in transgenic plants. We are applying this strategy to root parasites in the Orobanchaceae with the goal of identifying gene targets towards which robust resistances can be developed.

The thirty or so parasitic genera in the Orobanchaceae, including the *Striga* species, vary tremendously in their growth habits, life cycles and dependence on host derived resources.[3] The feature shared by these parasites is their ability to develop root-like haustoria that attach to and invade host roots (Chapter 4 and Ref. 4) Molecular phylogeny indicates a monophyletic placement of all Orobanchaceae in a common parasite clade distinct from non-parasitic relatives.[5] This indicates that the competence to develop haustoria originated one time in the evolution of this family and fundamental mechanisms for haustorium development are probably shared by all parasites. In contrast, mutations in genes encoding other parasite-associated phenotypes, such as the loss of photosynthesis or the use of host factors as germination stimulants, are observed in only some Orobanchaceae. Because haustorium development is common to all parasites, genetic resistances associated with haustorium development, invasion and maturation should be applicable across different parasite species.

The parasite lifestyle originated in some non-parasitic progenitor and the earliest parasites probably had growth habits and life cycles similar to those of facultative parasites. *Triphysaria* is a facultative parasite that grows as a common springtime annual throughout the Pacific Coast of North America.[6] *Triphysaria* can be grown to maturity in the absence of host plants and will develop viable seeds, but under these conditions few haustoria develop. The parasites develop haustoria in the presence of host plants that provide functional connections between host and parasitic members. Like other facultative parasites, *Triphysaria* has a broad host range that includes at least 27 families of angiosperms ranging from *Arabidopsis* to maize.[7,8] An intriguing exception is that *Triphysaria* do not parasitize other *Triphysaria.*[9] The mechanism of vegetative self-recognition in *Triphysaria* is not currently known but is an active area of investigation because of its obvious relevance to host genetic resistance.

Haustorium development in *Triphysaria* can be monitored *in vitro* by applying host roots, root exudates, or purified root factors to roots of aseptic seedlings.[10] In brief, *Triphysaria* seeds are surface sterilized and germinated on agar plates. One to two weeks after germination aseptic seedlings are transferred to square Petri dishes containing nutrient agar and incubated at a near vertical angle so that the *Triphysaria* roots grow down along the surface of the agar. Contact between host and parasite roots is made by laying roots of aseptic *Arabidopsis* seedlings across those of *Triphysaria.* Early haustorium development can be observed in most *Triphysaria* roots within twelve hours after contact with host. Most of the haustoria have attached and invaded *Arabidopsis* roots by twenty four hours.

Alternatively, haustoria can be induced in the absence of host roots by applying host root exudates or purified haustorium inducing factors to the roots. Several phenol derivatives were thus identified including simple phenolics, flavonoids, and quinones that induce haustorium formation in *Triphysaria.*[11] A similar set of molecules have been identified as haustorium inducing factors for *Striga* and *Agalinis.*[4] Many of these molecules are commonly found in root exudates and play critical signaling roles in the attraction and/or repulsion of rhizosphere populations.[12] The triggering of haustorium development by multiple allelochemicals suggests that there is a redundancy of signaling factors acting in the rhizosphere. This hypothesis is supported by our observation that inbred lines of *T. pusilla* selected for the lack of haustorium formation after exposure to a specific haustorium inducing molecule still form haustoria when exposed to complex root exudates.[10] In addition, there are multiple recognition alleles in the parasite receptor or receptors for host signals. The redundancy in host signal molecules is also consistent with the inability to identify host (*Arabidopsis*) mutants that do not induce haustoria in *Triphysaria* (unpublished results). An alternative possibility for the lack of *Arabidopsis* mutants that do not induce haustoria in *Triphysaria* is that haustoria inducing factors are critical to plant survival and mutations in such genes are lethal. We consider this less likely because mutations in phenylpropanoid biosynthesis that affect the production of certain haustorial inducing flavonoids are not lethal and would be recovered in these screens if only one class of inducer was active.

Early morphological events in haustorium development have been detailed for both obligate and facultative parasites. Haustorium development in *Triphysaria* is similar to that previously described for *Agalinis*.[13] The first response to haustorium inducing conditions is the rapid cessation of root elongation. Haustorial hairs begin to proliferate in the root elongation zone within five hours of contact with host root factors. Concomitantly, cortical cells underlying the haustorial hairs begin to expand and by twelve hours a hairy, swollen knob appears distal to the root tip. In the presence of a host root haustorial hairs attach firmly to the host root and haustorial intrusion cells penetrate host tissues via a combination of enzymatic activity and physical pressure.[14] In the absence of host root contact, the swelling and hair proliferation continue for about 24 hours, at which time the *Triphysaria* roots revert to their normal growth program. Haustorium development is highly synchronous; when several *Triphysaria* are simultaneously exposed to host factors, haustoria develop at the same location distal to the root tip. A time lapse animation of haustorium development can be seen at http://www.plantsciences.ucdavis.edu/yoder/lab/.

2. Identification of Haustorium Gene Candidates

We are interested in identifying genes that are selectively regulated in parasite roots during haustorium development with the goal of using these as gene targets for the engineering of resistant crop plants. Our ongoing strategy is to apply different haustorium-inducing treatments to *Triphysaria* roots and isolate RNA at different stages of haustorium development. The RNAs are reverse transcribed into cDNAs that are then enriched for regulated transcripts by suppressive subtractive hybridization.[15] The enriched cDNA pools are cloned, sequenced and characterized by bioinformatics to identify treatment specific transcripts.[16] cDNA arrays are also used to identify transcripts that are consistently responsive to particular treatments.[17] Using both sets of information we are selecting haustorium gene candidates that we hypothesize are necessary for haustorium development. These are being transformed into *Triphysaria* roots using *Agrobacterium*-based vectors designed to silence candidate gene expression.[18,19]

A "full-length" cDNA library was made from a pool of polyA RNAs isolated from *Triphysaria* roots before and after exposure to various haustorium-inducing treatments. The library was enriched for near full length transcripts using the BD Biosciences SMART technology.[20] Full length sequences are important for defining regulatory regions on haustorium genes and for determining protein function by expression in heterologous systems. They are also useful for homology comparisons among other parasitic and non-parasitic plants to identify swapped or frequently mutated domains.

The cDNAs were normalized for transcript abundance by duplex-specific nuclease normalization using the kamchatka crab nuclease to improve the frequency of novel gene recovery.[21] The library was constructed in the pDNR-LIB vector, which allows simple transfer of sequenced insertions into any Cre-Lox based acceptor vector (BD Biosciences).

A second set of libraries was made that were enriched by suppressive subtractive hybridization (SSH) for transcripts either up or down regulated by the treatments.[16,17] This PCR-based protocol includes cDNA-cDNA hybridizations in combination with suppression PCR to enrich differentially expressed transcripts and normalize transcript abundance.[15] Forward-subtracted probes were made using mRNA obtained from induced roots as target and water treated roots as driver; these probes are enriched for transcripts induced by the treatments. Reverse-subtracted probes were similarly prepared except that the target mRNA was derived from water treated roots and the driver mRNA from host treated roots; reverse libraries are enriched for down-regulated transcripts. Subtracted cDNAs were cloned into pCR8/GW/TOPO TA, which allows rapid subcloning into any Gateway compatible vector (Invitrogen, Carlsbad, CA)

Triphysaria roots were exposed to five different treatments prior to RNA isolation. One treatment was simply overlaying roots of aseptic *Arabidopsis* seedlings across those of *Triphysaria* growing in a square Petri dish. *Triphysaria* roots were dissected and frozen in liquid nitrogen at times ranging from immediately after contact until five hours later. Untreated samples were collected from *Triphysaria* roots exposed to media but not *Arabidopsis*. The library enriched for transcripts upregulated in response to host root contact is called the Host Forward (HF) library while the library enriched for transcripts downregulated following host contract is called Host Reverse (HR). A second set of forward and subtracted libraries were prepared from *Triphysaria* roots exposed to host root exudates collected from hydroponic cultures of *Arabidopsis*.[10] A third treatment used *Triphysaria* root extracts because we have observed that while *Triphysaria* root exudates do not induce haustoria, extracts of those same roots do. We are interested in comparing genes expressed after treatment with extracts to those expressed after treatment with exudates in order to identify those genes specifically associated with haustorium development.

Two additional SSH libraries were made from roots treated with chemical inducing factors. Peonidin is an anthocyanidin that induces haustorium development at concentrations between 1-1000 μM.[11] Peonidin is an antioxidant and is not phytotoxic to roots even at high concentrations.[22] In contrast, DMBQ (2,6-dimethoxybenzoquinone) is an active factor between 1-50 μM but at higher concentrations it is phytotoxic.[2,23] The comparison of transcript levels between DMBQ and peonidin treated roots is interesting because both induce haustoria while having dramatically different secondary effects.

3. The Parasitic Plant Sequence Database

To date we have sequenced approximately 40,000 ESTs from *Triphysaria* root tips treated as described above. These assemble into 7022 assemblies of at least 2 ESTs and 5656 singlets. 86% of the assemblies and 13% of the singlets have BLASTX hits in the *Arabidopsis* ATGC 08/04/2006/ database.

We submitted about 10,000 processed EST sequences in 2005 to GenBank's dbEST repository, which is accessible through the National Center for Biotechnology Information.[24] These ESTs were derived from the three libraries shown in Table 1. Proteins predicted to be encoded by the assemblies were annotated from the BLASTX reports comparing *Triphysaria* sequences to all proteins in GenBank or to all predicted proteins in *Arabidopsis* (ATH1.pep_cm_20040228). The BLAST reports, EST sequences and assemblies for individual libraries can be obtained from the Pscroph database (http://pscroph.ucdavis.edu/). The data are stored in a MySQL database that is available over the web using a phpMyAdmin interface on a server housed in the Plant Sciences Department at the University of California-Davis. BLAST reports can be accessed at the web site as full text files or by keyword searches of protein annotations. The keyword search function reports the best three hits obtained from either GenBank or the TAIR plant databases with e values $\leq 10^{-8}$. Each best hit is hyperlinked to the corresponding report page at NCBI or TAIR. The website also provides a BLAST function that allows homology searches against DNA or protein sequences in each or all libraries.

Table 1. Homologs to parasite genes are present in *Arabidopsis*.

Library treatment	Total ESTs	Unique transcripts	% with *Arabidopsis* homologs
Host forward	3386	1074	82%
Host reverse	3428	1344	80%
DMBQ	2216	1402	85%

Arabidopsis homologs were identified with BLASTX searches of *Arabidopsis* predicted proteins at an e value of less than 10^{-8}. The total number of transcripts shown in the table is higher than the number figured in the text because different libraries have overlapping sequences.

Genes that reside in large gene families can be distinguished using sequence specific probes. Such sequences are typically located in the untranslated regions of genes. We mapped the virtual translations of the SSH ESTs onto the most homologous protein in the plant protein database in order to determine the distribution of SSH products relative to the 3' and 5' ends of the encoding gene. With the length of the target ORF and the amino acid locations corresponding to the start and stop of the aligned region between the SSH and plant homologs, we estimated the number and length of non coding *Triphysaria* sequences. Depending on the library, from 34% to 62% of the *Triphysaria* sequences were predicted to include non-coding sequences; one to ten percent of the cDNAs included both 5' and 3' non-coding regions.[16] These regions provide good candidate sites for identifying gene-specific primers.

We used BLASTN to identify nucleotide sequences in common between the different libraries (PyMood Software, Allometra.com). About seventy percent of the sequences were specific to a single library. Seven percent of the assemblies were found in both Host Forward (HF) and DMBQ libraries but not Host Reverse (HR) library; these represent likely candidates for early haustorium development.[16]

BLASTX was used to assign putative protein functions to each library-specific assembly. Roughly 80% of the library specific sequences had homologies in the *Arabidopsis* protein database at an e value $\leq 10^{-8}$ (Table 1). The corresponding Gene Ontology (GO) annotations for each best hit were obtained through functions available at The *Arabidopsis* Information Resource.[25] GO annotations provide a uniform vocabulary to describe the roles of genes and gene products in all organisms,[26] and allowed us to categorize the putative functions of each translation product into one of nine general biological processes. The number of transcripts in each category of different libraries provided a way to determine which biological functions were over- or under-represented by different treatments. Three classes of transcripts were significantly ($p < 0.01$) more abundant in the HF than HR libraries; those involved in electron transport, those involved in stress responses, and those involved in cellular transport (Table 2). As previously observed, many of the transcripts in these classes putatatively function in xenobiotic detoxification and/or protection from reactive oxygen species.[17] The enrichment for transcripts putatatively associated with stress responses is consistent with the hypothesis that parasitic plants recruit defense related genes for host recognition.[17,27]

Table 2. Putative biological functions of parasite transcripts after contact with host roots.

Putative functions	Host forward	Host reverse	P
Total # transcripts	702	910	
electron transport	93	64	$p < 0.001$
response to stress	52	34	$p < 0.001$
other transport	157	153	$p < 0.01$
DNA or RNA metabolism	28	44	NS
cell organization and biogenesis	44	48	NS
protein metabolism	174	220	NS
signal transduction	28	35	NS
transcription	43	48	NS
response to abiotic or biotic stimulus	58	47	NS

Virtual translations of sequences from the host forward or host reverse libraries were grouped into functional classes using Gene Ontogeny descriptions. The total number of transcripts sequenced from each library is shown in the top line. Chi square was used to determine whether certain pathways were over or under represented in the total number of sequences.[16]

4. Identification of Haustorium Gene Candidates via cDNA Arrays

Hybridization to cDNA arrays can be used to identify genes transcriptionally regulated during development. A subset of these genes will be those critical for parasite success. Two criteria were used to select cDNAs for inclusion on the arrays. The first was differential expression in colony hybridizations. Seven thousand SSH colonies were picked and arrayed on nylon membranes. Membranes were then probed with forward and reverse subtracted probes from *Triphysaria* treated with host roots, host exudate, *Triphysaria* extract, peonidin and DMBQ. We also probed with tip specific SSH products (eleven probes total). About 1400 colonies produced differentially abundant spots with one or more probes. We selected 364 clones for further analysis based on the cDNAs being coordinately regulated by two or more treatments.

The second criterion for inclusion in the cDNA arrays were the annotations of putative functions. Annotations led us to select 2200 transcripts potentially associated with signal transduction, response to biotic stimulus, response to stress, quinine oxidoreduction, auxin transport, and disease resistance. cDNAs were amplified from 2564 SSH clones and printed in quadruplet on nylon membranes at a density of 43 spots per cm^2 using a Vicki Ultrahigh Density Array and Registration System. Arrays were probed with ^{32}P labeled cDNA from *Triphysaria* roots treated with either *Arabidopsis* roots or DMBQ for 30 min, 2h, 5h and 24h. Roots were also collected from untreated or mock treated *Triphysaria*. There were three biological replications for each time point. Images were quantified on a STORM scanner and analyzed using Phoretix Array software. The intensity of each spot was quantified and local background levels subtracted. Spots with thresholds two times over background were taken for further analysis. Each spot was then normalized to a set of six standard cDNAs that were previous shown by northern analyses to be unregulated. Approximately 300 genes were identified as being co-regulated by both *Arabidopsis* contact and DMBQ (Tomilov, Tomilova and Yoder, unpublished). Thirty candidate genes were used as northern probes to validate the arrays and 75% were consistent with the array results. A shortened list of representative candidate genes is shown in Table 3.

Table 3. Parasite gene candidates from cDNA arrays and sequence annotation.

Contig #	Annotation via NCBI NR	Putative functions
EDIT_011 HF_0184	pirin[17,28]	Induction is primary response in *Triphysaria* after treatment; interacts with G protein subunit to regulate development
EDIT_0330 HF_0231	glutathione transferase[29]	Binds flavonoids in the cytosol prior to deposition in vacuole
HF_0762	isoflavone reductase[30]	Secreted by border cells, metabolize flavonoids, associated with plant-rhizosphere interactions
HF_1112	Avr9/Cf-9 elicited protein[31]	Part of a receptor complex that recognizes the fungal Avr elicitor
EDIT_0030	calmodulin[32]	Calcium signaling and mechano-sensing in roots, involved in rhizosphere symbioses
EDIT_000X	TvQR1[33]	Induction is primary response in *Triphysaria* after treatment; encodes quinone oxidoreductase, a putative haustorium signal

5. Conclusions

The product of the work described here is a database containing sequences of RNA transcripts produced by parasite roots as they perceive and attack host plants. Comparative analyses between the cDNA libraries represented in these databases with other plant transcriptomes will define those genes and pathways that distinguish parasitic from non-parasitic plants. The role of these genes in parasitism will be determined by transforming vectors designed to silence gene expression into parasite roots (Chapter 14 and Ref. 18). The database also contains sequences of genes that are not parasite specific but essential for survival; these are targets for engineering *Striga* resistance. One approach to inactivating critical genes in parasites is to use the database to design gene silencing vectors that are specific to parasites and then transform the silencing vector into crop plants. If the silencing construction is parasite specific, there will be no effect of the transgene on the crop. However if the transgenic plant is infected with *Striga*, the silencing vector will be translocated across the haustoria connecting the host and parasitic partners, leading to ultimate death of the parasite. The sequence databases described here provide some of the requisite information needed to exploit biotechnological solutions for *Striga* control.

Acknowledgements

This work was supported by NSF grant #0236545. The authors gratefully acknowledge the excellent assistance and technical advice of Alex Kozik and Steve Edberg.

References

1. M. C. Press and J. D. Graves, Eds., *Parasitic Plants*, (Chapman and Hall, London, 1995).
2. E. M. Estabrook and J. I. Yoder, *Plant Physiol.* **116**, 1 (1998).
3. D. Nickrent, *Parasitic Plant Connection*, (Southern Illinois University at Carbondale, 2006), http://www.parasiticplants.siu.edu/
4. J. L. Riopel and M. P. Timko, in *Parasitic Plants*, Eds. Press, M. C. and Graves, J. D. (Chapman and Hall, London, 1995), p. 39.

5. C. W. dePamphilis, N. N. Young and A. D. Wolfe, *Proc. Natl. Acad. Sci. USA* **93**, 7367 (1997).
6. J. C. Hickman, *The Jepson Manual; Higher Plants of California,* (University of California Press, Berkeley, CA, 1993).
7. L. D. Thurman Ph.D. Thesis, University of California, Berkeley, 1966.
8. Y. Goldwasser, J. H. Westwood and J. I. Yoder, in *The Arabidopsis Book*, Eds. Somerville, C. and Meyerowitz, E. (American Society Plant Biologists, Rockville, MD, 2002).
9. J. I. Yoder, *Planta* **202**, 407 (1997).
10. D. S. Jamison and J. I. Yoder, *Plant Physiol.* **125**, 1870 (2001).
11. H. Albrecht, J. I. Yoder and D. A. Phillips, *Plant Physiol.* **119**, 585 (1999).
12. J. O. Siqueira, M. G. Nair, R. Hammerschmidt and G. R. Safir, *Crit Rev Plant Sci.* **10**, 63 (1991).
13. W. V. Baird and J. L. Riopel, *Am. J. Bot.* **71**, 803 (1984).
14. D. Losner-Goshen, V. H. Portnoy, A. M. Mayer and D. M. Joel, *Ann. Bot.* **81**, 319 (1998).
15. L. Diatchenko, Y. F. Lau, A. P. Campbell, A. Chenchik, F. Moqadam, B. Huang, S. Lukyanov, K. Lukyanov, N. Gurskaya, E. D. Sverdlov and P. D. Siebert, *Proc. Natl. Acad. Sci. USA* **93**, 6025 (1996).
16. M. J. Torres, A. A. Tomilov, N. Tomilova, R. L. Reagan and J. I. Yoder, *BMC Plant Biol.* **24**, 24 (2005).
17. M. Matvienko, M. J. Torres and J. I. Yoder, *Plant Physiol.* **127**, 272 (2001).
18. A. Tomilov, N. Tomilova and J. I. Yoder, *Planta* in press, (2007).
19. M. D. Curtis and U. Grossniklaus, *Plant Physiol.* **133**, 462 (2003).
20. Clontech, *Clontechniques* **13**, 12 (1998).
21. P. A. Zhulidov, E. A. Bogdanova, A. S. Shcheglov, L. L. Vagner, G. L. Khaspekov, V. B. Kozhemyako, M. V. Matz, E. Meleshkevitch, L. L. Moroz, S. A. Lukyanov and D. A. Shagin, *Nucl. Acids Res.* **32**, e37 (2004).
22. M. P. Kahkonen and M. Heinonen, *J. Agric. Food Chem.* **51**, 628 (2003).
23. M. Chang and D. G. Lynn, *J. Chem. Ecol.* **12**, 561 (1986).
24. NCBI, *National Center for Biotechnology Information*, (National Library of Medicine and National Institite of Health, 2006), www.ncbi.nlm.nih.gov
25. TAIR, *Gene Ontology at TAIR: The Arabidopsis Information Resource*, (NCGRI and Carnegie Institute of Washington, 2005), http://www.arabidopsis.org/tools/bulk/go/index.jsp
26. M. Ashbuner, C. A. Ball, J. A. Blake, D. Botstein, H. Butler, J. M. Cherry, A. P. Davis, K. Dolinski, S. S. Dwight, J. T. Eppig, M. A. Harris, D. P. Hill, L. Issel-Tarver, A. Kasarskis, S. Lewis, J. C. Matese, J. E. Richardson, M. Ringwald, G. M. Rubin and G. Sherlock, *Nat. Genet.* **25**, 25 (2000).
27. P. R. Atsatt, *Amer. Nat.* **107**, 502 (1973).
28. Y. R. Lapik and L. S. Kaufman, *Plant Cell* **15**, 1578 (2003).
29. D. Dixon, A. Lapthorn and R. Edwards, *Genome Biology*, **3** 1 (2002).
30. Q. Wu and H. VanEtten, *Mol Plant Microbe Interact*, **17** 798 (2004).

31. A. Heese, A. A. Ludwig and J. D. G. Jones, *Plant Physiol.* **138**, 2406 (2005).
32. L. Sanchez, S. Weidmann, C. Arnould, A. R. Bernard, S. Gianinazzi and V. Gianinazzi-Pearson, *Plant Physiol.* **139**, 1065 (2005).
33. M. Matvienko, A. Wojtowicz, R. Wrobel, D. Jamison, Y. Goldwasser and J. I. Yoder, *Plant J.* **25**, 375 (2001).

CHAPTER 4

GERMINATION OF *STRIGA* AND CHEMICAL SIGNALING INVOLVED: A TARGET FOR CONTROL METHODS

Zhongkui Sun[a,b], Radoslava Matusova[b,c] and Harro Bouwmeester[b,c,*]

[a]*Laboratory for Molecular Biology, Wageningen University, Dreijenlaan 3, 6703 HA Wageningen, The Netherlands*
[b]*Plant Research International, P.O. Box 16, 6700 AA Wageningen, The Netherlands*
[c]*Laboratory for Plant Physiology, Wageningen University, Arboretumlaan 4, 6703 BD Wageningen, The Netherlands*
[*]*E-mail: harro.bouwmeester@wur.nl*

Although there are some promising control methods of *Striga* there is a continuous need for new approaches to stay ahead of the parasite. The life cycle of *Striga* could be a suitable target for new control methods and particularly the chemical signals involved in its regulation. We focus on the germination stimulants, signaling molecules that are secreted by the host plant and regulate the first step in the *Striga* life cycle. We use information on the biosynthetic origin, distribution and sensitivity of seeds to germination stimulants, host specificity, and ecological significance, to describe a number of potential *Striga* control strategies. Some have been suggested before: suicidal germination agents, trap and catch crops and breeding for low germination stimulant formation. Some are proposed for the first time: molecular breeding to reduce germination stimulant formation, use of carotenoid biosynthesis inhibitors, use of knowledge about dormancy, use of arbuscular mycorrhizal fungi and phosphate, use of host specificity knowledge.

1. Introduction

Many research groups around the world continue to investigate the biology of *Striga* in the hope of finding the ultimate solution to this problem. One important aspect in the biology of *Striga* and other parasitic plants that may provide options to develop control methods is their requirement for a signaling molecule indicating the presence of a suitable host (Fig. 1). These molecules can induce the germination of the seeds of these parasites and are hence called germination stimulants. In this Chapter we will review the knowledge about the chemical signaling between *Striga* and its hosts with special focus on the germination stimulants and discuss the possibilities to explore the knowledge about their role in the biology of *Striga* for the development of new control methods.

2. Life Cycle and Chemical Signaling Between *Striga* and its Host

During the life cycle of *Striga* several steps have been shown or suggested to be regulated or affected by signaling molecules that are exchanged between *Striga* and its host (Fig. 1).[1-5] The chemical stimuli that are initiating the life cycle are called germination stimulants. They are secreted by the host root and trigger the germination of *Striga*.[2,6] Before *Striga* seeds can respond to these stimuli they require a pre-

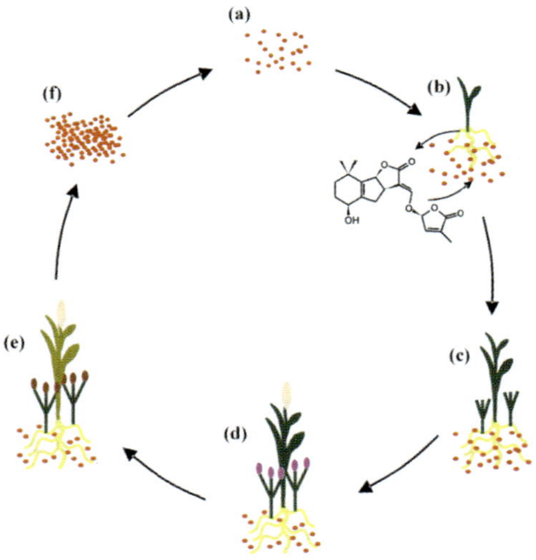

Figure 1. Life cycle of Striga *spp. (a) the seeds are buried in the soil; (b) become sensitive to germination stimulants exuded by host roots and may germinate; (c) form a haustorium and attach to the host root establishing a xylem connection, and the parasite then emerges; (d) parasitic plants flower; (e) produce seeds that enter the soil seed bank (f); in the next season the cycle starts again (a). Redrawn from Matsova and Bouwmeester[27] with permission of Springer Science and Business Media.*

treatment at a suitable temperature under moist conditions, a treatment known as preconditioning or warm stratification.[7,8] After preconditioning, the *Striga* seeds will germinate only if exposed to sufficiently high concentrations of germination stimulants hence assuring that germination only occurs in close vicinity of the host root. When the *Striga* seeds have germinated, the radicle must grow towards the host root. This is a process possibly directed by the concentration gradient of the germination stimulants.[9] The radicles attach to and penetrate the host root (see Chapter 2). After penetration the parasite forms a shoot, emerges above the soil, flowers and produces seeds after which the life cycle can start again (Fig. 1).

3. Germination Stimulants of *Striga* spp

3.1. *Origin and Distribution*

A number of different classes of secondary metabolites have been described to have germination stimulant activity: the benzoquinones (e.g. dihydrosorgoleone) (see Chapter 2), the strigolactones, the sesquiterpene lactones and an isoflavanone.[2,10-12] The best explored class of secondary metabolites with germination stimulant activity are the strigolactones. Up to now about eight natural strigolactone germination stimulants have been reported. The first germination stimulant strigol was isolated from the non-host cotton (Fig. 2).[13] Later, strigol was also detected in the exudates of the *Striga* hosts maize, proso millet, and sorghum.[14] Alectrol was identified in cowpea[15] and together with orobanchol also

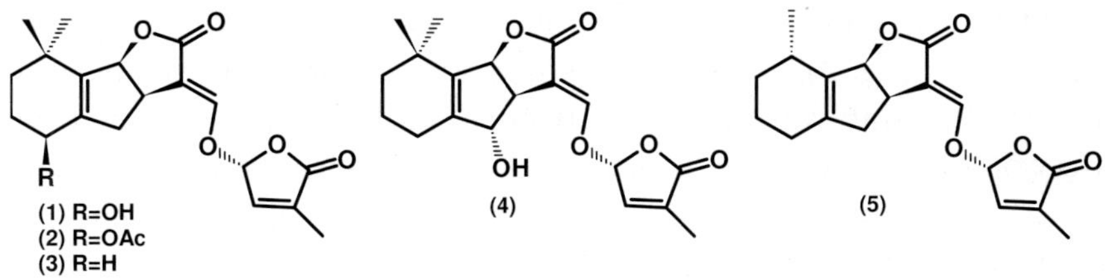

Figure 2. Structure of strigolactones: strigol (1), strigyl acetate (2), 5-deoxystrigol (3), orobanchol (4), sorgolactone (5). Structures of other strigolactones have not yet been unambiguously assigned.

isolated from the *Orobanche* host, red clover.[16] The strigol analogue sorgolactone was isolated from sorghum[17] and recently an isomer of strigol, named sorghumol, was detected in sorghum cultivars.[18] 5-Deoxystrigol, which was first isolated from *Lotus japonicus* root exudates as a branching factor for arbuscular mycorrhizal (AM) fungi[19] (see below) was later reported to be the major strigolactone present in maize, millet, and sorghum.[18] Surprisingly, sorgolactone was not present in the sorghum cultivars examined by these authors. Recently, several known but also new strigolactones were detected in the exudates of tomato, tobacco, spinach and white lupin by Yoneyama and coworkers[20-22] suggesting that the strigolactones are structurally diverse and are produced by many different plant species.

3.2. *Detection of Germination Stimulants by Germination Bioassay*

Plants are usually grown in medium that can be readily removed from the roots for easy collection of germination stimulants. Therefore, hydroponics, sand, perlite and vermiculite are good choices. The root exudates are collected in water or in a nutrient solution for several up to 24 hours. The exudates are then applied to preconditioned *Striga* seeds and two days later germination can be evaluated.[23] In addition to the use of root exudates, methods have been described using "live" cut roots.[24] Considering the reported large effect of ethylene on germination of *Striga*[25] and the risk that ethylene is released from decaying or wounded plant material we have tested whether this method is not generating too much ethylene. The commercial ethylene adsorbent Ethysorb considerably reduced the germination of *Striga* seeds in the presence of "live" sorghum roots (but not in the presence of GR24) in our experiments (Fig. 3). This suggests that at least part (but perhaps even all if the Ethysorb was not 100% effective) of the germination inducing effect of these "live" roots is due to ethylene and we therefore strongly advise against the use of this method for the selection of low germination stimulant producing germplasm.

3.3. *Perception of Germination Stimulants*

The availability of the synthetic germination stimulant GR24 sometimes has obscured the relevance of dormancy in parasitic plant seeds. In many cases, a standard preconditioning treatment to break dormancy has been combined with micromolar doses of GR24. However, if lower concentrations (~30 nM) of GR24 or natural germination stimulants are used, it becomes suddenly clear that preconditioning strongly affects the responsiveness of the seeds to the stimulants.[8,26] Preconditioning at an optimal temperature (e.g., about 30°C for *S. hermonthica*) releases dormancy within 2-3 weeks and increases the sensitivity to GR24 by several orders of magnitude. After reaching maximum sensitivity, prolonged preconditioning induces secondary dormancy, i.e., decreases sensitivity to GR24.[8] It is important to note that the rapid changes in sensitivity during prolonged preconditioning are only (or particularly) visible at low concentrations of GR24 (~30 nM). At a concentration of 3 μM or higher, GR24 almost always induces high germination, regardless of the preconditioning period. The changes in sensitivity to germination stimulants are suggestive of a safety mechanism that ensures that seeds can only respond to the germination stimulants produced by their host during a restricted period of the year. This is of ecological importance for the parasite as it requires a long enough period of time to grow and reproduce. Germination during the later stages of host development would not allow this. This ecological safety mechanism may however also have practical significance for *Striga* control as we will discuss below.

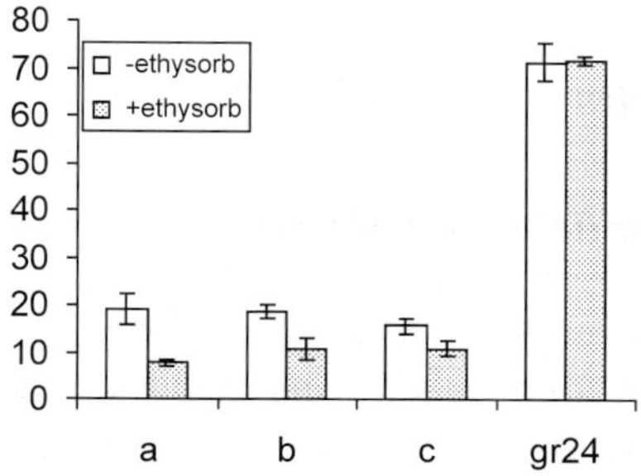

Figure 3. The ethylene adsorbent Ethysorb reduces the germination of Striga *seeds induced by cut sorghum roots (a, b, c) but not germination induced by3 μM GR24.*

3.4. *Implication for host specifity*

The survival of the parasitic *Striga* spp. fully depends on their ability to detect the presence of a host plant. Therefore the parasitic plants have evolved a mechanism to recognize host exuded chemical signals, ensuring that the roots of the host are in close vicinity. However, even though the recognition mechanism at the germination stage is a most crucial point of no return, it seems that it is not fully specific. The induction of germination of parasitic plant seeds by non-host plants is the most obvious example of a lack of specific recognition. Also the fact that germination of *Striga* and *Orobanche* seeds can be induced by the synthetic germination stimulant GR24, regardless of the parasitic plant species or its host history, is not suggestive of a strong host specificity during the germination phase. Conversely, there are several indications that the composition of root exudates does play a role in determining host specificity during the germination phase. For example, *Striga* seeds collected from maize and sorghum responded differentially to the germination stimulants present in the root exudates of maize (host), cowpea (nonhost) and GR24.[27] The novel evidence for the presence of several different strigolactones in root exudates of different host species e.g. sorghum and tomato or even varieties[18,22] and conversely the presence of the same strigolactones (e.g. strigol, 5-deoxystrigol, orobanchol) in the exudates of several different host species[14,18] may help (but also make it more difficult) to unravel the recognition of the germination stimulants by parasitic plants and the mechanism of host plant selectivity at the germination stage.

4. The Strigolactones

4.1. *Ecological Significance of Strigolactones*

A puzzling question that was asked when the strigolactone germination stimulants were first discovered was: why do plants produce these signaling molecules while they induce germination of one of their worst enemies? Akiyama and coworkers showed that the strigolactones are required by AM fungi for their host root colonization process.[19,28] One of the primary roles of AM fungi in the symbiotic relationship with plants is the delivery of mineral nutrients, and particularly phosphate.[29] The availability of phosphate is limiting plant growth in many areas of the world, not the least in the African continent. AM fungi can help to improve the uptake of phosphate and hence improve agricultural production in these areas.[30,31] In agreement with their role in the uptake of phosphate, it was shown that root exudates produced by phosphate-limited plants are more stimulatory to AM fungi.[32] Indeed, low phosphate conditions also stimulate the exudation of the strigolactone orobanchol by red clover[33] and we have shown that this is also true for hosts of *Striga* spp (Sun, Charnikhova, Bécard and Bouwmeester, unpublished results). This could well explain the dramatic increase of the *Striga* problem in areas with limited phosphate availability.[3] Interestingly, several groups have reported that colonization by AM fungi can reduce the infection of sorghum and maize by *Striga*.[34-36] Experiments with exudates of plants colonized by AM fungi show that this effect is, at least partly, caused by a down-regulation of strigolactone formation after colonization by AM fungi.[36,37]

4.2. *Biosynthetic Origin of the Strigolactones*

Strigolactones are exuded from the roots of host plants in extremely low concentrations and are often unstable which makes the isolation and characterization of these compounds quite complicated,[38] and the study of the biosynthesis of these compounds has been difficult. However, by using the germination of *Striga* seeds as the most sensitive assay available for the detection of strigolactones, we could analyze the production of germination stimulants by single plants.[23] Using this system we analyzed germination stimulant production in plants treated with specific inhibitors of isoprenoid biosynthetic pathways and in maize mutants. Our results showed that the germination stimulant(s) of *Striga* exuded by the roots of maize, cowpea and sorghum is (are) derived from the carotenoid pathway. The exact position in the carotenoid pathway where strigolactone biosynthesis branches off from the main pathway has not yet been identified, but it is clear that carotenoid cleavage must be involved in germination stimulant biosynthesis. Carotenoid cleavage commonly occurs in a number of biosynthetic pathways, for example in the production of other plant signaling molecules such as the plant hormone abscisic acid.[39] We have also postulated how, after carotenoid cleavage, further enzymatic conversions may lead to the production of all strigolactones known to date[23] and we are currently further characterizing this pathway.

5. Control Methods Using Knowledge of Germination Stimulants

5.1. *Control Through Enhanced Germination*

5.1.1. *Suicidal Germination Using Chemicals*

There has been a great interest in using the germination stimulants as a method for control of *Striga* in agricultural fields. Indeed, the introduction of a germinating agent before a crop is planted could potentially reduce *Striga* populations via suicidal germination (Fig. 4).[40,41] Work on synthetic germination stimulants in the group of B. Zwanenburg has led to the development of molecules that have potential as parasitic weed control agents through the induction of suicidal germination.[42,43] Limitations of this approach are that the synthetic stimulants should be inexpensive enough for farmers in the Developing World to be able to buy them. Also, the application of the chemicals to sufficient depth in the soil requires suitable equipment and possibly large amounts of water.

5.1.2. *Trap and Catch Crops*

Another control strategy, based on suicidal germination stimulants is the use of trap and catch crops in monoculture or in intercropping (Fig. 4). The crops used for this strategy produce germination stimulants, sometimes in high amounts, and hence induce massive germination of the parasite, but are resistant in a later stage of the parasite's life cycle (trap crops) or harvested before the seeds of the parasite are shed (catch crops).[44] The effectivity of catch and trap crops could possibly be increased if overproduction of germination stimulants can be achieved through selection or molecular breeding. The latter can be achieved by over expression of one or more (rate-limiting) genes from the strigolactone biosynthetic pathway (see below under 5.4). Over expression of strigolactone formation could possibly also improve colonization by AM fungi and hence benefit the trap/catch crop. The use of intercropping and rotation, particularly with legumes is also otherwise advantageous, because it improves soil fertility.

5.2. *Control Through Reduced Germination*

5.2.1. *Using Chemicals*

The results with the application of fluridone to maize, cowpea, and sorghum in our laboratory experiments (see above)[23] have inspired us to look at the possibility of using carotenoid biosynthesis inhibitors to reduce infection with parasitic plants *in situ* (Fig. 4). We found that treating rice with low doses of fluridone significantly reduced the number of germinated/attached Striga seeds even at very low concentrations of 0.001 to 0.1 μM (Sun, Bouwmeester *et al.*, unpublished results). Leaf bleaching did not occur at these low doses. These results clearly demonstrate that the unidentified rice germination stimulants are

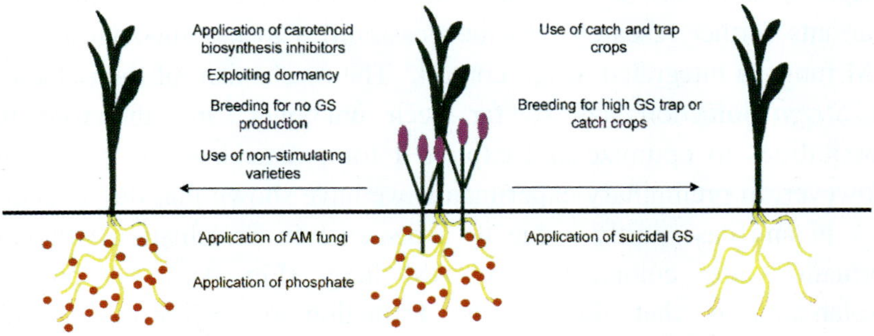

Figure 4. Diagram showing possible control methods based on the knowledge about germination stimulants.

also strigolactones. Our results show that herbicides that inhibit carotenoid biosynthesis can be used to significantly reduce the germination of parasitic seeds and that treating plants with low concentrations of such herbicides at one or more time intervals may be an effective and cheap method to reduce parasitic-weed induced yield losses of crop plants.

5.2.2. *Dormancy*

As described above, the dormancy of parasitic plant seeds is released during preconditioning and induced again upon prolonged preconditioning, and possibly this phenomenon can be used to control parasitic weeds. Indeed, there are several publications showing that a later crop sowing date strongly reduces infection by parasitic plants, for example of sunflower by *Orobanche cumana*[45] and of sorghum by *S. hermonthica*.[46] Although there is no direct proof that this is due to the re-induction of dormancy (= a decrease in sensitivity to germination stimulants), it seems worthwhile to investigate whether this plays a role in the positive effect of delayed sowing or transplanting and whether it could be developed into a control strategy, if shortage of water does not preclude the use of these strategies.

5.2.3. *AM Fungi and Phosphate*

The fact that several groups have reported that AM fungi can reduce *Striga* infection of sorghum and maize in pot and field experiments[34,35] warrants further research into the possibilities to use inoculation with AM fungi in integrated *Striga* control. The mechanism of the reduction in *Striga* infection has so far been unknown, and therefore the possibilities to optimize and exploit it for practical use were limited. However, in preliminary experiments, we have shown that the reduction is - in any case partly - due to a decrease in germination stimulant formation after colonization by AM fungi (Fig. 4).[36] A possible explanation is that due to the formation of mycorrhiza-specific apocarotenoids,[47] the formation of the *Striga* germination stimulant is reduced. Alternatively, colonization by AM fungi may directly down-regulate the strigolactone production pathway. Research could now be aimed at optimizing the use of AM fungi for controlling parasitic plants through reduced germination by selecting suitable AM fungus strain – host (variety) combinations. A factor that seems to be tightly linked to the effects of AM fungi on *Striga* infection is the effect of phosphate on the secretion of strigolactones. This definitely warrants more research on the possible direct positive effects of phosphate fertilization on *Striga* control.

5.3. *Control Using Host Specificity*

There is ample evidence that the composition of the mixture of germination stimulants that is exuded may vary between different crop species as well as between varieties of one crop species.[18] Detailed knowledge about the germination stimulant composition in the exudate of a crop variety to be sown and the effect of this on germination of a certain field population of *Striga* may help to choose or design (through breeding) a crop (variety) or combination of crops (varieties), for example in rotation, with the aim of reducing germination of *Striga* seeds from the local seed bank or to exhaust the seed bank as quickly as possible. Choosing the right varieties of one crop based on this knowledge may be useful especially in areas where a broad rotation of different crops is not possible or not attractive.

5.4. *Control Using Breeding*

In sorghum a selection program for low-germination stimulant formation has resulted in low-stimulant sorghum varieties with improved resistance to *Striga* (Chapter 7).[48,49] Once the strigolactone biosynthesis pathway has been elucidated, it may become feasible to make low-stimulant producing plants through the inactivation of one or more steps in the pathway. For the time being, enzymes of the primary carotenoid pathway could be suitable targets but preferably then these knock-outs should be done in an organ-specific and/or development-specific manner. In this way the inhibition of carotenoid/ABA biosynthesis is restricted to time and place necessary to obtain resistance against parasitic plants and this possibly also avoids a side-effect on colonization by AM fungi. Better targets would be the dedicated pathway enzymes, i.e. the postulated enzymes involved in cleavage and further conversion of the cleavage product to the strigolactones.[23] As an alternative to knocking out enzymes of the germination stimulant pathway, over expression of key-enzymes of competing pathways to channel away substrate can also be considered as a strategy to reduce germination stimulant formation. A possible candidate is the cleavage enzyme that is responsible for apocarotenoid formation upon colonization by AM fungi (Sun, Bouwmeester, Walter *et al.*, unpublished results).

6. Conclusions

Knowledge about the identity, biological function, and physiological and biochemical regulation of the germination stimulants has rapidly grown over the past five years. We have summarized this information and have used it to describe a number of potential control strategies, some of which have been suggested before and some of which are proposed for the first time (Fig. 4). Clearly the *Striga* problem is too big to be tackled by just one approach, but we sincerely believe that targeting germination, preferably in combination with other approaches such as, for example, post-germination resistance, herbicide seed dressing, and biological control - one day could lead to alleviation of the *Striga* problem.

Acknowledgements

We acknowledge funding by the Netherlands Foundation for the Advancement of Tropical Research (WOTRO) (to SZ); the Netherlands Organisation for Scientific Research (NWO) (Vici-fellowship to HJB and RM, and NATO visiting-scientist fellowships to RM); the European Commission (the FP5 EU project Improved Striga Control in Maize and Sorghum; INCO-DEV, ICA4-CT-2000-30012); the Dutch Ministry of Agriculture, Nature Management and Fisheries' North-South programme (to HJB) and an IAC fellowship (to RM). We wish to thank Francel Verstappen for his help with some of the work described here.

References

1. L. G. Butler, in: *Insights into Allelopathy,* Ed. K. Inderjit and F. A. Einhellig (ACS Books, Washington, 1995), p. 158.
2. H. J. Bouwmeester, R. Matusova, S. Zhongkui and M. H. Beale, *Curr. Opin. Plant Biol.* 6, 358 (2003).
3. H. J. Bouwmeester, C. Roux, J. A. Lopez-Raez and G. Bécard, *Trends Plant Sci.* in press, (2007).
4. H. Shen, W. Ye, L. Hong, H. Huang, Z. Wang, X. Deng, Q. Yang and Z. Xu, *Plant Biol.* 8, 175 (2006).
5. J. I. Yoder, *Curr. Opin. Plant Biol.* 4, 359 (2001).
6. M. C. Press and J. D. Graves, *Parasitic plants*, (Chapman and Hall, London, UK, 1995).
7. D. M. Joel, J. C. Steffens and D. M. Matthews, in: *Seed develpment and germination.,* Ed. J. Kigel and G. Galili (Marcel Dekker, Inc., New York, USA, 1995), p. 567.
8. R. Matusova, T. van Mourik and H. J. Bouwmeester, *Seed Sci. Res.* 14, 335 (2004).
9. M. M. P. Dube and A. A. Olivier, *Can. J. Bot.* 79, 1225 (2001).
10. N. H. Fischer, J. D. Weidenhamer and J. M. Bradow, *Phytochemistry* 28, 2315 (1989).
11. W. J. Keyes, J. V. Taylor, R. P. Apkarian and D. G. Lynn, *Plant Physiol.* 127, 1508 (2001).
12. M. K. Tsanuo, A. Hassanali, A. M. Hooper, Z. Khan, F. Kaberia, J. A. Pickett and L. J. Wadhams, *Phytochemistry* 64, 265 (2003).
13. C. E. Cook, L. P. Whichard, M. E. Wall, G. H. Egley, P. Coggon, P. A. Luhan and A. T. McPhail, *J. Am. Chem. Soc.* 94, 6198 (1972).
14. B. A. Siame, Y. Weerasuriya, K. Wood, G. Ejeta and L. G. Butler, *J. Agric. Food Chem.* 41, 1486 (1993).
15. S. Muller, C. Hauck and H. Schildknecht, *J. Plant Growth Regul.* 11, 77 (1992).
16. T. Yokota, H. Sakal, K. Okuno, K. Yoneyama and Y. Takeuchi, *Phytochemistry* 49, 1967 (1998).
17. C. Hauck, S. Muller and H. Schildknecht, *J. Plant Physiol.* 139, 474 (1992).
18. A. A. Awad, D. Sato, D. Kusumoto, H. Kamioka, Y. Takeuchi and K. Yoneyama, *Plant Growth Regul.* 48, 221 (2006).
19. K. Akiyama, K. Matsuzaki and H. Hayashi, *Nature* 435, 824 (2005).
20. X. N. Xie, H. Kamioka, D. Kusumoto, Y. Takeuchi and K. Yoneyama. in 11th IUPAC International Congress of Pesticide Chemistry, (Kobe, Japan, 2006).
21. K. Yoneyama, H. Sekimoto, Y. Takeuchi and K. Yoneyama. in 11th IUPAC International Congress of Pesticide Chemistry, (Kobe, Japan, 2006).
22. K. Yoneyama, Y. Takeuchi, D. Sato, H. Sekimoto and T. Yokoka. in 8th International Parasitic Weed Symposium, ed. D. M. Joel, (The International Parasitic Plant Society, Durban, South Africa, 2004), p. 9.
23. R. Matusova, K. Rani, F. W. Verstappen, M. C. Franssen, M. H. Beale and H. J. Bouwmeester, *Plant Physiol.* 139, 920 (2005).
24. D. K. Berner and O. A. Williams, *Plant Dis.* 82, 1242 (1998).
25. D. C. Logan and G. R. Stewart, *Plant Physiol.* 97, 1435 (1991).

26. M. J. Van Hezewijk, A. P. Van Beem, J. A. C. Verkleij and A. H. Pieterse, *Can. J. Bot.* 71, 786 (1993).
27. R. Matusova and H. J. Bouwmeester, in: *Chemical ecology: from gene to ecosystem,* Ed. M. Dicke and W. Takken (Springer, Wageningen, 2006), p. 39.
28. K. Akiyama and H. Hayashi, *Ann. Bot.* 97, 925 (2006).
29. V. Karandashov and M. Bucher, *Trends Plant Sci.* 10, 22 (2005).
30. J. F. Johansson, L. R. Paul and R. D. Finlay, *FEMS Microbiol. Ecol.* 48, 1 (2004).
31. M. Bagayoko, E. George, V. Romheld and A. Buerkert, *J. Agric. Sci.* 135, 399 (2000).
32. G. Nagahashi and D. D. Douds, *Mycol. Res.* 108, 1079 (2004).
33. K. Yoneyama, K. Yoneyama, Y. Takeuchi and H. Sekimoto, *Planta* in press, (2006).
34. N. A. Gworgwor and H. C. Weber, *Mycorrhiza* 13, 277 (2003).
35. V. W. Lendzemo, T. W. Kuyper, M. J. Kropff and A. van Ast, *Field Crops Res.* 91, 51 (2005).
36. V. W. Lendzemo, T. W. Kuyper, R. Matusova, C. H. de Vos, H. J. Bouwmeester, L. Narvaez and A. van Ast, *Plant Signal. Behavior* in press, (2006).
37. V. W. Lendzemo, PhD thesis in Department of Soil Quality, pp 112 (Wageningen University, Wageningen, 2004).
38. D. Sato, A. A. Awad, Y. Takeuchi and K. Yoneyama, *Biosci. Biotechnol. Biochem.* 69, 98 (2005).
39. M. Seo and T. Koshiba, *Trends in Plant Science. 2002;* 7, 41 (2002).
40. R. E. Eplee, *Crop Prot.* 1, 3 (1992).
41. B. Zwanenburg and J. W. J. F. Thuring, *Pure and Applied Chemistry* 69, 651 (1997).
42. A. S. Mwakaboko, PhD thesis in Organic Chemistry, pp 203 (Radboud University, Nijmegen, The Netherlands, 2003).
43. S. C. M. Wigchert and B. Zwanenburg, *J. Agric. Food Chem.* 47, 1320 (1999).
44. B. M. Chittapur, C. S. Hunshal and H. Shenoy, *Allelopath. J.* 8, 147 (2001).
45. H. Eizenberg, D. Plakhine, J. Hershenhorn, Y. Kleifeld and B. Rubin, *J. Exp. Bot.* 54, 1305 (2003).
46. G. Gbehounou, E. Adango, J. C. Hinvi and R. Nonfon, *Crop Prot.* 23, 379 (2004).
47. D. Strack and T. Fester, *New Phytol.* 172, 22 (2006).
48. G. Ejeta. in International Congress: In the Wake of the Helix: From the Green Revolution to the Gene Revolution, (Bologna, Italy, 2003), p. 239.
49. A. Mohamed, P. J. Rich, T. L. Housley and G. Ejeta. in 7th International Parasitic Weed Symposium, eds. A. Fer, P. Thalouarn, L. J. Musselman, C. Parker and J. A. C. Verkleij, (Nantes, France, 2001), p. 96.

CHAPTER 5

CHEMICALS INVOLVED IN POST-GERMINATION INHIBITION OF *STRIGA* BY *DESMODIUM*: OPPORTUNITIES FOR UTILIZING THE ASSOCIATED ALLELOPATHIC TRAITS

John A. Pickett[1,*], Zeyaur R. Khan[2], Ahmed Hassanali[2] and Antony M. Hooper[1]

[1]*Rothamsted Research, Harpenden, Hertfordshire, AL5 2JQ, UK*
[2]*The International Centre of Insect Physiology and Ecology (icipe), PO Box 30772, Nairobi, Kenya*
[*]*E-mail: john.pickett@bbsrc.ac.uk*

A new and highly effective intervention against *Striga*, particularly *S. hermonthica* in cereals, involves intercropping with the cattle fodder legumes, *Desmodium* spp. We now sufficiently understand the mechanism by which *Desmodium* suppresses *Striga*, including the identification of secondary metabolites involved, to propose ways to develop the biochemical trait involved in edible beans and possibly cereal crops themselves. This will provide greater flexibility to farmers, particularly those without livestock, for controlling *Striga* and will contribute even more to stabilising and improving cereal production in the poorest farming regions. Here we explain how *Desmodium* is used to control *Striga*, the mechanism by which it acts, as is understood so far, and our approach to developing this trait in edible beans and cereals.

1. Introduction

During investigations into the control of insect damage to maize crops in subsistence farming in Kenya, which involved intercropping with repellent plants,[1,2] the fodder legumes silverleaf *Desmodium uncinatum*, and greenleaf *D. intortum*, dramatically reduced the infestation of maize by *Striga*, specifically *S. hermonthica*.[3,4] This effect was confirmed by further field testing and was significantly greater than that observed with other legumes, e.g. cowpea, as were the concomitant yield increases.[5] The mechanism was investigated, and although soil shading and addition of nitrogen fertiliser showed some benefits against *Striga* infestation, an allelopathic mechanism was implicated. A highly significant reduction in *Striga* infestation was obtained in screenhouse studies when an aqueous solution, eluting from pots in which *Desmodium* plants were growing, was used to irrigate pots of maize planted in soil seeded with high levels of *Striga*. Growth of *Striga* was almost completely suppressed, whereas extensive infestation occurred with the control eluate. Water-soluble chemical components exuded from cleaned *Desmodium* roots contained both a germination stimulant for *Striga* as well as an allelopathic inhibitor of *Striga* attachment to the cereal roots and vascular system. The latter phenomenon was measured by the impairment of radicle elongation, although other aspects of inhibitory action may also be involved.[5] The bioassays are described in Tsanuo *et al.*, 2003.[6] However, new studies on the mechanism by which *Desmodium* inhibits *Striga* will provide more refined bioassays in the future.[7]

A number of candidate structural variants have been isolated from *Desmodium* root exudates, comprising novel substituted isoflavonoids with different effects on *Striga*.[6] More recently, from a more water soluble and inhibitory fraction, a di-*C*-glycoside of the flavone, apigenin has been identified with high activity in inhibiting *Striga* radicle growth in laboratory assays but without apparently causing adverse effects on cereals (unpublished). Although other compounds may contribute to the inhibitory mechanism, this flavone di-C-glycoside accounts for a major part of the inhibitory activity and can be used as a target for the types of biotechnological development elaborated in this paper. Food legumes such as cowpea, beans, soybean and other pulses, etc., share the flavone/isoflavonoid metabolic pathways with *Desmodium*. Other legumes also produce *Striga* germination stimulants, but demonstrate no significant post-germination allelopathic effects. This suggests close similarity between the two groups of legumes differentiated by a lack of specific tailoring enzymes, e.g. *C*-glycosyl transferases, that convert common precursors, i.e. apigenin, to the highly active post-germination inhibitors. There is now, therefore, the need to identify specific genes that will convert those precursors already present in edible legumes and cereals into the same agents that are released from roots of *Desmodium* species, and that inhibit *Striga* development so efficiently on farm.

2. Immediate Prospects for Breeding

The current understanding of the mechanism by which *Desmodium* prevents *Striga* infestation, as indicated in Section 1, is that there is both germination stimulation and a post-germination inhibitory effect as measured by interference with radicle elongation. Most legumes probably have sufficient inherent germination stimulation capacity for this,[8] but further investigations into potential target legumes, including cowpea, have to be compared to that caused by *Desmodium*. The extent to which *Desmodium* induces germination also needs to be quantified for these comparative studies to be meaningful.

It is unlikely that direct crosses could be made with *Desmodium* and other legume genera but there may be the prospect of selection in these for the trait if present vestigially. Further studies on other *Desmodium* spp. will also be made, for example the Botanic Garden at Jena, Germany, cultivates many *Desmodium* species, and from colleagues in Sudan, where a number of wild species are extant.[9] One objective is to create an edible bean active against *Striga* in cereals. Some early investigations into adverse human toxicological aspects of *Desmodium* need to be made. Although any new breeding lines would need to be studied, initially this could be done on *Desmodium* seeds themselves. We now have a considerable history of feeding *Desmodium* directly to cattle and other ruminants, which, besides indicating no toxic effects, has high nutritional value.[2]

Simple discriminatory tests are being developed so that the effects on *Striga* seed germination and radicle length inhibition can be measured simultaneously for a range of legumes, particularly cowpea cultivars, including those showing some resistance to *Striga gesnerioides*, which mainly attacks legumes. In order to maintain minimum costs and to economise on the use of advanced techniques, only those plants showing promising levels of radicle length inhibition will be investigated to see if the chemistry identified from *Desmodium* is responsible. In the latter part of these studies, and particularly for cowpea cultivars, seed will be sought from around the world, particularly from IITA in Nigeria. Thus, from this work, the prospect for a conventional breeding programme in cowpea or other bean plants, e.g. *Phaseolus* species, depending on the consensus from target farmers, can be initiated. Comparative studies will also be made between West African legumes that have co-evolved with *S. gesnerioides* and legumes from elsewhere to try to establish the evolutionary origin of *Striga* resistance in *Desmodium*.

3. New Chemical Studies

3.1. *Germination Stimulants*

The precursors for the isoflavonoid germination stimulants are commonly found in legumes (Table 1).[10] These compounds arise *via* the isoflavone synthase (IFS) (Fig. 1) and, as stated in Sec. 2., are expected to be present at sufficiently high levels in legume species chosen for breeding programmes. However, if the specific compounds found in *Desmodium* are essential in terms of the germination stimulation component, then further breeding for these traits would be required. The incorporation of the isoprenyl transferase genes necessary to isoprenylate isoflavonoids such as genistein (Fig. 1)[6] is an example. Initially, older bioassays[6] or newer ones[7] may be appropriate.

Table 1. Isoflavonoids of Fabaceae (= Leguminosae).

Soybean *Glycine max*	Licorice *Glycyrrhiza pallidiflora*
Daidzein	Pallidiflorin
Genistein	Di-*O*-methyldiadzein
Biochanin	Calycosin
Afrormosin	Prunetin
	Afrormosin
Alfalfa or lucerne *Medicago sativa*	*Glycyrrhiza uralensis*
Alfalone	Semilicoisoflavone B
	Gancaonins
	Lupiwighteone
	Isoangustone A

3.2. *Inhibitory C-Glycosylated Flavones*

Characterization of the inhibitory principles from *Desmodium* active against *Striga* requires either *de novo* synthesis or larger samples of compounds involved, potentially from root cultures referred to in Section 4.2. The di-*C*-glycosylated flavones being identified here are formed from flavones prior to IFS in Fig. 1 and are likely to employ the potential precursors apigenin and the mono-*C*-linked glycosides vitexin or isomollupentin (Fig. 2).[11] Thus, chemical studies should concentrate

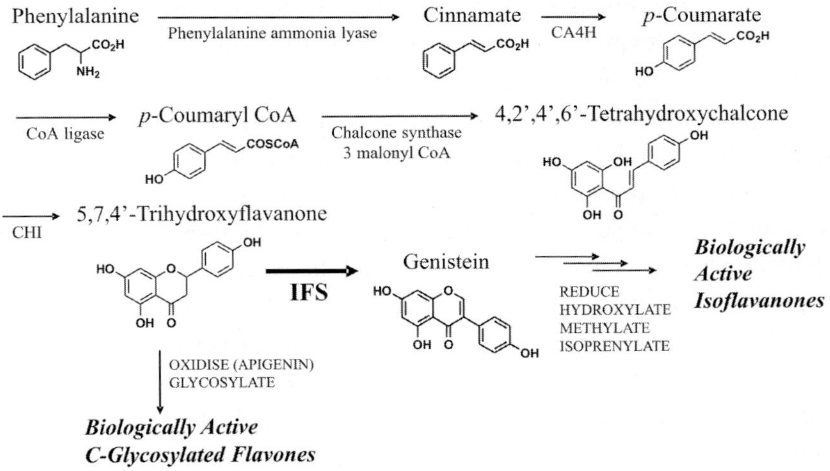

Figure 1. Putative biochemical pathway of Desmodium-*derived allelochemicals.*

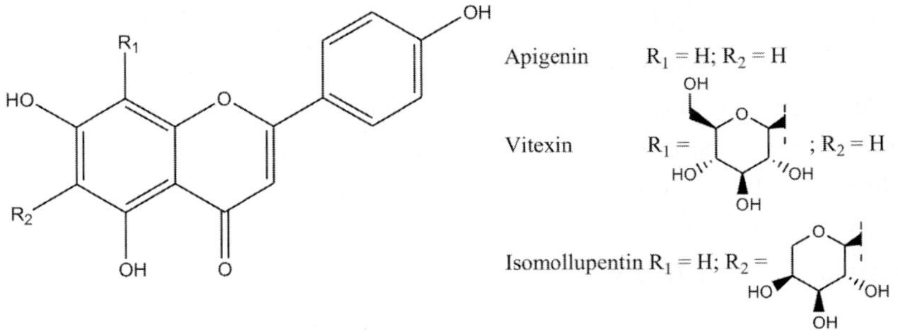

Figure 2. Chemical structures of apigenin, vitexin and isomollupentin.

on identifying apigenin, vitexin or isomollupentin[12] in cultivars of legumes that can be parents for breeding in, or upregulating, the second *C*-glycosylation step from vitexin or isomollupentin to the inhibitory di-*C*-glycosides found in *Desmodium*.

4. Opportunities Through Biotechnology

4.1. *Locating the Allelochemical Genes in Model and Crop Legume and Cereal Plants*

As part of the chemical investigations in Section 3, legumes with extensive EST (expressed sequence tag) libraries and where full genomic sequences will be available in the near future, particularly lotus (*Lotus japonica*) and medicago (*Medicago truncatula*), will also be searched for signs of the chemistry relating to *Striga* for stimulatory and inhibitory effects. If the appropriate chemistry is present, then this could greatly facilitate the identification of genes involved in the biosynthesis of other allelochemicals. This information could be used to initiate breeding programmes or even heterologous transferral of the biosynthesis genes from *Desmodium* to other legume species. The prospect of transferring such genes into cereals can also be considered. Certainly, apigenin and vitexin are important components of pearl millet (*Pennisetum*) spp., and fonio millet (*Digitaria exilis*),[13] and recently found in wheat.[14] The situation with cereals could be more difficult compared to legumes because, in cereals, expression of the earlier parts of the inhibitory pathway are not directed to the roots, as in legumes. Nonetheless, a number of potentially useful root-specific promoters are emerging that could be applied to solving this problem.

As a longer and more expensive lead time will be necessary for the approaches here, the main effort will, however, be on the inhibitory pathway. These could take two approaches. EST databases for *L. japonicus* and *M. truncatula* will be searched, using sequence information from *O*-glycosyl transferases[15] in the event that we detect *C*-glycosyl transferase activity, which would be evidenced by the existence of the mono-*C*-linked glycosides. There may, in some of the sequences, be motifs specific to *O*-glycosyl transferases but also common to those present in *C*-glycosyl transferases. In all probability, the enzyme binds and orientates a similar reactive electrophilic glycosyl species. The nature of the nucleophilic donor is the only criterion by which the two reactions differ.

The feasibility of using a proteomic approach for the identification of this genetic activity, linked to studies in Sec. 4.2., will also be assessed based on an *in vitro* bioassay for the glycosylation of flavonoid intermediates. The other approach will be to use sequence data from known *C*-glycosyl transferases acting on polyphenolic substrates, e.g. from *Streptomyces* species,[16,17] to search the full cowpea genome database[18] or the *L. japonicus* and *M. truncatula* EST databases. Although there will be substantial differences among the sequences encoding these known genes and those in the higher plants, new bioinformatics approaches to creating searches using algorithms based on functional structural features could now be used.[19] The evidence for *C*-glycosyl transferases being in wheat[14] and maize[20] may also allow the associated genes to be identified. This could greatly facilitate the generation of cereal cultivars directly expressing allelochemicals inhibiting *Striga*.

4.2. *Isolating the Genes from Desmodium*

Professor John Hamill, Monash University, Australia, has created hairy root cultures of *D. uncinatum* and *D. intortum* (personal communication). Currently, these are being fed apigenin so that analysis can be made with and without the addition of such substrates, to establish how far the pathway is represented in these cultures and to what extent primary substrates can be incorporated. If the latter is successful, then we will add vitexin or isomullopentin (Fig. 2) and this, if further *C*-glycosylated to inhibitory material, could facilitate, using various biochemical and molecular genetic approaches, identification of the necessary *C*-glycosyl transferases in *Desmodium*. Various approaches could then be adopted to incorporate these into edible bean species. Although in the current political and social climate there is by no means universal support for a transgenic approach involving heterologous gene expression, by the time this aspect of the work would be in place, there may be a greater enthusiasm for these technologies, driven by considerably increased world demand for food. For this approach, and indeed some of the others, metabolite engineering, for example to give increased levels of the potential substrate apigenin, may need to be investigated and this would be facilitated by the already existing wide body of knowledge on the underpinning molecular biology.[21,22]

5. Conclusions and Outlook

The way is now clear to develop edible bean crop plants by feasible breeding programmes, with contributions from biotechnological approaches where appropriate. These would be suitable for intercropping into maize and other cereals in poor farmer communities. They could be bred or transgenically introgessed into cereal crops themselves, which would incorporate the powerful *Striga* controlling properties of the cattle forage legume *Desmodium*. The various options described above all rely on scientific contributions from analytical chemistry through to plant molecular genetics. A continuation of this integrated approach, taking on board all options, would be the most promising course for translating both the science base and the current practical use of the *Desmodium* intercropping system into new farming practices.

Acknowledgments

Rothamsted Research receives grant-aided support from the Biotechnology and Biological Sciences Research Council (BBSRC), UK, with additional funding provided under the Biological Interactions in the Root Environment (BIRE) initiative. *icipe* appreciates the long-standing core support from the Governments of Sweden, Switzerland, Denmark, Norway, Finland, France and Kenya. The push-pull project was primarily funded by The Gatsby Foundation, with additional support from The Rockefeller Foundation.

References

1. Z. R. Khan, K. Ampong-Nyarko, P. Chiliswa, A. Hassanali, S. Kimani, W. Lwande, W. A. Overholt, J. A. Pickett, L. E. Smart, L. J. Wadhams and C. M. Woodcock, *Nature* **388**, 631 (1997).
2. A. Hassanali, H. Herren, Z. R. Khan and J. A. Pickett, in *Philosophical Transactions of the Royal Society of London B, Special Issue on Sustainable Agriculture*, Eds. I. Crute, H. Dalton, C. Leaver and J. Pretty. (Royal Society, London, in press).
3. Z. R. Khan, A. Hassanali, T. M. Khamis, J. A. Pickett and L. J. Wadhams, Proceedings of the BCPC Conference – Weeds 2001, Brighton, UK, 12-15 November 2001, 895 (2001).
4. Z. R. Khan, C. A. O. Midega, A. Hassanali and J. A. Pickett, in *Integrating New Technologies for Striga Control: Towards Ending the Witch-hunt,* Chapter 18, this volume.
5. Z. R. Khan, A. Hassanali, W. Overholt, T. M. Khamis, A. M. Hooper, J. A. Pickett, L. J. Wadhams and C. M. Woodcock, *J. Chem. Ecol.* **28**, 1871 (2002).
6. M. K. Tsanuo, A. Hassanali, A. M. Hooper, Z. R. Khan, F. Kaberia, J. A. Pickett and L. J. Wadhams, *Phytochemistry* **64**, 265 (2003).
7. BBSRC Research Grant, Rothamsted Research and the University of Sheffield, "The mechanism of suppression by the legume *Desmodium uncinatum* of the parasitic weed *Striga*" (awarded 2007).
8. D. Musambasi, O. A. Chivinge and I. K. Mariga, *African Crop. Sci. J.* **10**, 163 (2002).
9. F. W. Andrews. The Flowering Plants of the Anglo-Egyptian Sudan. Volume II. (Sterculiaceae – Dipsacaceae). (Published for the Sudan Government by T. Buncle & Co. Ltd., Arbroath, Scotland, 1952).
10. J. B. Harborne, Ed., *The Flavonoids: Advances in research since 1986.* (Chapman and Hall, London, 1993).
11. G. Rath, A. Toure, M. Nianga, J. L. Wolfender and K. Hostettmann, *Chromatographia* **41,** 332 (1995).
12. E. Ayanoglu, A. Ulubelen, T. J. Mabry, G. Dellamonica and J. Chopin, *Phytochemistry* **21**, 799 (1982).
13. H. Sartelet, S. Serghat, A. Lobstein, Y. Ingenbleek, R. Anton, E. Petitfrere, G. AguieAguie, L. Martiny and B. Haye, *Nutrition* **12**, 100 (1996).
14. I. Cummins, M. Brazier-Hicks, M. Stobiecki, R. Franski and R. Edwards, *Phytochemistry* **67**, 1722 (2006).
15. E-K. Lim, C. J. Doucet, Y. Li, L. Elias, D. Worrall, S. P. Spencer, J. Ross and D. J. Bowles, *J. Biol. Chem.* **277**, 586 (2002).
16. B. Faust, D. Hoffmeister, G. Weitnauer, L. Westrich, S. Haag, P. Schneider, H. Decker, E. Künzel, J. Rohr and A. Bechthold, *Microbiology* **146**, 147 (2000).
17. K. Ichinose, M. Ozawa, K. Itou, K. Kuneida and Y. Ebizuka, *Microbiology* **149**, 1633 (2003).
18. http://cowpeagenomics.med.virginia.edu/.

19. J-J. Zhou, W. Huang, G-A. Zhang, J. A. Pickett and L. M. Field, *Gene* 327, 117 (2004).
20. M. D. McMullen, P. F. Byrne, M. E. Snook, B. R. Wiseman, E. A. Lee, N. W. Widstrom and E. H. Coe, *Proc. Natl. Acad. Sci. USA* **95**, 1996 (1998).
21. R. A. Dixon and C. L. Steele, *Trends Plant Sci.* **4**, 1360 (1999).
22. R. A. Dixon and D. Strack, *Phytochemistry* **62**, 815 (2003).

CHAPTER 6

GENETIC DIVERSITY OF *STRIGA* AND IMPLICATIONS FOR CONTROL AND MODELING FUTURE DISTRIBUTIONS

Kamal I. Mohamed[a,*], Jay F. Bolin[b], Lytton J. Musselman[b] and A. Townsend Peterson[c]

[a]*Oswego State University of New York, Oswego, NY, USA*
[b]*Old Dominion University, Norfolk, Virginia, USA*
[c]*University of Kansas, Lawrence, KS, USA*
[*]*E-mail: Mohamed@oswego.edu*

The current knowledge of genetic diversity of *Striga asiatica, S. hermonthica,* and *S. gesnerioides* is reviewed. The genetic variability of these species has not been sufficiently evaluated relative to their wide distributions. Genetic diversity is a result of hybridization, clinal variation, local adaptation, and frequent colonization events. Colonization events of autogamous species formed genetically uniform populations. There is a general correlation between geographic distance and genetic distance and evidence of host specific *Striga* populations. The genetic diversity inherent in *Striga* is extremely important for modeling its future dispersal in light of global climate change. Under present day climate conditions, ecological niche models predict *Striga* species as serious agronomic threats to tropical and subtropical regions including the Western Hemisphere. Future climate change scenarios may result in an overall reduction in spread of *Striga* species in tropical and subtropical habitats with modest expansion in temperate regions.

1. Introduction

The genus *Striga* (witchweed) along with other parasitic genera once placed in the Scrophulariaceae are now considered part of the Orobanchaceae.[1,2] Crops with some measure of resistance are being integrated into *Striga* management programs. However, new *Striga* resistant crops are immediately challenged by the *Striga* seed bank. The massive seed bank precludes cropping in some areas,[3,4] is structured temporally, and represents several generations of plant parasites.[5] In addition crop breeders must cope with the diversity of species within the genus.[6,7] Although *S. hermonthica, S. asiatica,* and *S. gesnerioides* may constitute the greatest economic threat to agriculture, other species should not be ignored because they act as a reserve of genes via hybridization, as documented in fertile *S. aspera* X *S. hermonthica* crosses.[8]

The spread of witchweed throughout much of Africa as well as other parts of the world shows that rapid movement and gene flow are the norm. These dispersal events are agricultural in origin with the transport of contaminated crop seed or via livestock.[9] Economically important *Striga* species have broad distributions across Africa and Asia, setting conditions for genetically structured populations based on geographic clines. Locally adapted *Striga* races have long been observed that specialize on particular crops.[10] Thus, several factors contribute to genetic diversity in *Striga*: (1) a persistent seed bank of several generations of witchweed populations; (2) hybridization; (3) broad geographic distributions; (4) long distance dispersal; (5) and locally adapted host races. Consequently, a *Striga* resistant crop must be able to cope with the great potential genetic diversity within each *Striga* species, a condition difficult to address.

Since the advent of techniques to estimate genetic diversity, workers have focused on delineating morphotypes, hybrids, local races, and general genetic diversity within the genus. Allozyme electrophorisis was the first method of choice for investigating genetic diversity in *Striga* a few decades ago.[11] Unlike some PCR based techniques allozyme markers are co-dominant, and thus heritability can be inferred. However, allozymes generally underestimate genetic variability because less than 50% of nucleotide substitutions result in polymorphic loci.[12] Allozyme markers have largely been supplanted by polymerase chain reaction (PCR) based fingerprinting techniques. Allozyme and recent PCR based techniques are usually coherent if not directly comparable.[12,13] A variety of PCR based techniques have been applied to investigations of *Striga*, including randomly amplified fragment length polymorphism (RFLP) and amplified fragment length polymorphism (AFLP). Again issues of correlation between different PCR based fingerprinting techniques have been raised, particularly for reproducibility across time and space. Nevertheless, comparisons of RAPD, RFLP, AFLP, and other PCR based results correlate for estimates of genetic distances and variability within and among populations.[14]

The primary goal of this paper is to summarize the genetic variability of *S. asiatica, S. gesnerioides,* and *S. hermonthica.* Initially each species will be discussed separately, considering what is known of within and among population genetic diversity, relationships between geographic and genetic distance, methods used to measure genetic diversity, and the study areas relative to the overall distribution of the species. Then we present a synthesis of our understanding of the genetic diversity underlying the *Striga* plague in Africa in the context of predicting and modeling new *Striga* infestations in the face of global climate change. The invasive potential of *Striga* into new areas as inferred from ecological niche modeling is also presented.

2. Genetic Studies

2.1. *Striga Asiatica*

Striga asiatica is located in the African agroecosystems and natural habitats from portions of southern (including Madagascar), central, and western Africa.[7] *Striga asiatica* is also found across India and southeast Asia.[15] The taxonomic and phylogenetic relationships between plants known as *S. asiatica* in Asia and Africa need to be studied because that name has been applied to a broad geographic range and variable taxa. Populations have been found in areas outside of its typical range, for example the disjunct Mediterranean population in the Nile Delta, mediated by movement of contaminated grains. *Striga asiatica* is reported to be mainly autogamous, this is noteworthy because breeding system can strongly influence genetic structure.[16] The first study to use allozymes to investigate genetic diversity within *Striga* was for a *S. asiatica* introduction in the southeastern United States. All individuals sampled using 18 enzymes and 32 loci were monomorphic, suggesting that the entire US population was the result of a single colonization event.[11]

An AFLP study of 14 populations of *S. asiatica* in Benin, indicated genetic structure within and among populations with genetic distances of 0.028–0.038 and 0.019–0.088, respectively.[17] This is one of the most thorough studies of *Striga* genetic diversity to date. A significant regression was present (R^2=.61) between geographic and genetic distance.[17] Both findings are congruent with expected genetic structure for autogamous plants.[16] The results of Botanga *et al.*[17] support the notion of locally adapted *Striga* ecotypes based on their analyses of geographically distant populations and floral morphotypes.

AFLP was used to examine genetic diversity in 17 coastal populations of *S. asiatica* in Kenya.[18] Unlike the previous study, they found little evidence of within or among population structure, genetic distances for populations of *S. asiatica* ranged from 0.009 to 0.116 (mean of 0.032). Moreover, no relationship was observed between geographic distance and genetic distance suggesting high levels of gene flow with the more recent spread of contaminated crops.[18]

2.2. *Striga Gesnerioides*

Striga gesnerioides has the greatest distribution of all *Striga* species across Africa with extensions to Arabia and Asia between 33°10'N and 32°15'S.[7] It is an important pest of cowpea and other dicotyledons. As an autogamous species it is no surprise that several host specific strains of *S. gesnerioides* have been recognized, but they lack morphological discontinuity.[19] Allozyme techniques were first applied to *S. gesnerioides* to investigate host specific partitioning of parasite genotypes after sowing a single Niger seed (*Guizotia abyssinica*) source on two lines that had been growing in cowpea fields.[20] Significantly different ranges of parasite genotypes were observed on each cowpea line, showing selection for virulence.[20] In the only molecular genetic study on *S. gesnerioides* AFLP markers were used to examine the genetic diversity and parasite/host interaction of four populations of introduced *S. gesnerioides* parasitic on *Indigofera hirsuta* in central Florida.[21] These were compared to *S. gesnerioides* parasitic on *I. hirsuta* and cowpea from West Africa.[21] There was a high degree of genetic uniformity for the introduced *S. gesnerioides* population of central Florida, all but one of the 71 plants sampled were identical (genetic distances 0.000–0.067), suggesting a single introduction of *S. gesnerioides* in the United States or a host driven selection. The Florida strain and the West African strain parasitic on indigo were more closely related to one another (genetic distances 0.214–0.274) relative to the Florida strain and the West African strain parasitic on cowpea (genetic distances range 0.320–0.390). Remarkably, the Florida *S. gesnerioides* was stimulated to germinate by root exudates from cowpea varieties known to be susceptible to *S. gesnerioides* in West Africa but the *Striga* failed to attach.[21,22] These results suggest that *S. gesnerioides* is presently an unlikely agronomic pest in the United States.[21] Because it is a weed of disturbed areas it is not unlikely that this strain could show a shift in host preference and spread to agroecosystems. Questions of how many genes separate strains of *S. gesnerioides* specific to agronomic versus wild hosts and how long ago strain divergence occurred remain to be answered. Currently, no studies have described the relationship between genetic distance and geographic distance in *S. gesnerioides*. However, the large number of host specific strains of *S. gesnerioides* (Chapter 9), its wide geographic range, and the findings of the studies above[20,21] indicate that genetic structure differences are quite probable across the continent.

2.3. *Striga Hermonthica*

Striga hermonthica is mainly distributed from Senegal to Ethiopia and south to Tanzania. Collections have been made in many other areas of Africa including the Nile Delta and Namibia, and likely represent more recent introductions.[7] *Striga hermonthica* is an obligate outcrosser [23] and its hybridization events with other *Striga* species have caused some taxonomic confusion.[8] The first study of genetic diversity in *S. hermonthica* used allozyme electrophoresis (9 loci coding 8 enzymes) on samples from two populations in Burkina Faso, one adapted to pearl millet and one adapted to sorghum and one population adapted to sorghum from the Sudan.[24,25] There was a high heterozygosity within each population (H=0.261-0.365).[24,25] Within population variability was larger than the mean values for other obligately outcrossing species.[16,26] Nevertheless, Bharathalakshmi *et al.*[24,25] suggested that the extremely high fecundity/seed set in *S. hermonthica* may be a contributing factor. Their data also showed that geographic distance played a more important role in genetic differentiation of *S. hermonthica* populations than host specialization.[24,25]

Gel electrophoresis (2 DNA loci) was used to study genetic diversity and host specificity in 14 populations of *S. hermonthica* parasitizing sorghum, pearl millet, maize, and wild grasses in Burkina Faso (9 populations), Mali (4 populations), and Niger (1 population).[27,28] The results indicated low allelic divergence within populations, suggesting that the outcrossing populations were in Hardy-Weinberg equilibrium for most populations. Allelic frequencies were expected to remain constant from generation to generation in these populations. There were slight geographic distance effects and little or no host specificity effects on genetic variability, indicating low selectivity for hosts may be the trend in *S. hermonthica*.[27,28] However, the low number of loci investigated undermines any strong conclusions.

Contrasting results were presented using gel electrophoresis (14 loci in 8 enzyme systems. High levels of genetic diversity were apparent among six West African (Benin, Mali, and Burkina Faso) and nine Kenyan populations of *S. hermonthica*.[29] Again geographic distance was the primary driver of genetic differentiation with no differentiation by host.[29] RAPD markers showed higher levels of genetic diversity within *S. hermonthica* relative to *S. aspera* and their hybrids.[8] The low similarity between *S. hermonthica* and *S. aspera* (55% similarity) as measured by RAPD clearly delimits the two species.[8]

Koyama[30] conducted the first study to combine allozyme electrophoresis (47 loci in 10 enzyme systems) and RAPD markers (33 loci with five primers) to investigate genetic diversity of *S. hermonthica*. She surveyed populations from two sites in Mali and one site each from Nigeria and Kenya. Using cluster analyses with both methods showed high levels of genetic distance between geographic locations, with allozyme variance estimates of between 3.908–6.882 and RAPD variance estimates of 5.725–8.789.[30] Unfortunately, these results must be interpreted with caution. *Striga* plants were not sampled from their respective populations *in situ*, but were reared from bulked seed (from each population) sown on potted *Sorghum* in a controlled experiment.[30] Thus, the results reported do not reflect actual population genetic diversity, but genetic diversity within the individuals selected for by the strain of *Sorghum* used in the experiment and the experimental conditions applied.

This oversight is surprising because in a related study, Koyama[31] applied the same allozyme and RAPD markers to demonstrate strain specific forms of *S. hermonthica* on five *Sorghum* cultivars.[31] Finally, an AFLP analysis of genetic diversity for 24 populations of *S. hermonthica* from Kenya showed genetic distance values range from 0.007-0.025, very low genetic diversity, and no geographic distance to genetic distance relationship was detected.[18] The observed homogeneity of the Kenyan populations of *S. hermonthica* may be in part due to colonization (a founder event) from the Lake Victoria basin east into Kenya and its allogamous breeding system.[18]

2.4. *Synthesis of Genetic Diversity Studies*

Colonization events, linkage with agroecosystems/hosts, geographic clines, and hybridization are the central drivers of genetic diversity in *Striga*. Studies of *S. asiatica*[11] and *S. gesnerioides*[21] colonization events in the United States both showed genetic uniformity in introduced populations, suggesting single successful colonization events. This is consistent with the low genetic diversity in the relatively recently introduced Kenyan *S. asiatica* populations,[18] which is particularly remarkable for an autogamous species. Of the studies reviewed many did not demonstrate strong correlations of allozyme or PCR based markers with host-specific *Striga* strains.[24,27,29] However by combining pot studies and higher resolution of AFLP techniques, Botanga *et al.*[17] showed host specialization of *S. asiatica* in Benin. Moreover with the same combination of techniques Botanga and Timko[21] demonstrated convincingly that the introduced strain of *S. gesnerioides* in Florida (USA) is unable to effectively parasitize cowpea. Taken as a whole this suggests that allozyme markers were insufficiently variable at the scale used to identify host specific genotypes relative to AFLP.

The studies reviewed indicated a relationship between geographic distance and genetic distance.[17,21,24,30] Exceptions are attributable to either insufficient markers/loci to detect differences[27] or sampling of a relatively small geographic area, or a recent parasite introduction.[18] However with adequate markers significant correlations between geographic and genetic distances were observed in an area as small as the Republic of Benin for *S. asiatica.*[17] With a total of 30 or more species of *Striga* in Africa the storage of virulence genes in 'wild' *Striga* congeners is very real danger as evidenced by the RAPD and breeding study of the *S. aspera* and *S. hermonthica* hybrids.[8]

The genetic variability of *Striga* species has not been evaluated in depth relative to their total current distributions. Practical issues of cost and accessibility have prevented continent wide studies of genetic diversity of *Striga.* However, it should be evident that crop breeding efforts towards obtaining resistant cultivars must take the view that *Striga* species are diverse at the intraspecific level.[31] Future matching of resistant crops with resident *Striga* strains must be considered with directed quarantine efforts to prevent movement of virulent strains of *Striga.* We also consider that the genetic diversity inherent in *Striga* may be extremely important for modeling of future dispersal events in light of global climate change. Maximum and minimum germination and flowering temperatures need be recorded for *Striga* ecotypes particularly at the climatic extremes of their ranges. These basic data are clearly lacking to effectively predict the worst case yet unlikely scenarios of dispersal events.

3. Ecological Niche Modeling and Invasive Potential of *Striga*

We used ecological niche modeling to predict the invasive potential of three *Striga* species which constitute the major agronomic threats.[32] The software used to generate the models was the Genetic Algorithm for Rule-Set Prediction (GARP).[33,34] Under current climate conditions, the ecological niche models predicted great invasive potential of *Striga* species that extends to tropical and subtropical regions worldwide including the Western Hemisphere (Fig. 1). The rainforest climatic conditions fall within the range favorable to *Striga* germination and development. However, the deep Amazon Basin rainforest and other similar communities are excluded because climatic conditions are not favorable for germination, as witchweeds are shade intolerant and

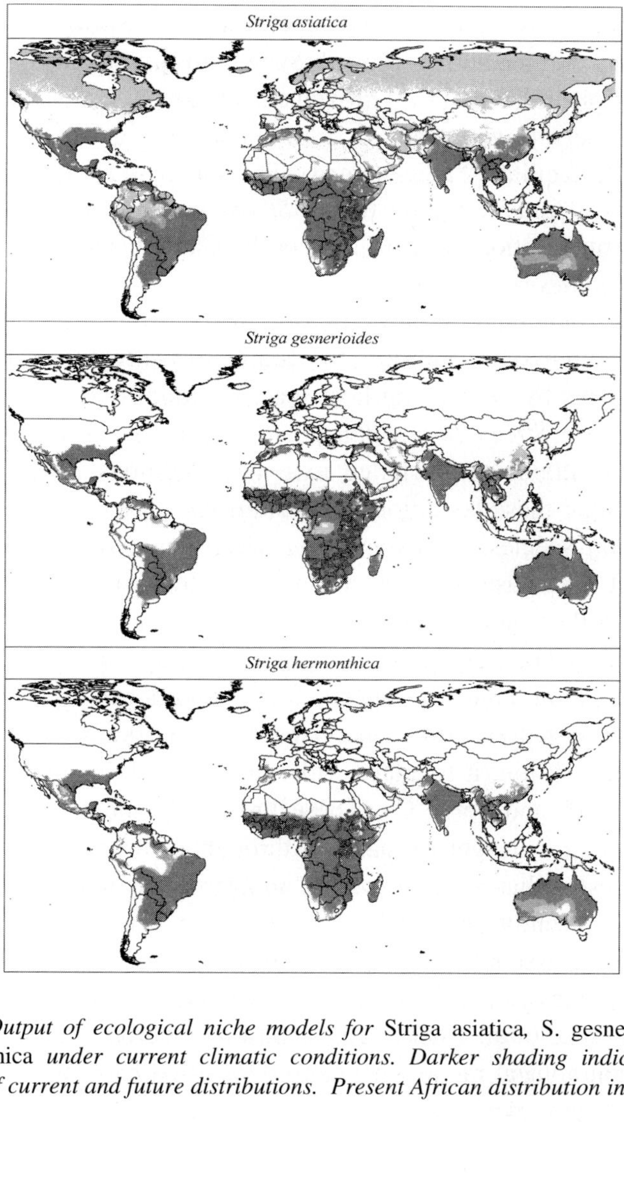

Figure 1. Output of ecological niche models for Striga asiatica, S. gesnerioides, *and* S. hermonthica *under current climatic conditions. Darker shading indicates higher likelihood of current and future distributions. Present African distribution indicated with dots.*

germination is retarded in wet and poorly aerated soils (wet dormancy). The invasive potential of *Striga* will likely increase in tropical Western Hemisphere with increasing disturbance, logging activity, and expansion of soybean farms and other potential hosts in this region. Recently, Brazil witnessed a huge expansion in farming at the expense of natural habitats increasing the risk of *Striga* infestation.

Our models showed that *Striga* should exhibit a worldwide expansion in savannas dominated by typical *Striga* hosts such as grasses and herbaceous plants (Fig. 1).[32] Movement of goods, people, other weed species, farmers' saved seeds in addition to the "informal" crop seed market could facilitate *Striga* spread. The southeastern United States is predicted as a suitable region for all *Striga* species consistent with the accidental introduction of *S. asiatica* in the Carolinas and more recently of *S. gesnerioides* in Florida.

Striga hermonthica, *S. gesnerioides*, and *S. asiatica* are well known for their impressive abilities to adapt to different habitats and agroecosystems by developing host-specific strains and ecotypes across their ranges.[15] *Striga hermonthica* and *S. gesnerioides* have evolved host-specific strains that tolerate extreme conditions in the semi-arid regions. Under these conditions, *S. gesnerioides* has evolved specificity to *Euphorbia* species.[35] *Striga hermonthica* has evolved and attacked pearl millet. Because of their adaptation to drought, these two species range the farthest north among *Striga* in Africa and can cause severe damage as their hosts are already stressed. *Striga hermonthica* can attain 50% germination and was successfully conditioned and germinated under conditions described as permanent wilting points for most other plants.[36] In addition, it tolerates wide ranges of day/night temperatures between 40/30° and 25/15°C. These broad climatic tolerances render *S. hermonthica* a dangerous parasite throughout its range. Aigbokhan *et al.*[8] suggested that *S. hermonthica*, an aggressive agroecosystem pest, is a species recently derived from *S. aspera* which is most commonly restricted to grassland savanna. Moreover *S. aspera* itself has been reported to attack rice and maize in Ivory Coast.[37] *Striga asiatica* is the most widespread of all witchweeds.[15] Based on herbarium studies, it has a wider geographical range with more diversified habitats and a greater host range than previously thought. Unfortunately, the taxonomy of the

S. asatica complex has been confused for some time.[7] It is essential that workers always deposit voucher specimens in accessible herbaria so that the plant identity can be verified. More basic taxonomic work remains to be done in the *S. asiatica* complex because of its widespread transcontinental distribution.

The potential presence of witchweeds in temperate regions is greatly reduced by the inability of their seeds to germinate, successfully attach, or reach maturity and set seeds under the climatic conditions in these ecosystems. Optimum temperatures for seed germination of most *Striga* spp. are 30-35°C.[38] In most studied species, germination percentages were very low at or below 20°C, even when the conditioning period was prolonged.[39] For example, under lab conditions, the germination percentage in the American *S. asiatica* was only 0.5% in seeds conditioned for 15 days at day/night temperatures of 20/14°C, compared to 37% germination after 2 days of conditioning at a day/night temperature of 32/26°C.[39] The minimum day/night temperature under which the American *S. asiatica* infecting maize can successfully flower is 29/23°C. The climatic conditions in the midwestern USA Corn Belt fall within the range tolerable to witchweeds. However, the day/night temperatures in the northern USA Corn Belt States are below that required for germination/flowering of witchweed.[39] These findings were consistent with the predictions of the ecological niche models (Fig. 1).

Future climate change may have a profound effect on geographic distribution and invasive potential of many plant species including root parasites. Early projections suggested that many plants may have broad geographic potential for invasion.[40] This idea has not been quantitatively tested. One study however indicated broader invasive potential in changing climates[41] but another model[42] predicted overall reduction in potential distributional area of invasive species with the potential for some regional expansions. Our preliminary predictions for *Striga* invasive potential under future climate change scenarios support the notion of Roura-Pascual *et al.*[42] for possible overall reduction in potential distribution and spread of *Striga* species (Mohamed and Peterson, unpublished). Ecological niche models indicate a loss in potential distributional areas for *Striga* in tropical and subtropical habitats with modest expansion in temperate regions, especially in North America.

These are welcoming results for badly impacted regions in Africa though it could be too late then.

In conclusion, our genetic algorithm based models suggest that changing climate will play a major role in determining geographic distributions of *Striga* directly by affecting germination, growth, and development, or indirectly through its hosts. The problems of *Striga* mostly affect small hold farmers in the developing world as they are unable to adopt expensive chemical control or use modern agricultural practices and because they depend on precisely those crops hardest hit by these parasites. In dry regions of the developing world, parasitic weeds take a large toll because of the limited number of crops that can be cultivated. Eradication programs require significant commitments of labor and financial resources over a long period of time and work only with limited infestations. For example, the United States took over 50 years and >$250 million to contain/eradicate *S. asiatica*.[43] This was a small investment compared to potential losses in corn production if *Striga* were to spread to the Corn Belt. In the United States, crops threatened by witchweeds are valued at $20 billion annually. The American experience is indeed a model for containment/eradication of parasitic weeds. It involved many logical steps that culminated in containment and eventually eradication. For an excellent review of the problems of invasives and containment see references 32, 44, and 45. Again, problems with witchweeds could be compounded by climate change, which may result in new invasions in regions anticipated to have temperatures and moisture within the ranges tolerated by witchweeds. Genetic diversity studies of *Striga* species while still not comprehensive or continent wide in scope still suggest locally adapted and host specific genotypes in some African agroecosystems. In light of changing global climate these data should be warning enough to underscore the differential invasive potential of certain genotypes within a *Striga* species. We suggest that the genotypes with the greatest potential for invasion into new systems need to be identified and tested empirically under simulated current and projected climatic conditions. We hope that this may allow us to more finely predict and marshal energy against future invasions.

Acknowledgments

Work reported here is based, in part, on research sponsored by USDA grant 58-319R-5-0006 and AID DHR 5600-G-00-1021-00. We also acknowledge the Mary Payne Hogan fund of Old Dominion University and support from SUNY Oswego.

References

1. R. G. Olmstead, C. W. dePamphilis, A. D. Wolfe, N. D. Young, W. J. Elisens and P. A. Reeves, *Am. J. Bot.* **88**, 348 (2001).
2. J. R. Bennett and S. Mathews, *Am. J. Bot.* **93**, 1039 (2006).
3. A. G. Carson, *Proc. OAU/FAO Workshop on Striga,* (FAO, Yaounde, Cameroon, 1985), p. 37.
4. S. Njinyam, *Proc. OAU/FAO Workshop on Striga,* (FAO, Yaounde, Cameroon, 1985), p. 20.
5. F. F. Bebawi, R. E. Eplee, C. E. Harris and R. S. Norris, *Weed Sci.* **32**, 494 (1984).
6. L. J. Musselman and E. S. Ayensu, in *Striga Biology and Control,* Ed. E. S. Ayensu,. H. Doggett, R. D. Keynes, J. Marton-Lefevre, L. J. Musselman, C. Parker and A. Pickering (ICSU Press, Paris, 1984), p. 35.
7. K. I. Mohamed, L. J. Musselman and C. R. Riches, *Ann. Mo. Bot. Gard.* **88**, 60 (2001).
8. E. I. Aigbokhan, D. K. Berner, L. J. Musselman and H. D. Mignouna, *Weed Res.* **40**, 375 (2000).
9. D. K. Berner, K. F. Cardwell, B. O. Faturoti, F. O. Ikie and O. A. Williams, *Plant Dis.* **78**, 402 (1994).
10. H. Doggett, Annual Report for the Botanist Ukiriguru, for the year 1950, Tanganyika Dep. Agric., Dar es Salaam, Tanzania, Annual Report, p. 221 (1952).
11. C. R. Werth, J. L. Riopel and N. W. Gillespie, *Weed Sci.* **32**, 645 (1984).
12. K. Weising, H. Nybom, K. Wolff and G. Kahl, Eds., *DNA fingerprinting in plants: Principles, methods, and applications*, (CRC Press, Boca Raton, FL., 2005).
13. N. J. Ouborg, Y. Piquot and J. M. Van Groenendael, *J. Ecol.* **87**, 551 (1999).
14. H. Nybom, *Mol. Ecol.* **13**, 1143 (2004).
15. V. Cochrane and M. C. Press, *J. Trop. Ecol.* **13**, 371 (1997).
16. M. D. Loveless and J. L. Hamrick, *Annu. Rev. Ecol. Syst.* **15**, 65 (1984).
17. C. J. Botanga, J. G. Kling, D. K. Berner and M. P. Timko, *Euphytica* **128**, 375 (2002).
18. J. G. Gethi, M. E. Smith, S. E. Mitchell and S. Kresovich, *Weed Res.* **45**, 64 (2005).
19. K. I. Mohamed, *Ph.D. Dissertation*, Old Dominion University, Norfolk (1994).
20. K. Shawe and M. Ingrouille, in *Proc. Brighton Crop Protection Conf. Weeds*, (BCPC, Brighton, UK, 1993), p. 919.
21. C. J. Botanga and M. P. Timko, *Phytopathology* **95**, 1166 (2005).
22. L. J. Musselman and C. Parker, *Weed Sci.* **29**, 594 (1981).
23. S. B. Safa, B. M. G. Jones and L. J. Musselman, *New Phytol.* **96**, 299 (1984).

24. Bharathalakshmi, C. R. Werth and L. J. Musselman, *Plant Syst. Evol.* **172**, 1 (1990).
25. L. J. Musselman, Bharathalakshmi, S. B. Safa, D. A. Knepper, K. I. Mohammed and C. L. White, in *Proc. Int. Workshop, Combating Striga in Africa,* Ed. S. K. Kim (IITA, Ibadan, Nigeria, 1991), p. 31.
26. J. L. Hamrick, Y. B. Linhart and J. B. Mitton, *Annu. Rev. Ecol. Syst.* **10**, 173 (1979).
27. A. Olivier, J.-C. Glaszmann, C. Lanaud and G. D. Leroux, *Plant Syst. Evol.* **209**, 33 (1998).
28. A. Olivier, J.-C. Glaszmann, C. Lanaud, G. Salle and G. D. Leroux, in *Advances in Parasitic Plant Research*, Ed. M. T. Moreno and J. I. Cubero (Junta de Andalucia, Cordoba, Spain, 1996), p. 114.
29. E. Kuiper, P. L. M. Koevoets, J. A. C. Verkleij and A. H. Pieterse, in *Advances in Parasitic Plant Research*, Ed. M. T. Moreno and J. I. Cubero (Junta de Andalucia, Cordoba, Spain, 1996), p. 124.
30. M. L. Koyama, in *Breeding for Striga resistance in cereals*, Ed. B. I. G. Haussmann, D. E. Hess, M. L. Koyama, L. Grivet, H. F. W. Rattunde and H. H. Geiger (Margraf Verlag, Weikersheim, Germany, 2000), p. 227.
31. M. L. Koyama, in *Breeding for Striga resistance in cereals*, Ed. B. I. G. Haussmann, D. E. Hess, M. L. Koyama, L. Grivet, H. F. W. Rattunde and H. H. Geiger (Margraf Verlag, Weikersheim, Germany, 2000), p. 247.
32. K. I. Mohamed, M. Papes, B. Williams, B. W. Benz and A. T. Peterson, *Ambio* **35**, 1 (2006).
33. D. R. B. Stockwell and I. R. Noble, *Math. Comp Simul.* **33**, 385 (1992).
34. D. R. B. Stockwell and D. P. Peters, *Int. J. Geogr. Info. Systems.* **13**, 143 (1999).
35. L. J. Musselman and F. N. Hepper, *Kew Bull.* **41**, 205 (1986).
36. D. Dawood, and J. Sauerborn, *Exp. Agric.* **30**, 249 (1994).
37. D. E. Johonson, C. Riches, R. Diallo and M. J. Jones, *Crop Prot.* **16**, 153 (1997).
38. D. C. Reid and C. Parker, in *Proc. 2nd Int. Sym. Parasitic Weed*, Ed. L. J. Musselman, A. D. Worsham and R. E. Eplee (Raleigh, NC State University, 1979), p. 202.
39. D. T. Patterson, R. L. Musser, E. P. Flint and R. E. Eplee, *Weed Sci.* **30**, 87 (1982).
40. J. S. Dukes and H. A. Mooney, *Trends Ecol. Evol.* **14**, 135 (1999).
41. T. L. Benning, D. LaPointe, C. T. Atkinson, and P. M. Vitousek, *Proc. Natl. Acad. Sci.*, USA, **99,** 14246 (2002).
42. N. Roura-Pacual, A. Suarez, C. Gómez, Y. Pons and A. L. Touyama, in *Proc. R. Soc. London Ser.* **B271**, 2527 (2005).
43. Westbrooks, R. G. and Eplee, R. E. (1999). in *Invasive Species and Biodiversity Management*, Ed., O. T. Sandlund, P. J. Schei, and A. Viken (Kluwer, Dordrecht, 1999), p. 283.
44. R. E. Eplee, *Plant Dis.* **65**, 951 (1981).
45. C. R. Veitch and M. N. Clout, Eds., *Turning the tide: the eradication of invasive species.* Invasive Species Specialist Group, (IUCN, Gland, Switzerland and Cambridge, UK., 2002).

Section III

Knowledge-based Breeding — Translating Information to Products

CHAPTER 7

DISSECTING A COMPLEX TRAIT TO SIMPLER COMPONENTS FOR EFFECTIVE BREEDING OF SORGHUM WITH A HIGH LEVEL OF *STRIGA* RESISTANCE

Gebisa Ejeta[a,*], Patrick J. Rich[a] and Abdalla Mohamed[b]

[a]Department of Agronomy, Purdue University, West Lafayette, IN, USA
[b]Gezira Research Station, Wad Medani, Sudan
[*]E-mail: gejeta@purdue.edu

We have made considerable progress in advancing sorghum breeding for *Striga* resistant cultivars by employing a knowledge-based approach that exploits the intricate biological association between the host and the parasite. This approach has provided a greater understanding of the host-parasite biology as we made key observations at individual components amenable to intervention. We developed laboratory assays that facilitated isolating unique genetic variants and elucidated the nature of signals exchanged between host and parasite. Significant advances were made in the isolation of key compounds essential for *Striga* germination, in conducting sound genetic analyses that yielded vital information on the mode of inheritance as well as in the characterization of the specific mechanisms involved in *Striga* resistance. More significantly, valuable genes for *Striga* resistance were introgressed to selected genotypes using this approach in a variety of strategies. This has led to the development and release of sorghum cultivars with high level of *Striga* resistance. These cultivars have been widely deployed in a number of African countries where they have had significant impact as cultivars per se, or as a central component of an integrated *Striga* management program.

1. Introduction

Host plant resistance has been advocated as a practical and economically feasible option for *Striga* control in Africa.[1,2] When effectively deployed, genetic control offers many benefits without a significant increase in cost, as the technology is embedded in the genetics of the crop cultivar planted.[3] Adequate genetic variation and availability of effective selection tools are essential for successful plant breeding efforts. Sorghum landraces with varying levels of *Striga* resistance have been found in our global genetic resources. Significant effort has been directed to breeding for *Striga* resistance in sorghum, though these efforts have been erratic and not sustained in any one program for considerable length of time. As a result, progress made from conventional breeding for *Striga* resistance has not been considerable. Empirical selection methods that worked well for improving other desirable crop traits have not operated at the same efficiency in *Striga* resistance breeding. The genetics of *Striga* resistance is a complex quantitatively inherited trait that is replete with a large genotype x environment interaction that limits selection efficiency. Plant breeding approaches were needed that consider the ever-growing knowledge of the interactive host-parasite biology, minimize environmental variation, and promise an enhanced efficiency for achieving crop cultivars with resistance to *Striga*. We therefore proposed a novel approach for breeding for *Striga* resistance by dissecting this seemingly complex trait into simpler components.[4,5] Laboratory assays were designed to dissect the expression of host resistance to specific points of parasitic establishment. The expression of potential resistance was narrowed to a specific point in the parasitic life cycle. The expression of these resistance reactions could be masked in a field setting by confounding environmental factors influencing parasite emergence and host crop performance. Hence, the development of carefully designed laboratory assays has been a key to this approach. This paper summarizes the progress made over the last two decades in a sustained effort to increase our understanding of host-parasite biology and exploit this increased knowledge towards developing *Striga* resistant sorghum cultivars.[2-7,12,19]

2. Materials and Methods

We developed an approach to *Striga* resistance breeding based on dissecting the trait to simpler components on the basis of the intricate biological relationships between the parasite and its hosts at each stage of the parasitic life cycle (Chapter 2, Fig. 1). Attaining a good understanding of the key events in the life cycle of *Striga* and the array of signal exchanges was crucial for the eventual establishment of a direct connection of the parasite to the host, an essential requisite to genetic exploitation. The rationale and premise behind this approach has been detailed in an earlier publication.[4] Briefly, our premise was based on the fact that the life cycle of *Striga* is intimately linked to that of its hosts and that, at each of these stages, there is good potential for genetic intervention leading to host plant resistance. Conventional selection for *Striga* resistance is difficult, because each of the discrete interactive events between host and parasite are unobservable in field grown plants. Each of these events is probably influenced by environmental conditions, albeit in a somewhat limited way. Laboratory methods that permitted observation of each of the early events in the developmental association between the host and parasite were needed. We hypothesized that genetic variation for each of these discrete events is likely to be found in nature or to be induced artificially, and that host plant resistance derived by disruption in any one of these critical stages may well be simply inherited, easy to select and transfer to other cultivars through breeding.

Bioassays were developed that target specific signal exchanges at the early stages of the parasitic process. We first developed an *in vitro* laboratory procedure, the agar gel assay that separated sorghum genotypes based on their capacity to produce the exudates required for *Striga* germination.[6] We subsequently developed two other *in vitro* assays: the extended agar gel assay and the paper roll assay that targeted both the pre- and post-attachment stages of parasitic development, respectively.[7] The extended agar gel assay distinguishes host genotypes on the basis of their ability to induce haustoria formation. The paper roll assay was developed for observation of the early stages of *Striga* attachment following germination and development of haustoria as the organ of attachment to host roots. We recently developed another procedure, sand-packed titer plate assay that allows visual, pictorial, and microscopic monitoring over a period of time from early stages of host parasite association for a long duration (unpublished). Our ability to systematically assemble, evaluate, and exploit genetic resources for *Striga* resistance has been enhanced by using these assays. The bioassays have provided insights into the interactive biological processes between *Striga* and the roots of host plants. They permit observation of discrete events during the early stages of the infection process. Identifying genetic variants that disrupt these interactions allows genetic control of *Striga* through development of resistant sorghum cultivars with single or multiple interventions at key stages in the parasitic life cycle. This is a powerful tool in pyramiding multiple mechanisms for a more durable resistance to *Striga*.

3. Results and Discussion

Efforts devoted to developing a thorough understanding of the basic biology involved in the signal exchanges between the *Striga* and its hosts have greatly facilitated the relative ease by which we have been able to breed sorghums with high level of *Striga* resistance. Increased knowledge of host-parasite biology has been useful in developing appropriate assays. These assays were used to identify sorghum germplasm with unique sources of *Striga* resistance and for characterizing and ascribing defined mechanisms of *Striga* resistance to each of these variants. The same assays have also been used for conducting genetic analyses to determine the mode of inheritance of each of the more discrete components. Selection for resistance in breeding populations was then practiced and methodologies were developed for efficient transfer of genes for *Striga* resistance from source genotypes to improved sorghum cultivars with enhanced levels of *Striga* resistance. As a result, there appears to be far more progress made in breeding sorghum for *Striga* resistance than for other crops.

3.1. *Characterization of Mechanisms of Resistance*

Empirical selection, as conventionally practiced through direct evaluation of genetic populations in *Striga* infested fields, overlooks

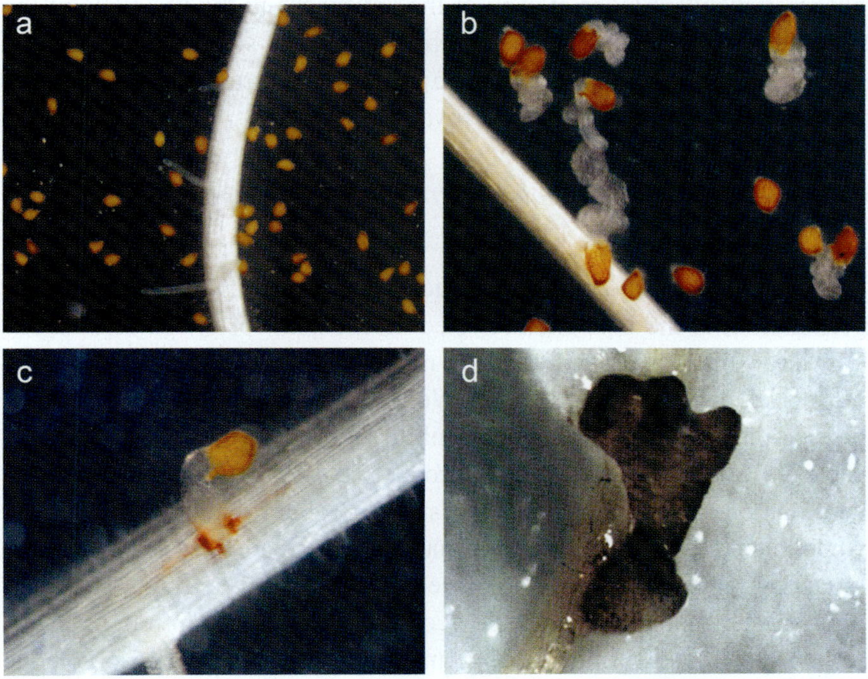

Figure 1. Four mechanisms of Striga *resistance in sorghum. A.* Striga *seeds do not germinate near the root of a low germination stimulant producing sorghum cultivar; B. Low production of the haustorial initiation factor by a wild sorghum apparently prevents* Striga *radicle apex differentiation to attachment structures; C. A sorghum root expresses a hypersensitive response in cells immediately surrounding the attachment site of* Striga*; D. An attached* Striga *stops developing and dies on the root of a sorghum expressing an incompatible response.*

some of the potential biological variations during key events in the life cycle of the parasite. That approach does not lead to increasing knowledge about the actual defenses that discourage parasitic growth and establishment. Successful exploitation of host plant resistance requires an understanding of the physiological and genetic mechanisms that govern parasitism. In this obligate parasite, both metabolic and developmental processes are needed to bridge connections between the parasite and its host, leading to its eventual survival. Paired comparative observations were made at each of the key stages in the life cycle of the parasite between known *Striga* resistant and susceptible sorghum variants to characterize their specific mechanism of resistance, as defined by their unique reaction to *Striga* invasion (Fig. 1).

3.1.1. *Resistance Based on Low Germination Stimulant (lgs) Production*

Low production of crop root exudates that are essential for *Striga* germination is the best characterized mechanism for *Striga* resistance.[8] We have fully exploited the *lgs* mechanism of *Striga* resistance in our sorghum research program. We developed the agar gel assay for phenotyping resistance and susceptibility to *Striga* on the basis of the capacity of host genotypes to produce these exudates required for germination.[6] Not all *Striga* resistant sorghum genotypes are low stimulant producers, as other mechanisms can lead to resistance. Yet, all susceptible genotypes we have phenotyped were high stimulant producers.[7] We identified the key compounds in sorghum root exudates responsible for eliciting *Striga* germination, namely dihydrosorgoleone and (sorgolactone), a strigolactone.[9,10] Though several classes of chemicals elicit *Striga* seed germination, the strigolactones appear to be the most active and correlate well with *Striga* resistance expressed in infested crop fields.[11] We established that low stimulant production in sorghum is inherited as a single recessive gene.[12] The bioassay developed for this character and the genetic information generated have been exploited in breeding *Striga* resistant sorghum cultivars.[2]

3.1.2. *Resistance Based on Low Production of the Haustorial Initiation Factor (LHF)*

Striga seeds that germinate near the roots of sorghum lines possessing resistance based on low production of the haustorial initiation factor, normally do not form haustoria and eventually die from their inability to attach to their potential host. A variety of phenolic compounds function as haustoria initiators in *Striga*, but the active signals from host roots have not yet been identified. A simple quinone, 2,6-dimethoxy-*p*-

benzoquinone (DMBQ), though not found in root exudates, acts as a strong haustorial initiation factor.[8] We developed a modified procedure, the extended agar gel assay for qualitatively sorting host genotypes on the basis of their ability to induce haustorial formation.[7] In this assay, the presence of haustoria can be microscopically detected around the growing host root at two days after ethylene treatment. We have not found any cultivated sorghum lines with LHF among sorghum germplasm we studied to date. However, we recently found wild sorghum lines that rarely developed haustoria.[13] This observation was confirmed through repeated assaying of an array of these genotypes.

3.1.3. *Resistance Based on the Hypersensitive Response (HR)*

Necrotic areas appear on roots at the site of *Striga* attachment in some sorghum genotypes. These red necrotic lesions start become brownish with time. They may be large, spreading up to 2mm from the center of attachment but most remain more localized. The hypersensitive response is also characterized by slowing the further advance of attached *Striga*, which does not develop normally and eventually dies on the host. Both cultivated and wild sorghum lines with powerful HR responses were developed utilizing yet another assay, the paper roll assay developed in our laboratory.[7] This phenomenon has been observed in sorghum lines Framida, CK32, and KP33, although the response appears graded depending on the background of the germplasm. A single infected root may show reddening around most, but not necessarily at all haustorial attachment sites. The overall character of lines possessing hypersensitive response, however, is a greatly reduced percentage of *Striga* complete attachments with relative to susceptible cultivars. Hypersensitive responses against attaching parasites have been reported in resistant cowpeas and vetch.[14] Although not called hypersensitive response, earlier reports describe reactions in sorghum with some similarities, particularly the release of colored phenolics at the attachment interface with *Striga*.[15,16]

3.1.4. *Resistance Based on an Incompatible Response (IR)*

We are the first to describe an incompatible relationship between both *S. hermonthica* and *S. asiatica* to both wild and cultivated sorghums.[17] A similar response was later reported in *Tripsacum dactyloides*, a wild race of maize.[18] An incompatible response is characterized by retarded growth and development of attached parasites even though vascular connections are sometimes established. There is no apparent necrosis in host root tissue surrounding the attachment site. In resistance based on this mechanism, *Striga* seedlings that penetrated into host tissue may not develop beyond the emergence of the first leaf primordia. Some *Striga* appear to develop normally at first, but show signs of stunted growth. This is a response similar to that observed when *Striga* unsuccessfully infests non-host plants, thus the use of the term incompatible response. Similar incompatible relationships with resistant hosts have also been reported for *Orobanche cumana* on sunflower,[19] and *O. crenata* growing on legumes.[20]

3.2. *Genetic Analyses of Resistance* to Striga

Knowledge about the inheritance of a trait is crucial for its successful exploitation in a breeding program. Information on the genetics of *Striga* resistance in crop plants has been generally scant. This is perhaps attributable to the paucity of germplasm of crop plants with a high level of resistance to *Striga*, and the lack of reliable methods for phenotyping described earlier as the rationale for knowledge-based breeding effort undertaken by our program. The genetics of low germination stimulant production was studied in populations of sorghum derived from the resistant cultivar SRN39.[12] The agar gel assay was employed to determine the inheritance of low stimulant production in progenies of SRN39 and three susceptible lines, Shanqui Red, P954063, and IS4225. Segregation ratios suggested that this trait was inherited as a single, nuclear, recessive gene with largely additive gene action. The gene symbol *lgs* was proposed. The same approach was employed to study the inheritance of the other two additional mechanisms of *Striga* resistance, the low production of the haustorial factor and the hypersensitive response but using the extended agar gel assay.[3] Analysis of progenies derived from a cross of *Striga* susceptible lines and a wild sorghum accession P78 with the low haustorial factor, suggests inheritance of the trait through a dominant allele of a single gene. Analysis of $F_{2:3}$ progenies from crosses between hypersensitive response expressers CK32 and KP33 and susceptible lines TX430 and TX2737 resulted in a segregation of progenies for the presence or absence of necrosis at the point of attachment at a ratio that reflected the presence of one dominant allele from either of two genes. The mode of inheritance of the incompatible response mechanism of *Striga* resistance has not been clearly established. However, we have determined that incompatible response is independently inherited from low germination stimulant production mechanism of *Striga* resistance.[7]

3.3. *Development and Deployment of* Striga*-Resistant Cultivars*

We developed and tested *Striga*-resistant sorghum cultivars for wide geographical distribution.[2,3] Early releases were based on the mechanism on low germination stimulant production alone. A bioassay specifically developed for this character has been exploited in developing *Striga*-resistant sorghum cultivars. The nature of induction of these genes is now known, although the relationship between the activity of these genes and the formation of germination stimulants has not yet been clearly established.[21] Powerful laboratory methods were also used to screen wild and cultivated sorghums for the ability to cause haustorial initiation on germinated *S. asiatica*. Wild accessions of sorghum were found that showed reduced haustorial formation.[13] The same assays have also been used in directed introgression of genes for *Striga* resistance into target cultivars. Recipient parents were either improved sorghum cultivars or landraces susceptible to *Striga* but with otherwise desirable sets of attributes. We developed and released 11 high-yielding *Striga*-resistant sorghum cultivars that have been widely distributed for use in *Striga* endemic areas in several African countries[3]. The breeding of many of these lines was accomplished through laboratory-mediated selection in early generation populations, followed by confirmatory field evaluation in *Striga* infested fields. A list of these lines and the local names ascribed to these selections in the respective countries where they were release can be obtained in a recent report.[3] Some of these are also described in Chapters 15 and 19. The released lines have been grown extensively as cultivars *per se* or as a component of an integrated *Striga* management package in some of these countries.

3.4. *Resistant Cultivars as Components of Integrated* Striga *Control*

We developed and deployed an integrated *Striga* management (ISM) protocol to promote the adoption of the *Striga* resistant releases and to enhance the benefit to farmers, as a pilot project in three African countries, namely Ethiopia, Eritrea, and Tanzania (Chapters 15 and 19). We evaluated the combined effects of *Striga* resistant cultivars, soil fertility management, and moisture conservation practices on *Striga* control and grain yield enhancement. The ISM technology significantly increased grain yield of sorghum and reduced *Striga* infestation, as is well described throughout this book. The synergistic effect of combining the component amendments in the ISM packages resulted in very dramatic yield responses and *Striga* control. In addition to reliable crop harvest, sustained use of the ISM practice would likely lead to a significantly reduced *Striga* seed inoculum in the soil for a major long-term benefit.

4. Conclusions

A paradigm for breeding *Striga* resistance in sorghum that is based on an enhanced understanding of the biological basis of host-parasite association and minimizes the effect of environmental influence on the genetic basis of *Striga* resistance has paid dividends in our sorghum research program. The research focused on essential signals exchanged between host and parasite that determine potential sites for intervention. The novel bioassays developed for use in the identification of unique sources of genetic variants in host plant germplasm, to elucidate the mode of inheritance of these variants, to characterize the specific mechanisms of *Striga* resistance involved, and to develop sorghum cultivars combining *Striga* resistance with other desirable agronomic attributes clearly paid off by providing rather resistant material. The multiple mechanisms of resistance that were pyramided together was deployed in a high yielding sorghum cultivar with decreased likelihood of breakdown of resistance genes. The impact of *Striga* resistant sorghum cultivars was synergized when used in combination with other agronomic interventions in an integrated *Striga* management program. Based on this experience, we promote research targeting the evaluation of synergistic effects of the combined use and integration of different sciences and approaches towards the ending the *Striga* menace in African agriculture.

Acknowledgements

Financial support for research that led to the results documented herein has been primarily obtained through a series of grants by the United States Agency for International Development (through INTSORMIL) and by the Rockefeller Foundation. Basic research was undertaken as graduate research dissertation of several M.Sc. and Ph.D. students at Purdue University. The more applied component of varietal testing and deployment was accomplished jointly with collaborating scientists from the national agricultural research services (NARS) in several countries.

References

1. C. Parker and C. R. Riches, in *Parasitic weeds of the world: Biology and control*, Eds. C. Parker and C. R. Riches (CAB International, Wallingford, Oxon, U.K. 1993).
2. G. Ejeta, L. G. Butler, D. E. Hess, T. Obilana and B.V. Reddy, in *International Conference on genetic improvement of sorghum and pearl millet,* Ed. D. T. Rosenow, Intsormil/Icrisat Conference Proceeding, (Lubbock, TX, USA, 1997), p. 504.
3. G. Ejeta, in *In the wake of the double helix: From the green revolution to the gene revolution*, Eds. R. Tuberosa, R .L. Phillips and M. Gale (Edizioni Avenue Media, Bologna, 2005), p. 239.
4. G. Ejeta and L. G. Butler, *Afr. Crop Sci.*, **1**, 75 (1993).
5. G. Ejeta, L. G. Butler and A. G. Babiker, Eds., *New approaches to the control of Striga* (Agricultural Experiment Station, Purdue University, Lafayette IN, 1998).
6. D. E. Hess, G. Ejeta and L.G. Butler, *Phytochemistry* **31**, 493 (1992).

7. G. Ejeta, A. H. Mohamed, P .J. Rich, A. Melake-Berhan, T. L. Housley and D. E. Hess, in *Breeding for Striga Resistance in Cereals* Eds. B. I. G. Haussmann, D. E. Hess, M. L. Koyama, L. Grivet, H. F. W. Rattunde and H. H. Geiger (Margraf Verlag, Weikersheim 2000), p. 29.
8. D. G. Lynn and M .Chang., *Ann. Rev. Plant Physiol. Plant Mol. Biol.* **41**, 497 (1990).
9. D. H. Netzly, J. L. Riopel, G. Ejeta and L. G. Butler, *Weed Sci.* **36**, 441 (1988).
10. B.A. Siame, Y. Weerasuriya, K. Wood, G. Ejeta and L. G. Butler, *J. Agric. Food Chem.* **41**, 1486 (1993).
11. C. Hauck, H. Schildknecht and S. Muller, *J. Plant Physiol.* **139**, 474 (1992).
12. R. K. Vogler, G. Ejeta and L. G. Butler, *Crop Sci.* **36**, 1185 (1996).
13. P. J. Rich, C. Grenier and G. Ejeta, *Crop Sci.* **44**, 2221 (2004).
14. J. A. Lane, T. H. M. Moore, J. Steel, R. F. Mithen and J. A. Bailey, in Biology and Management of Orobanche, Proceedings of the 3rd International Workshop on Orobanche and Related Striga Research Conference Proceeding (1994), p. 356.
15. M.-C. Arnaud, C. Veronesi and P. Thalouarn, *Aust. J. Plant Physiol.* **26**, 63 (1999).
16. U. Neumann, B. Vian, H .C. Weber and G. Sallé, *Protoplasma* **207**, 84 (1999).
17. C. Grenier, P. J. Rich, A. Mohamed, A. Ellicott, C. Shaner and G. Ejeta, in *7th International Parasitic Weed Symposium,* Eds. A. Fer, P. Thalouarn, D.M. Joel, L.J. Musselman, C. Parker and J. A. C. Verkleij Conference Proceeding, (Nantes, France, 2001), p. 220.
18. A. L. Gurney, D. Grimanelli, F. Kanampiu, D. Hoisington, J. D. Scholes and M. C. Press, *New Phytol.* **160**, 557 (2003).
19. P. Labrousse, M.-C. Arnaud, H. Serieys, A. Berville and P. Thalouarn, *Ann. Bot.* **88**, 859 (2001).
20. A. Pérez-De-Luque, D. Rubiales, J. I. Cubero, M. C. Press, J. Scholes, K. Yoneyama, Y. Takeuchi, D. Plakhine and D. M. Joel, *Ann. Bot.* **95**, 935 (2005).
21. H. Bouwmeester, R. Matusova, S. Zhongkui and M. Beale, *Curr. Opin. Plant Biol.* **6**, 358 (2003).

CHAPTER 8

BREEDING MAIZE FOR BROAD-BASED RESISTANCE TO *STRIGA HERMONTHICA*

Abebe Menkir[a,*], Baffour Badu-Apraku[a], Chabi G. Yallou[b], Alpha Y. Kamara[a] and Gebisa Ejeta[c]

[a]*International Institute of Tropical Agriculture, PMB 5320, Ibadan, Nigeria;*
[b]*Institut National des Recherches Agricoles du Benin, B.P. 884, Cotonou, Republic of Benin*
[c]*Agronomy Department, Purdue University, West Lafayette, IN 47907, USA*
[*]*E-mail: a.menkir@CGIAR.ORG*

Recurrent selection under artificial *S. hermonthica* infestation has significantly reduced the number of emerged *Striga* plants and increased grain yield under *Striga* infestation in broad-based populations. These populations have been sources of varieties and inbred lines with consistently high levels of resistance to *S. hermonthica* across locations and seasons. The number of parasites attached to the roots of diverse lines was significantly correlated with the number of emerged parasites in the screenhouse and in the field as well as with a reduction in grain yield due to *Striga*. AFLP and SSR markers clearly separated 41 *Striga* resistant inbred lines from four populations into groups according to their source populations. The consistent ranking of the general combining ability effects of selected inbred lines across locations and seasons also indicates that the inbred lines had a stable genetic basis that controlled the *Striga* resistance traits. Unraveling the complex mechanisms of resistance to *S. hermonthica* using rapid and efficient screening tools can facilitate the improvement of maize for resistance to different populations of the parasite.

1. Introduction

Resistance to *S. hermonthica* is an important trait for maize varieties specifically developed for the savannas. Exploiting host genetic variability to increase the level of resistance to the parasite can be a major component of an integrated approach to minimize yield losses from *S. hermonthica* in farmers' fields.

S. hermonthica has a high degree of genetic diversity due to being an obligate out-crossing species.[1,2] Our breeding strategy for maize germplasm development has thus focused on utilization of diverse sources of genetic materials against the parasite. Screenhouse and field inoculation techniques were developed and refined over the years to increase the uniformity and severity of *S. hermonthica* infestation.[3] Systematic screening of diverse maize germplasm using these screening techniques followed by repeated evaluation of potential sources of resistance in multiple locations over seasons yielded promising genetic materials with consistent expression of resistance or tolerance to *S. hermonthica*.[3,4]

The complex nature of the mode of inheritance of traits associated with resistance to *S. hermonthica*[5-7] prompted us to employ recurrent selection for increasing the levels of resistance to the parasite in populations and composites with diverse genetic backgrounds, maturities, and grain colors.[3] An advantage of recurrent selection is that new combinations of resistance alleles can come together through continual recombination in each cycle of selection, leading to the development of open-pollinated varieties with a high degree of genetic diversity that can impart polygenic resistance to different populations of *S. hermonthica*. The populations and composites undergoing improvement have also been good sources of inbred lines with accumulation of different combinations of resistance alleles that can be used as parents for developing synthetics and hybrids with high levels of polygenic resistance to *S. hermonthica*. Our recent progress is described in developing maize germplasm with broad-based resistance to *S. hermonthica*.

2. Current Breeding Strategies and Progress

2.1. *Accumulating Resistance Alleles in Broad-Based Populations and Composites*

The presence of adequate genetic variation is an important prerequisite for efficient selection of resistance to *S. hermonthica*. *Striga* damage rating, number of emerged *Striga* plants and grain yield under *Striga* infestation are important traits for defining the degree of resistance of genotypes to *S. hermonthica* in our breeding program. Significant genetic variation for the three *Striga* resistance traits has been detected among elite maize germplasm, landrace accessions and wild relatives.[3,7] These traits had moderate to high heritability estimates in broad-based and bi-parental populations (Table 1), providing scope for their improvement under *S. hermonthica* infestation. Polygenic resistance can be obtained in breeding populations by accumulating resistance genes of small effect derived from different sources.[8] Maize genotypes of diverse origin with proven moderate levels of resistance to *S. hermonthica* were thus selected as parents and crossed to adapted germplasm for developing several broad-based populations and composites in the late 1980s and the 1990s.[3] Selfed progeny and full-sib family selection schemes have been used for continual accumulation and increase in frequency of resistance alleles in these breeding populations under artificial *S. hermonthica* infestation. The best 11 to 30% of the lines or families that combined higher yield and ear number under *Striga* infestation with lower *Striga* damage rating and fewer emerged *Striga* plants were selected in each selection cycle using a base-index and were inter-crossed to form the genetic material for each new cycle of selection.

Table 1. Moderate to high heritability of three traits recorded under artificial *Striga* infestation in two bi-parental populations at three environments in 1999 and 2000 and in a broad-based population (TZLCOMP1-Y) at two test environments in 2003.

Traits	TZLCOMP1-Y (121 S1 lines)	607/1393 (290 F3 families)	91-5-2/1393 (280 F3 families)
		Heritability estimates	
Grain yield, infested[a]	0.51±0.144	0.89±0.083	0.70±0.086
Striga damage[b]	0.57±0.141	0.82±0.084	0.43±0.091
Striga emergence	0.63±0.138	0.71±0.085	0.65±0.087

The populations and composites were subjected to at least two cycles of selection for improved performance under artificial *Striga* infestation. Independent trials consisting of the original and advanced cycles of selection of two populations were conducted with and without artificial *S. hermonthica* infestation at two locations over seasons to determine the effectiveness of recurrent selection for the improvement of the three *Striga* resistance traits. The susceptible check in each trial had about 80% yield loss, sustained the highest *Striga* damage and supported the most emerged parasites (Table 2), indicating that the level of infestation was severe during trial evaluation. Selection increased grain yield by 14% per cycle in the intermediate and by 26% per cycle in the late populations (Table 2), while at the same time it significantly reduced *Striga* damage in the two populations. Although the reduction in the number of emerged *Striga* was significant only in the late population, it decreased with selection in the intermediate population. The response in grain yield under non-infested condition was either non-significant or was significant and positive when selection for improved performance was done under *S. hermonthica* infestation (Table 2). Recurrent selection increased grain yields in the two populations to levels that were comparable to the yield potential of a tolerant hybrid control. Conversely, the two populations sustained significantly lower *Striga* damage and one of them also supported significantly fewer emerged *Striga* (Table 2) in comparison to the tolerant hybrid control. The latest cycle of selection of the intermediate population can be classified tolerant while that of the late maturing population can be classified resistant, as the number of emerged *Striga* is considered to be a good indicator of resistance. The significant gains from recurrent selection across diverse environments provide evidence that genotype effects of selection were more important than the genotype x environment interaction effects. Such significant genetic gains for at least two of the three *Striga* resistance traits in the two populations could result from the presence of adequate genetic variation for these traits as well as the high selection intensity and effectiveness of the recurrent selection schemes used.

Polygenic resistance is often difficult to breed because several genes must be manipulated at the same time and its improvement takes a long time. Our studies illustrate the potential usefulness of full sib and S_1 recurrent selection schemes for attaining substantial and rapid progress from selection for polygenic resistance to *S. hermonthica* in adapted breeding populations. The level of resistance conferred by polygenes can be very high in some cases and may not be distinguishable from major gene resistance.[9]

Table 2. Recurrent selection resulted in enhanced performance of two populations evaluated at Abuja and Mokwa in Nigeria for at least two years.

	Grain yield		*Striga* damage rating	Emerged *Striga* plants
	Infested	non-infested	10 weeks	10 weeks
Cycles	(kg/ha)		(1-9)[a]	(number per plant)
	Intermediate maturing (2003, 2004 and 2005)			
Susceptible check	698	3344	8.0	7.3
Tolerant check	2304	4309	5.5	5.1
C_0	1983	4401	6.0	5.9
C_2	2546	4204	4.0	5.5
Standard error	178	206	0.2	0.7
Gain per cycle (%)	14.2**	-2.2	-12.3**	-3.5
	Late maturing (2003 and 2004)			
Susceptible check	579	3086	8	7.5
Tolerant check	2304	4309	5.5	5.1
C_0	1045	3628	6.4	6.3
C_6	2673	4449	4.3	2.6
Standard error	228	400	0.4	1.0
Gain per cycle (%)	26.0**	3.8**	-5.5**	-9.7**

**Significantly different from C_0 at $p<0.01$ level using a paired t-test.
[a]*Striga* damage rating : 1 = no damage symptoms and 9 = severe damage.

The lack of significant changes in the number of emerged parasitic plants in the intermediate population underscores the need to incorporate new genetic variation to significantly shift gene frequencies for this trait. Elite inbred lines with resistance to *S. hermonthica* available in our breeding program[3] can be used as sources of resistance to enhance performance in this breeding population. Further progress in improving the performance of the population and the composite under *Striga* infestation should thus be possible using a base index with appropriate adjustment in weights assigned to reduce both *Striga* damage and emergence and increase grain yield under *Striga* infestation. The improved populations and composites may be used as open-pollinated varieties and as a source of germplasm for developing inbred lines and hybrids with high levels of resistance to *S. hermonthica*. Because of its complex genetic basis, polygenic resistance derived from different sources may act as a buffer against the different populations of *S. hermonthica*.

2.2. *Evaluating Consistency of Resistance to* **S. hermonthica** *in Multiple Locations*

The different populations and composites of maize undergoing continual improvement for resistance to *S. hermonthica* have been the source of open-pollinated varieties for further testing in multiple locations. Such testing exposes the genetic materials to the diversity of *S. hermonthica* populations, which may be encountered in field production. During evaluation of varieties over location and years under *S. hermonthica* infestation, genotype x environment interaction assumes prominence because varieties can show varying levels of resistance under different environments, while the parasite may exert different levels of aggressiveness in different environments. Therefore, changes in parasite aggressiveness can result in changes in resistance ranking of the varieties.

Genotype x environment interaction can affect grain yield under *Striga* infestation. This will be illustrated by two examples. In a recent performance trial involving 10 late-maturing maize varieties bred from improved source populations evaluated at two locations for three years, the variety x environment interaction was significant for all the traits recorded in this trial (Table 3). Further analysis using Kendall's[10] coefficient of concordance of the communality of ranks of the varieties in six environments was significant ($p<0.01$) for grain yield under *Striga* infestation ($W = 0.71$), *Striga* damage rating ($W = 0.76$) and number of emerged parasites ($W = 0.74$). Twelve early-maturing open-pollinated varieties derived from improved source populations were evaluated at two locations for two years in a second performance trial. A significant variety x environment interaction was found only for grain yield under infested and non-infested condition (Table 4). Kendall's[10] coefficient of concordance for grain yield under *Striga* infestation in four environments was found to be significant (W = 0.64, $p<0.01$). These results suggest that the observed significant variety x environment interaction for the different traits recorded in each trial represented a non-crossover type of interaction and thus the relative ranking of the varieties and the checks for these traits was consistent across environments.

Table 3. Late-maturing *Striga* resistant varieties perform far better than susceptible varieties in trials conducted at Abuja and Mokwa in Nigeria in 2003 to 2005.

Variety	Grain yield (kg/ha) Infested	Grain yield (kg/ha) Non-infested	*Striga* damage rating 10 weeks (1-9)[a]	Emerged *Striga* plants 10 weeks (per 7.5 m^2)
8338-1 (Susceptible)	698	3344	8	288
TZB-SR (Common)	1569	3481	6	250
9022-13STR (Tolerant)	2304	4309	6	208
TZLCOMP1SYNW-1	3204	3545	4	40
Acr.97 TZL Comp.1-W	3061	4173	4	98
TZLCOMP1-W C6	2727	3813	4	84
ZEA DIPLO BC4 W C3	2673	3329	4	80
TZLCOMP1SYNY-1	2667	3677	5	85
ZEA DIPLO BC4 Y C32	2559	3950	5	158
IWD C2 SYN F2	2546	4204	4	219
ZEADIPLOSYNW-1	2457	3168	4	83
EV IWD STR C0	1983	4401	6	236
MID-ALTITSTRSYN2	1765	2820	5	80
Mean	2270	3451	4	125
SE	178	206	0.2	29.7
CV (%)	29	23	17	48
F probability for variety	***	***	***	***
F probability for VAR x ENV	**	*	***	***

[a]*Striga* damage rating: 1 = no damage symptoms and 9 = severe damage.

The variation among varieties was significant for the three *Striga* resistance parameters in each performance trial. Among the varieties included in these trials, eight late and three early maturing varieties were significantly less damaged by *Striga*, supported significantly fewer emerged parasites and produced significantly higher yields than the respective susceptible check (Tables 3 and 4). These varieties can be classified as resistant based on the definition of

Table 4. Early-maturing *Striga* resistant varieties performed better than a susceptible variety in trials conducted at Abuja and Mokwa in Nigeria in 2004 and 2005.

Variety	Grain yield		*Striga* damage	Emerged *Striga* plants
	Infested	non-infested	10 weeks	10 weeks
	(kg/ha)		(1-9)[a]	(per 7.5 m^2)
TZE Comp.4 C3 (Susceptible)	841	2539	7	219
TZE Comp.5-W C7 F2	2612	4028	4	155
Acr.94 TZE Comp.5-W	2211	3675	5	180
Acr.94 TZE Comp.5-Y	2154	3272	5	166
TZE-W POP/LD SYN (A)	1987	3178	6	172
TZE-W POP/1368STR SYN-A	1849	3186	5	176
TZE-W POP/1368STR SYN-B	1845	3227	6	148
TZE-Y POP CO SYN	1810	3471	6	194
TZE-W POP/LD SYN-B	1676	3518	6	183
EARLY STR-SYN-1	1572	3341	6	209
EARLY STR-SYN-2	1569	3847	6	257
ACR 94 POOL16 DT STR	1430	3889	7	274
TZE-W POP Co SYN)	1296	3271	7	259
Mean	1796	3423	6	198
SE	209	235	0.2	21
CV (%)	38	22	14	39
F probability for variety	**	*	***	**
F probability for VAR x ENV	*	*	ns	ns

[a]*Striga* damage rating: 1 = no damage symptoms and 9 = severe damage.

Ejeta *et al.*[11] The remaining varieties in each maturity group that supported as many *Striga* plants as the respective susceptible check and with significantly higher grain yields than the susceptible check under *Striga* infestation could be classified as tolerant varieties (Tables 3 and 4). Both the resistant and tolerant varieties from each maturity group had incomplete resistance to *S. hermonthica*. As these varieties and synthetics are heterogeneous, each has the potential to possess an array of resistance genes, gene combinations and resistance mechanisms. The consistently high levels of polygenic resistance expressed in these varieties across locations and seasons suggest that they can be suitable candidate varieties for use in rotation with legumes that elicit suicidal germination of *S. hermonthica* as well as other control methods, including appropriate rate of fertilizer, herbicides and biocontrol, for more effective control of the parasite on subsistence farmers' fields. Also the incomplete resistance in combination with other control measures provides good prospects for durable protection of the maize crop against *S. hermonthica.*

2.3. *Developing and Identifying Superior Parental Lines*

The source populations and composites undergoing improvement constantly create new genotypes through recombination. Inbreeding fixes the combinations of new resistance allele complexes in individual lines and thus facilitates the development of less-related lines that maintain favorable linkage blocks intact for exploitation in breeding programs.[12] Inbreeding can also eliminate deleterious recessive alleles and increases the sensitivity of lines to *S. hermonthica* infection allowing more effective selection. We have, therefore, repeatedly screened selfed families or lines selected from advanced selection cycles of populations in the field and in the screenhouse under artificial *Striga* infestation to fix resistance alleles. Several promising maize inbred lines with consistently few emerged parasites, low *Striga* damage and high grain yield under *S. hermonthica* infestation at two locations over seasons have been bred from diverse source populations.

Pot, screenhouse and field experiments were conducted to determine whether the observed field resistance of the diverse inbred lines to *S. hermonthica* was related to the number of root-attached parasites.[13] These studies were important because the number of emerged *Striga* plants, which has been used as a major selection criterion during inbred line development, could only represent 10 to 30% of the actual number of attached parasites underground in severely infested areas.[14,15] The inbred lines had significant differences between the number of underground attachments in pots, and the emerged *Striga* plants. The results in the screenhouse were consistent across seasons. Significant differences were detected among inbred lines for *Striga* damage and emergence in the field.[13] Multivariate analysis revealed that high parasite attachment was significantly correlated in the field with a large reduction in grain yield and plant height, prolonged delay in tasseling, high *Striga* damage and emergence and poor ear aspect scores.[13] The regression coefficient of the numbers of attached parasites to the roots on the first principal component axis scores was positive and significant,[13] suggesting that simultaneous selection for a combination of traits in the field may increase the probability of identifying maize inbred lines supporting fewer attached and emerged parasites.

Several *Striga* resistant inbred lines were derived from different populations that share common sources of resistance to *S. hermonthica* in their genetic backgrounds. We thus conducted diversity assessment studies of 41 *Striga* resistant inbred lines derived from four populations with AFLP and SSR markers to examine the genetic structure and extent of diversity of the lines.[16] These results should be useful for efficient selection of parental genotypes for crossing,[17] to develop new hybrids and for accumulating resistance alleles in elite germplasm. Accurate diversity assessment of these inbred lines might be useful to ensure long-term and sustained gain from selection for resistance to *S. hermonthica*. The inbred lines from each source population had a broad range of genetic similarity with the two types of markers. Both AFLPs and SSRs revealed similar levels of within population genetic variation for all source populations. Cluster and principal component analysis of genetic similarity with the two markers revealed clear differentiation of the *Striga* resistant inbred lines into groups according to their source.[16] The occurrence of significant changes in allelic frequencies in different directions during intensive screening of the source populations for resistance to *Striga* at the various cycles of recurrent selection could lead to such differentiation of the source populations. In general, genetically unrelated inbred lines are likely to have fewer resistance genes in common than closely related inbred lines. Therefore, the inbred lines that originated from different source populations may have different genes for resistance to *S. hermonthica*. Accumulation of complementary resistance alleles through utilization of inbred lines with a broad genetic base can enhance the stability of resistance across populations of *S. hermonthica*. These lines may, be used to develop source populations with low *S. hermonthica* infection or crossed with other adapted germplasm to increase the frequency of resistance alleles to the parasite. They may also be used as potential candidate genotypes for studies to elucidate the mechanism of resistance to *S. hermonthica* and their genetic basis in maize.

2.4. *Assessing Consistency of Genetic Resistance in Multiple Locations*

The expression of resistance to *S. hermonthica* depends both on the genetics of the host as well as that of the parasite interacting with the environment. Inbred lines with stable expression of resistance to *S. hermonthica* across locations and seasons are ideal for studying the consistency of the genetic basis of polygenic resistance across environments. Genetic analysis with such inbred lines could pinpoint superior parental materials with high breeding value for use in breeding maize for broad-based resistance to *S. hermonthica*. Five new inbred lines derived from a source population containing *Zea diploperennis* as a donor parent, one new inbred line derived from a tropical composite and four old inbred lines with varying levels of resistance to *S. hermonthica* were crossed in a diallele mating scheme to generate 45 single-cross hybrids. The hybrids were evaluated with and without artificial *Striga* infestation at two locations each in Nigeria and the Republic of Benin for three years to examine the consistency of the combining ability of the *Striga* resistant inbred lines.

The combined analysis of variance over twelve environments showed significant general combining ability (GCA) for all *Striga* resistance parameters (Table 5). The variance due to specific combining ability (SCA) was significant for grain yield under *Striga* infestation but not for *Striga* damage and number of emerged parasites. There was significant GCA x environment interaction for all *Striga* resistance parameters in this study. Examination of the consistency of relative ranking of the GCA effects of the inbred lines in 12 environments using Kendall's[10] coefficient of concordance found significant correlations for grain yield under *Striga* infestation (W = 0.53, p<0.01), *Striga* damage rating (W = 0.64, p<0.01) and number of emerged *Striga* plants (W = 0.63, p<0.01). The SCA x environment interaction was significant for grain yield under *Striga* infestation and damage. This significant interaction was also reflected in weak Kendall's[10] coefficient of concordance of the consistency of ranks of SCA effects for grain yield under *Striga* infestation (W = 0.17) and *Striga* damage rating (W = 0.12) recorded in the 12 environments.

Table 5. Combined analysis of variance revealed significant or non-significant general combining ability (GCA) and specific combining ability (SCA) for three traits recorded at two locations each in Benin and Nigeria in 2000 to 2002.

		Grain yield		*Striga* damage rating[b]	*Striga* Emergence[c]
Source	DF	Infested	non-infested	10 weeks	10-weeks
Environment (ENV)	11	238824841**	362905477**	163.3**	816516**
REP (ENV)	24	3817281**	4221077**	6.9**	10701**
GCA	9	50189967**	7492809**	86.6**	136666**
SCA	35	1398719**	1863032**	1.3	2819
GCA*ENV	99	4423763**	1638453**	5.0**	15495**
SCA*ENV	385	593746**	698922	0.9**	2670
Error	1056	478515	705384	0.7	2652

*, ** Significantly different from zero at p<0.05 and p<0.01 levels, respectively.

The ratio of GCA to SCA sums of squares was 9 for grain yield under *Striga* infestation, 17 for *Striga* damage rating and 12 for number of emerged *Striga* plants, indicating the predominance of genes with additive effects controlling these traits. Similarly, Gethi and Smith[6] found additive genetic effects being more important than non-additive genetic effects in controlling resistance parameters for *S. hermonthica.* The new inbred lines combined the desirable positive GCA effects for gain yield under *Striga* infestations with favorable negative GCA effects for *Striga* damage and number of emerged *Striga* plants (Table 6). Conversely, three of the four old inbred lines combined negative GCA effects for grain yield under *Striga* infestation with positive GCA effects for *Striga* damage rating and number of emerged *Striga* plants. Among the new inbred lines, four had significant GCA effects for grain yield under *Striga* infestation, two new inbred lines had significant and negative GCA effects for *Striga* damage, and one had significant and negative GCA effects for number of emerged *Striga* plants (Table 6). Some inbred lines were also identified in broad-based source populations with better combining ability and resistance to *S. hermonthica* and *S. asiatica* than the adapted local inbred lines.[6] The consistent ranking of the GCA effects in the new inbred lines across locations and seasons suggests that the genetic factors controlling resistance are stable across potentially different populations of *S. hermonthica* in Benin and Nigeria. These lines may thus combine well with other maize inbred lines for resistance to *S. hermonthica* and can be used as parental materials to create suitable breeding populations for improving resistance to *S. hermonthica*, as well as possessing good agronomic traits.

Table 6. New inbred lines show desirable general combining ability effects for three traits recorded in 10 inbred lines evaluated at two locations each in Benin and Nigeria in 2000 to 2002.

	Grain yield		*Striga* damage rating	Emerged *Striga* plants
Lines	Infested	non-infested	10 weeks	10 weeks
	General combining ability, relative units			
Zd 282 (New)	170	-283	-0.25	-13
Zd 290 (New)	319	57	-0.45	-8
Zd 467 (New)	314	42	-0.29	-4
Zd 472 (New)	272	24	-0.29	-11
Zd 551 (New)	314	-53	-0.55	-28
TZL TC 87 (New)	231	21	-0.31	-17
TZi 25 (Old)	-26	167	0.12	-10
TZi 4 (Old)	-468	53	0.51	25
TZi 10 (Old)	-202	225	0.29	30
TZi 11 (Old)	-923	-253	1.23	35
SE	134	163	0.16	10

3. Distributing *S. Hermonthica*-Resistant Maize Germplasm

Regional trials have been used as major vehicles for channeling extra-early, early, intermediate and late maturing *Striga* resistant open-pollinated varieties, hybrids and inbred lines to collaborators in and outside west and central Africa (WCA). Seeds of *Striga* resistant varieties were supplied to collaborators for on-farm testing and eventual release. In collaboration with the national agricultural extension systems, two *Striga* resistant maize varieties of early (ACR 94 TZE, COMP.5-W) and late (ACR 97TZL COMP.1-W) maturity were introduced to farmers at the Federal Capital Territory around Abuja in Nigeria in 1999. A total of 153 on-farm trials were conducted for three years to assess the performance of the *Striga* resistant varieties under diverse growing conditions and to expose these varieties to farmers. The average grain yields of the early (1594 kg ha^{-1}) and the late (1863 kg ha^{-1}) maturing *Striga* resistant open-pollinated varieties were higher than that of the farmers' variety (887 kg ha^{-1}). In addition, the *Striga* damage rating was 26% less for the early and 37% less for the late maturing *Striga* resistant varieties compared with the farmers' variety. The early and late maturing *Striga* resistant varieties also supported 42% and 48% less *Striga* than the farmers' varieties, respectively. Extensive on-farm trials were also conducted in the northern Guinea savanna of Nigeria to promote the use of one of the *Striga* resistant varieties (ACR 94 TZE, COMP.5-W) in rotation with legumes that cause suicidal germination of *S. hermonthica* seeds.[18,19] This integrated approach increased crop productivity by an average of 88%.[19] The use of the resistant maize variety after a legume crop resulted in a consistent doubled net benefit over farmers' practices across seasons.[18,19] The use of the *Striga* resistant variety alone or in rotation with legumes also reduced *Striga* seed density in the soil by 29 to 50%.[19] These integrated approaches spread within and beyond the villages where on-farm trials were conducted through farmer-to-farmer diffusion.[18,19]

On-farm trials of *Striga*-resistant maize varieties and other integrated control methods have been pursued in many countries of west and central Africa. The Semi-Arid Africa Agricultural Research and Development of the African Union (AU/SAFGRAD) has also promoted the delivery of *Striga* resistant/tolerant varieties along with improved cultural practices to farmers through farmer managed on-farm demonstration trials in Benin, Burkina Faso, Cameroon, Cote d'Ivoire, Ghana, Mali and Nigeria from 2002 to 2004. This *Striga* control project involved more than 5000 farmers in technology evaluation, demonstration and dissemination. Grain yields of the *Striga* resistant maize varieties averaged 2131 kg ha^{-1} while the farmers' maize varieties produced an average of 1517 kg ha^{-1}. The use of the *Striga* resistant maize varieties either in rotation or intercropping with legumes resulted in 30 to 41% greater grain yield and 40% fewer emerged *Striga* plants compared to the farmers' practices under natural infestation in farmers' fields. This superiority of the integrated approach over the local practice was irrespective of the level of *Striga* infestation in the field. These activities have promoted the adoption of *Striga* resistant varieties in rotation with legume cultivars selected for efficacy in causing higher levels of suicidal germination of *S. hermonthica*. Some of the *Striga* resistant varieties are already in the hands of farmers in Nigeria and Benin.

References

1. J. Bharathalakshmi, C. R. Werth and L. J. Musselman, *Plant Syst. Evol.* **17**, 1. (1990).
2. M. L. Koyama, In B. I. G. Haussmann, *et al.*, Eds., Breeding for *Striga* Resistance in Cereals. (Margraf Verlag, Weikersheim, Germany, 2000), p. 227.
3. J. G. Kling, J. M. Fajemisin, B. Badu-Apraku, A. Diallo, A. Menkir and A. Melake-Berhan, In B. I. G. Haussmann, *et al.*, Eds., Breeding for *Striga* Resistance in Cereals. (Margraf Verlag, Weikersheim, Germany, 2000), p. 103.
4. D. K. Berner, J. G, Kling and B. B. Singh, *Plant Dis.* **79,** 652 (1995).
5. L. Akanvou, E. V. Dou and J. Kling, *Africa Crop Sci. J.* **5**, 1 (1997).
6. J. G. Gethi and M. E. Smith, *Crop Sci.* **44**, 2068 (2004).
7. S. K. Kim, *Crop Sci.* **34**, 900 (1994).
8. J. E. Parlevliet and J. C. Zadoks, *Euphytica* **26**, 5 (1977).
9. P. W. Wilkins, *Ann. Appl. Biol.,* **81**, 257 (1975).
10. M. G. Kendall, Rank correlation methods. (3rd ed., Griffin, London, 1962).
11. G. Ejeta, L. G. Butler, D. E. Hess and R. K. Vogler, In J. K. Ransom *et al.*, Eds., *Proc. Fifth Int. Symp. Parasitic Weeds*, CIMMYT, Nairobi, Kenya, p. 539 (1991).
12. A. R. Hallauer, and S. Pandey, In L. R. Lamkery and M. Lee, Eds., Plant breeding: (Blackwell Publishing, Oxford, UK), p. 73 (2006).
13. A. Menkir, *Plant Breed.* **125**, 131 (2006).
14. H. Doggett, *Agric Sci. J.,* **65**, 183 (1965).
15. B. I. G. Haussmann, D. E. Hess, H. G. Welz and H. H. Geiger, *Field Crops Res.* **66**, 195 (2000).
16. A. Menkir, J. G. Kling, B. Badu-Apraku and I. Ingelbrecht, *Theor. Appl. Genet.* **110**, 1145 (2005).
17. J. P. Russell, J. D. Fuller, M. Macaulay, B. G. Hatz, A. Jahoor, W. Powell and R. Waugh, *Theor. Appl. Genet.* **95**, 714 (1997).
18. J. Ellis-Jones, S. Schulz, B. Douthwaite, M. A. Hussaini, B. D. Oyewole, A. S. Olanrewaju and R. White, *Expl. Agric.* **40**, 353 (2004).

19. A. C. Franke, J. Ellis-Jones, G. Tarawali, S. Schulz, M. A. Haussini, I. Kureh, R. White, D. Chikoye, B. Douthwaite, B. D. Oyewole and A. S. Olanrewaju, *Crop Prot.* **25**, 868 (2006).

CHAPTER 9

MOLECULAR MARKERS FOR ANALYSIS OF RESISTANCE TO *STRIGA GESNERIOIDES* IN COWPEA

Michael P. Timko[a,*], Bhavani S. Gowda[a], Jeremy Ouedraogo[b] and Boukar Ousmane[c]

[a]*Dept. of Biology, University of Virginia, Charlottesville, VA 22904 USA,*
[b]*Institut de l'Environnement et de Recherches Agricoles (INERA), 01 BP 476 Ouagadougou 01, Burkina Faso*
[c]*Institut de la Recherche Agricole pour le Développement (IRAD), Cameroon*
[*]*E-mail: mpt9g@virginia.edu*

Cowpea is one of the most important food and forage legumes in the semi-arid tropics. Losses due to attack by the root hemiparasitic weeds *Striga gesnerioides* (witchweed) and *Alectra vogelii* are among the major constraints to cowpea production in West and Central Africa. At least seven races of *S. gesnerioides* have been identified within the cowpea-growing regions of West Africa based on host differential response and genetic diversity analysis. Race-specific resistance genes have been identified and mapped to two linkage groups (LG1 and LG6) of the cowpea genetic map. Molecular markers have been identified that are associated with specific resistance genes, and at present two markers have been developed as sequence-confirmed amplified regions and are available for germplasm evaluation and efficacy testing on field populations. Marker–assisted selection has yet to be implemented in cowpea but the groundwork has now been laid for its development.

1. Introduction

Cowpea (*Vigna unguiculata*) is one of the most important food legumes in the semi-arid tropics covering Asia, Africa, Southern Europe, Southern United States and Central and South America.[1,2] It serves a critical role in the lives of millions of people in Africa and other parts of the developing world, where it is a major source of dietary protein that nutritionally complements staple low-protein cereal and tuber crops.[3,4] In addition to its nutritional value, cowpea is a valuable and dependable commodity that produces income for farmers and helps to restore soil fertility for succeeding cereal crops grown in rotation with it.[5-7]

Yields of cowpea grain are reduced by a variety of biotic and abiotic constraints of which attack by two root parasitic angiosperms, *Striga gesnerioides* and *Alectra vogelii*, are a major concern.[8] *S. gesnerioides* causes extensive damage to cowpea in the Sudano-Sahelian belt of West and Central Africa.[9] Annual yield losses range from slight to moderate in most regions, however, total crop loss is not unusual in some parts of Nigeria, Niger, and Burkina Faso.[10,11] *A. vogelii* also infects a number of grain legume crops in an agroecological range extending from the northern agricultural regions of South Africa and Swaziland, through Central Africa to Burkina Faso and Mali in the west and Kenya in the east. In addition to cowpea, soybean, bambara groundnut (*Vigna subterranea*), common bean, mung bean (*Phaseolus radiata*) and many legume fodder crops, including *Lablab purpureus*, siratro (*Macroptilium atropurpureum*) and velvet bean (*Mucuna pruriens*) are also parasitized by these two noxious pests.[1,12,13]

Several control strategies have been developed for parasitic weeds including improved cultural practices, breeding using wild and cultivated germplasm as sources of resistance, and the use of chemical control.[1,8] These topics are covered elsewhere in this volume.

The breeding of improved cultivars with pyramided desirable agronomic traits and multiple disease and pest resistances requires effective screening and selection procedures. Cowpea is considered to have been domesticated in Africa and is likely to have co-evolved with *Striga* and *Alectra*. Thus, it may have many of the requisite genes for resistance. Resistance against most parasitic weeds, including *S. gesnerioides* and *A. vogelii* is often difficult to assess due to numerous confounding factors in the field, including parasite variability, unpredictable environmental influences, and imprecise selection criteria. Despite these difficulties, significant success has been achieved in the identification of heritable sources of resistance to both *S. gesnerioides* and *A. vogelii*, and the inclusion of germplasm having these traits into cowpea selection and breeding programs.[1-3]

2. Variation in Host Preference

There is considerable variation in host specificity among isolates of *S. gesnerioides*. In addition to cowpea, members of the wild legume genera *Alysicarpus, Indigofera,* and *Tephrosia*, and non-legumes such as *Ipomea, Jaquemontia, Merremia, Euphorbia,* and *Nicotiana* are among the known hosts of *S. gesnerioides*.[9,14] Strains of the parasite growing on cowpea, *Indigofera* spp., *Tephrosia* spp. and *Jacquemontia* spp. would only attack and emerge on the host species from which they had been collected[15]. A strain of *S. gesnerioides* parasitic on tobacco in South Africa and Zimbabwe germinates in the presence of other potential hosts but is only capable of completing its life cycle on tobacco.[16] Similarly, *S. gesnerioides* parasitic on *Indigofera* species will germinate in the presence of cowpea roots, but is not capable of parasitizing this host.[8,17]

Evidence for the existence of distinct races of *S. gesnerioides* that attack cowpea is also based upon the observation that some cowpea cultivars are differentially resistant to various geographic isolates of the parasite. It was proposed that there are five distinct races of *S. gesnerioides* in west and central Africa based on their ability to differentially parasitize different cowpea lines.[18-20] A broader collection of *S. gesnerioides* isolates from this region was analyzed using genotypic profiling with molecular markers and host differential resistance response studies, and at least seven distinct races of the parasite were recognized.[21] The races were designated as follows: SG1 (Burkina Faso), SG2 (Mali), SG3 (Nigeria and Niger), SG4 (Benin), SG4z (localized to the Zakpota region of Benin), SG5 (Cameroon), and SG6 (Sénégal). SG1 and SG5 are the most closely related, while SG4 and SG3 are the most diverged. SG6, one of the new races of the parasite identified in Sénégal, was genetically most similar to SG4. The hypervirulent isolate of *S. gesnerioides* from Zakpota (SG4z) is genotypically distinct from other populations of SG4 located in this region and elsewhere in Benin.

Geographic variation in host preference has also been observed in *A. vogelii*. *A. vogelii* populations from West Africa and Cameroon attack cowpea and groundnut.[22] Isolates from eastern Botswana and northern portions of South Africa parasitize cowpea, groundnut, and mung bean, while those from the eastern portions of South Africa, Kenya, Malawi and Zimbabwe parasitize cowpea, groundnut, mung bean, and bambara groundnut. *A. vogelii* also has distinct races that differentially parasitize cowpea.[21,23,24] For example, the cowpea landrace B301 is resistant to *A. vogelii* in Kenya, but susceptible to isolates from Malawi, Botswana, and some areas of South Africa[22] and the cowpea breeding line IT81D-994 is resistant to *A. vogelii* in Nigeria, but susceptible to isolates from Malawi (C.J. Botanga, N. Skizim, and M.P. Timko, unpublished).

3. Mechanisms of Resistance

At least two mechanisms of resistance to *S. gesnerioides* have been described.[25,26] Neither type is resistance due to reduced parasite germination or failed haustorial formation as the parasite succeeds in attaching to the potential host and initiates penetration of the host tissue. Penetration of the resistant cowpea cultivar 58-57 from Sénégal by *Striga* was associated with rapid necrosis of the host cells around the point of infection, leading to the death of the parasite in 3 to 4 days. This mechanism of resistance is analogous to the hypersensitive response shown in plant-pathogen interactions.[27] The response in the host was specific with rapid death of cowpea tissue localized to the sites of parasite invasion.[28]

The second type of resistance mechanism was observed in cultivars B301 and IT81D-994, where resistance to *S. gesnerioides* parasitism was not as dramatic. In these interactions, the majority of *Striga* seedlings penetrated the cortex and reached the host stele. Although tubercles began to develop on the host root surface, these did not enlarge, remaining less than 0.5 mm in diameter (on B301), or failing to expand their cotyledons (on IT81D-994). In these same studies, the host resistance response was also dependent on which race of *S. gesnerioides* was used.[28,29] Tubercle arrest is also seen during interactions of *Striga* strains adapted for growth on one host species, when attempting to parasitize a non-host.[17] In these cases, careful examination has revealed that the parasite neither forms vascular bundles (no xylem-xylem connections are evident with the host) nor develops proper internal organization.

4. Genetics of Resistance to Parasitic Plants in Cowpea

The genetic basis of resistance to *S. gesnerioides* and *A. vogelii* parasitism has been examined by a few laboratories. Monogenic dominant inheritance has been demonstrated in the progeny of the crosses between Suvita 2, 58-57 or B301 and susceptible lines.[30-32] The region where the study was conducted and the strain of *S. gesnerioides* involved is not usually specified when the results of inheritance are mentioned, which limits the interpretation of these results.[33]

Single dominant genes confer resistance to SG1, SG2 and SG3 in the cultivar B301[34] (Table 1). Resistance to SG3 in the cultivars B301 and IT82D-849 may be conferred by different alleles at the same locus or tightly linked genes, as two types of resistance response are manifested[35]. Contradictory to other reports, resistance to *S. gesnerioides* race SG3 in Niger was conferred by a single recessive gene in IT82D-849.[34] Prior studies indicating the presence of a recessive gene for resistance to *S. hermonthica* and *S. asiatica* in sorghum were cited in support of their interpretation.[34] These results could indicate that more than one race of *S. gesnerioides* is present in Niger or that the response to SG3 in Niger is influenced the by level of parasite infestation, or environmental factors.

The inheritance of resistance to SG1 in Burkina Faso, was studied using two resistant cowpea varieties, HTR (from Niger) and Wango-1 (from Burkina Faso).[36] Resistance in HTR was controlled by one or two dominant genes that are nonallelic and independent of the resistance gene active against SG1 in IT82D-849 and B301 but possibly linked to the SG1 resistance gene in IT81D-994. Resistance to SG1 in Wango-1 is conferred by a single dominant gene probably allelic to the resistance gene in Gorom, and possibly linked to the resistance gene in IT81D-994.[36] Unfortunately, no supporting data are provided for these findings. However, more compelling data are available for SG1 resistance in the cowpea cultivar IT81D-994 conferred by a single dominant gene.[37]

Table 1. Inheritance of resistance to *S. gesnerioides* in some cowpea lines.

Cultivar/line	Inheritance	Race of *S. gesnerioides*	Ref.
B301	Single dominant	SG1 (Burkina Faso), SG2 (Mali)	26,34,36
		SG3 (Niger), SG3 (Nigeria)	32,34,35,43
IT82D-849	Single dominant	SG1 (Burkina Faso); SG3 (Nigeria)	34,35
		SG2 (Mali)	34
	Single recessive	SG3 (Niger)	34
Suvita 2	Single dominant	SG1 (Burkina Faso), SG2 (Mali)	34,35
IT81D-994	Single dominant	SG1 (Burkina Faso), SG2 (Mali)	37,53
HTR	1 or 2 dominant genes	SG1 (Burkina Faso)	36
Wango-1	Single dominant	SG1 (Burkina Faso)	36

Approximately 650 local cowpea varieties and exotic accessions were screened for resistance to *A. vogelii.* Landraces B301 and B359 from Botswana were among the most resistant genotypes.[38,39] The superiority of B359 as a source of resistance for southern Africa was demonstrated when it was shown to remain completely resistant to isolates of the parasite from Malawi, while B301, IT90K-59 and IT90K-76 (two lines derived from B301 as parent), all supported the emergence of parasites of a population from Malawi.[41] B359 was resistant in pot trials to isolates of *A. vogelii* from different locations in east, southern and west Africa, including Botswana, Cameroon, Mali, Malawi, Nigeria and South Africa.[22,39-41]

A number of cowpea lines were screened for resistance to *S. gesnerioides* and *A. vogelii* and the landrace B301 was resistant to both parasites.[32,35,42,43] In contrast, line IT82D-849 is resistant to *S. gesnerioides* but susceptible to *A. vogelii*, Suvita-2 (Gorom local) is resistant only to *S. gesnerioides* in Burkina Faso but susceptible elsewhere and susceptible to *A. vogelii*, and IT81D-994 is moderately resistant to *S. gesnerioides* as well as *A. vogelii*. While resistance to *S. gesnerioides* in B301 is controlled by a single dominant gene designated *Rsg*1, resistance to *A. vogelii* in this cultivar is controlled by duplicate genes, *Rav*1 and *Rav*2.[32,35] The data also indicate that the genes conferring resistance to *A. vogelii* in B301 are non-allelic and independent of each other and not linked to the *Striga* resistance gene *Rsg*1.[42] In subsequent studies the duplicate dominant genes for *Alectra* resistance in B301 were determined to be nonallelic to a single gene for resistance found in IT81D-994. *Rav*1 and *Rav*2 are used to designate the genes for resistance to *A. vogelii* in B301 and *Rav*3 for the resistance gene in IT81D-994.[35,43]

Table 2. Agronomic growth habit and disease and pest resistance trait loci on the current cowpea genetic map.[a]

Trait	Locus	Linkage Group
Pod pigmentation	P	LG1
S. gesnerioides SG1 resistance	*Rsg*2-1	LG1
S. gesnerioides SG3 resistance	*Rsg*4-3, *Rsg*1-1	LG1
Meloidogyne incognita resistance	*Rk*	LG1
Nodes to 1st Flower (D1301a)	NTF	LG2
Dehydrin protein	*Dhy*	LG2
Resistance to cowpea mosaic virus	CPMV	LG2
Resistance gene analog[e]	RGA-438	LG2
Resistance gene analog[e]	RGA-468	LG2
Resistance gene analog[e]	RGA-490	LG2
Fusarium oxysporum resistance	*Fus*R	LG3
Cowpea severe mosaic virus resistance	CPSMV (ims)	LG3
Cowpea mosaic virus resistance	CPMV	LG3
Resistance gene analog[e]	RLRR3-4B	LG3
General flower color factor	C	LG4
Seed weight (OB6a)	SW	LG5
Resistance gene analog[e]	RGA-434	LG5
Southern bean mosaic virus resistance	SBMV(*sbc-1, 2*)	LG6
S. gesnerioides SG1 resistance	*Rsg*3-1, *Rsg*-994	LG6
Blackeye cowpea mosaic virus resistance	BlCMV	LG8
Resistance gene analog[a]	RLRR3-4T	LG9
Previously mapped traits not placed on the current map[b]		
Cowpea aphid (*Aphis craccivora*) resistance[c]	*Rac1*	
50% Flowering[d]	50%FL	
Seed weight[d]	SW	
Plant height[d]	HT	
Pod number per plant[d]	PodN	

[a]Adapted from Ouédraogo *et al.*[44]
[b]Insufficient marker data is available to allow placement on current map
[c]From Myers *et al.*[55]
[d]From Fatokun *et al.*[54]
[e]Function has not yet been determined[52]

5. Genetic Mapping of *Striga* Resistance Genes

The most complete genetic map currently available was drawn by Ouédraogo, *et al.*[44] It is based on segregation in 94 recombinant inbreds derived from a cross between IT84S-2049, an advanced breeding line of African origin (Nigeria), and 524B, a Blackeye type which encompasses the genetic variability available in cowpea cultivars in California.[45] The cowpea genetic map consists of 11 linkage groups (LGs) spanning a total of 2670 cM, with an average distance of ca. 6 cM between markers. It includes 242 AFLP and 18 disease or pest-resistance-related markers,[44,52] plus 133 RAPD, 39 RFLP, and 25 AFLP markers from the map of Menéndez, *et al.*,[45] for a total of 441 markers, of which 432 were assigned to a specific LG. The various agronomic and disease resistance trait loci that have now been placed on the cowpea genetic map are listed in Table 2.

Three AFLP markers are linked to *Rsg*2-1, a gene that confers resistance to SG1 present in Burkina Faso, and six AFLP markers linked to gene Rsg4-3, a gene that provides resistance to SG3 from Nigeria (Fig. 1).[37] Two of the AFLPs were associated with both *Rsg*2-1 and *Rsg*4-3.[37] Two AFLP markers are closely linked to *Rsg*1-1, a gene that also confers resistance to SG3 in Nigeria.[50] Five markers were subsequently found linked to the *Rsg*994-1 gene on LG6 that also confers resistance to SG1.[46]

The *Striga* resistance genes mapped thus far cluster in two locations in the cowpea genome (Fig. 1.). Markers linked to the *S. gesnerioides* race SG1 and SG3 resistance genes (*Rsg*2-1, *Rsg*1-1 and *Rsg*4-3) present in the resistant cowpea lines B301, IT82D-849 and Tvu 14676, respectively map to LG1, whereas markers linked to the *S. gesnerioides* race SG1 resistance genes *Rsg*3-1 and *Rsg*994-1 present in Suvita-2 and IT91D-994, respectively, were located to LG6.[37,44,46]

6. Molecular Markers and Marker-Assisted Selection

Marker-assisted selection (MAS) is the identification of DNA sequences located near genes that can be tracked to help in the selection of traits that are difficult to observe. In practice, MAS is a tool to more efficiently assemble alleles of interest into an improved cultivar and thereby increase the overall efficiency and effectiveness of crop

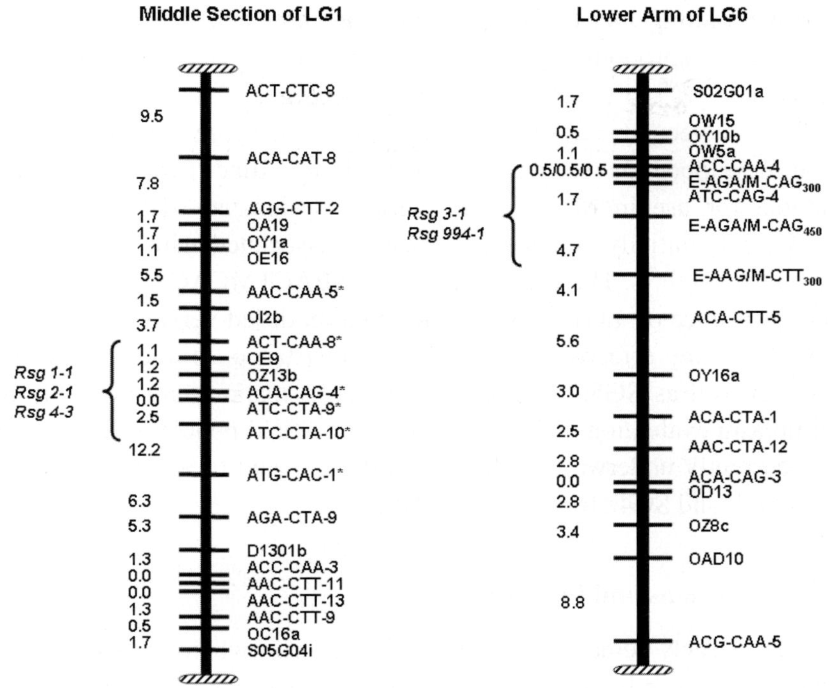

Figure 1. Location of molecular markers linked to S. gesnerioides *race-specific resistance genes in cowpea. Portions are shown of linkage groups 1 (LG1) and 6 (LG6) of the cowpea genetic map developed by Ouédraogo et al.*[44] *indicating the location of AFLP, RAPD, and other markers linked to resistance to* S. gesnerioides *race 1 (SG1) (*Rsg2-1 *and* Rsg1-1*) and race 3 (SG3) (*Rsg4-3*) (*left side*) and* S. gesnerioides *race 1 (SG1) (*Rsg3-1 *and* Rsg994-1*) (*right side*). The relative map distances are given in centimorgans (cM). AFLP markers indicated by an asterisk are being used to develop sequence confirmed amplified regions (SCARs).*

improvement programs.[47] In some cases, MAS can allow smaller populations to be used, reduce the number of generations needed to reach a goal, or increase the accuracy of evaluations.[48] MAS offers the only practical method to combine multiple resistance genes into one cultivar to provide more durable resistance.[49]

MAS has yet to be implemented in cowpea, but some of the groundwork has been laid for its development by constructing a genetic map pinpointing loci controlling important pest and disease resistance genes and agronomic traits including genes for *S. gesnerioides* resistance.[37,44,46]

Two molecular markers linked to *S. gesnerioides* resistance have been developed as sequence confirmed amplified regions (SCARs) suitable for use in MAS. One marker, designated 61R (E-ACT/M-CAA), was initially isolated as a marker associated with resistance to SG1 on LG1.[37,44] The second SCAR is SEACTMCAC83/85 linked to SG3 resistance on LG1.[50] Both 61R and a modified version of it termed MahSE2[51] are effective in identifying resistance to races SG1 and SG3, as well as SG5. At present, these two markers are available for germplasm evaluation and efficacy testing on field populations. Work is also currently underway to identify markers linked to resistance to SG2 from Mali and SG4z from Zakpota, Benin.

7. Conclusions and Perspective

Cowpea largely remains an underexploited crop where relatively large genetic gains can be made with only modest investments in both applied plant breeding and molecular genetics. One of the major goals of cowpea improvement programs is to combine resistances to numerous pests and diseases and other desirable traits (such as those governing maturity, photoperiod sensitivity, plant type, and seed quality) in agroecologically adapted cultivars. Landraces and local cultivars with many of the desired disease and pest resistance traits (e.g., resistance to cowpea weevil, cowpea aphid, bacterial blight, CABMV, root knot nematodes) and resistance to one or more of the defined races of *A. vogelii* and *S. gesnerioides*, have been identified and are presently being integrated in various cowpea breeding programs around the world.[1,2] A decade, more or less, is needed to breed a superior improved line using traditional selection and hybridization strategies depending on the source of the trait being introgressed.

The current focus in applied breeding is leveraging biotechnological tools to develop more and better markers linked to important disease and pest resistance traits and the establishment of breeder friendly protocols that will allow marker-assisted selection (MAS) and marker-assisted breeding (MAB) to be readily employed. The hope is that MAS and MAB will complement and extend conventional breeding efforts in cowpea and speed up the delivery of improved cultivars to the farmer. To date, however, progress in marker development and delivery of useful markers has been slow. With well-defined race-specific markers it should be possible to breed cultivars with resistance to all currently defined races of *S. gesnerioides* and *A. vogelii*.

It is also hoped that the application of knowledge being gained from basic genomic research on other crop plants and "model species" will also contribute to more rapid cowpea improvement. As information on genome structure and composition becomes available from a wide variety of legumes, comparative genomics can be employed for gene/trait identification in cowpea where existing bioassays may not be readily available or are too difficult to conduct. Understanding syntenic relationships is one of the many research areas that will have cross cutting impact on breeding in all legumes.

The integration of genetic engineering and transgenic crops into traditional breeding programs is another issue that needs to be considered. At present, the ability to transform cowpea and generate transgenic lines containing desired resistance and agronomic traits is limited.[2] Without improved selection technologies, it is likely to take as long to introgress a molecularly engineered trait into an improved cultivar as it takes for a natural gene variant, if one is there to be found. Finding recessive mutants is also highly unlikely. So far, low stimulant and low attachment mutants have not been found, as they have with sorghum (Chapter 7). The challenge facing us in the near future is to demonstrate that biotechnologically-based alternative methods can generate knowledge and cost-effective tools that enable germplasm enhancement and product development opportunities that are either complementary or superior to those currently in use. The limitation is how rapidly refinements and changes to plant breeding methodology can be made available to the breeder. We are clearly still at the first of many steps in this long process.

Acknowledgements

We wish to thank Dr. Christopher Botanga for his comments and input into these studies and the members of the Timko laboratory especially Jianxiong Li, Edmond Chipumuro, Aaron Asare, and Josesph Opuko for their suggestions. We also wish to thank our collaborators on this work, Drs. Moctar Wade, Ndiaga Cisse, and Festo Massawe. This work was supported in part by grants from the Kirkhouse Trust, Generation Challenge Program, and National Science Foundation awarded to MPT.

References

1. B. B. Singh, in *Genetic Resources, Chromosome Engineering and Crop Improvement. Volume 1*, Eds. R. J. Singh and P. P. Jauhar (CRC Press, Boca Raton, FL, 2005), p. 117.
2. M. P. Timko, J. D. Ehlers, and P.A. Roberts (2007) in, *Genome Mapping and Molecular Breeding in Plants, Volume 3 Pulses, Sugar and Tuber Crops*, Ed. C. Kole (Springer-Verlag, Berlin, Heidelberg, 2007), p. 49.
3. B. B. Singh, (2002) in, *Challenges and Opportunities for Enhancing Sustainable Cowpea Production,* Eds. C. A. Fatokun, S. A. Tarawali, B. B. Singh, P. M. Kormawa, and M. Tamo (International Institute of Tropical Agriculture, Ibadan, Nigeria, 2002), p. 3.
4. S. Langyintuo, J. Lowenberg-DeBoer, M. Faye, D. Lambert, G. Ibro, B. Moussa, A. Kergna, S. Kushwaha, S. Musa, and G. Ntoukam, G. *Field Crops Res.* **82**, 215 (2003).
5. R. J. Carsky, B. Vanlauwe and O. Lyasse, in *Challenges and Opportunities for Enhancing Sustainable Cowpea Production,* Eds. C.A. Fatokun, S.A. Tarawali, B. B. Singh, P. M. Kormawa, and M. Tamo (International Institute of Tropical Agriculture, Ibadan, Nigeria, 2002), p. 252.
6. S. A. Tarawali, B .B. Singh, S. C. Gupta, R. Tabo, F. Harris, S. Nokoe, S. Fernández-Rivera, A. Bationo, V. M. Manyong, K. Makinde and E. C. Odion, in, *Challenges and Opportunities for Enhancing Sustainable Cowpea Production,* Eds. C. A. Fatokun, S. A. Tarawali, B. B. Singh, P. M. Kormawa, and M. Tamo (International Institute of Tropical Agriculture, Ibadan, Nigeria, 2002), p. 233.
7. N. Sanginga, K. E. Dashiell, J. Diels, B. Vanlauwe, O. Lyasse, R.J. Carsky, S. Tarawali, B. Asafo-Adjei, A. Menkir, S. Schulz, B.B. Singh, D. Chikoye, D. Keatinge and R. Ortiz, *Agric. Ecosyst. Environ.* **100**, 305 (2003).
8. C. Parker and C. R. Riches, *Parasitic Weeds of the World: Biology and Control* (CAB International, Wallingford, UK, 1993).
9. K. I. Mohamed, L. J. Musselman and C. R. Riches, *Annals Missouri Bot. Gard.* **88**, 60 (2001).
10. V. D. Aggarwal and J. T. Ouédraogo, *Trop. Agric.* **66**, 91 (1989).

11. M. Emechebe, B. B. Singh, O. I. Leleji, I. D. K. Atokple and J. K. Adu, in, *Combating Striga in Africa*, Ed. S. K. Kim (IITA, Ibadan, Nigeria, 1991), p. 18.
12. H. Bagnall-Oakley, V. Gibberd and T. E. Nyongesa, in, *Proceedings of the 5th International Symposium on Parasitic Weeds*, Eds, J. K. Ransom, L. J. Musselman, A. D.Worsham and C. Parker (CYMMIT, Nairobi, Kenya, 1991).
13. M. Mbwaga, J. Kaswende and E. Shayo, in, *A Reference Manual on Striga Distribution and Control in Tanzania* (Ilnga Agricultural Research Institute, Kilosa, Tanzania, 2000).
14. J. A. Lane, T. H. M. Moore, D. V. Child and K. F. Cardwell, *Plant Dis.* **80**, 299 (1996).
15. L. J. Musselman and C. Parker, *Weed Sci.* **29**, 594 (1981).
16. H. Wild, *Rhodesia Agric. J.* **45**, 208 (1948).
17. C. J. Botanga and M. P. Timko, *Phytopathology* **95**, 1166 (2005).
18. J. A. Lane, T. H. M. Moore, D. V. Child, K. F. Cardwell Lane, B. B. Singh and J. A. Bailey, *Euphytica* **72**, 183 (1994).
19. J. A. Lane, D. V. Child, G. C. Reiss, V. Entcheva and J. A. Bailey, in *The Gene-for-Gene Relationship in Plant-Parasite Interactions,* Eds. I. R. Crute, E. B. Holub, and J.J . Burdon (CAB, Wallingford, UK 1997), p. 81.
20. J. A. Lane, T. H. M. Moore, D. V. Child and J. A. Bailey, in, *Advances in Cowpea Research*, Eds. B. B. Singh, D. R. Mohan Raj, K. E. Dashiell & L. E. N. Jackai (International Institute for Tropical Agriculture and Japan Agricultural Research Centre for Agricultural Sciences, Ibadan, Nigeria, 1997), p. 225.
21. C. J. Botanga and M. P. Timko, *Genome* In Press (2007).
22. C. R. Riches, K. Hamilton and C. Parker, *Ann. Appl. Biol.* **37**, 361 (1992).
23. T. I. Polniaszek, C. Parker and C. R. Riches, *Trop. Pest Manage.* **37**, 152 (1991).
24. B. B. Singh, A. M. Emechebe and I. D. K. Atokple, *Crop Sci.* **33**, 70 (1993).
25. J. A. Lane, M. J. Kershaw, T. H. M. Moore, D. V. Child, P. J. Terry and J. A. Bailey, in, *Proceedings of the Fifth International Symposium on Parasitic Weeds*, Eds., J. K. Ransom, L. J. Musselman, A. D. Worsham and C. Parker (CIMMYT, Nairobi, Kenya, 1991) pp. 237.
26. J. A. Lane and J. A. Bailey, *Euphytica* **63**, 85 (1992).
27. M. C. Heath, *Plant Mol. Biol.* **44,** 321 (2000).
28. J. A. Lane, J. A., Bailey, R. C. Butler and P. J. Terry, *New Phytologist* **125**, 405 (1993).
29. J. A. Lane, T. H. M., Moore, D. V. Child, K. F. Cardwell, B. B. Singh and J. A. Bailley, *Euphytica* **72**, 183 (1993).
30. V. D. Aggarwal, N. Muleba, I. Drabo, J. Souma, and M. Mbewe, in, *Proceeding of the Third International Symposium on Parasitic Weeds*, Eds., C. Parker *et al.* (ICARDA-IPSPRG, Aleppo, Syria 1984).
31. V. D. Aggarwal, in, *Combatting Striga in Africa*, Ed. S. K. Kim (IITA, Ibadan,1991), p. 90.
32. B. B. Singh and A. M. Emechebe, *Crop Sci.* **33**, 879 (1990).
33. M. P. Dubé and A. Olivier, *Can. J. Bot.* **79**, 1225 (2001).
34. M. Touré, A. Olivier, B. R. Ntare, J. A. Lane and C. -A. St-Pierre, *Euphytica* **94**, 273 (1997).

35. D. K. Atokple, B. B. Singh, and A. M. Emechebe, *J. Hered.* **86**, 45 (1995).
36. M. P. Dubé, F. Belzile, J. B. Tignégré and A. Olivier, in, *Proceedings of the 7th International Parasitic Weed Symposium*, Eds. A Fer, P.Thalouarn, D M. Joel, L. J. Musselman, C. Parker, and J. A. C. Verkleij (Imprimerie Parenthèses, Nantes, France, 2001), p. 243.
37. J. T. Ouédraogo, V. Maheshwari, D. Berner, C.-A. St-Pierre, F. Belzile and M. P. Timko, *Theor. Appl. Genet.* **102**, 1029 (2001).
38. C. R. Riches, in, *Parasitic Flowering Plants*, Eds., H. Chr. Weber and W. Forstreuter (Phillips University, Marburg, Germany, 1987), p. 701.
39. C. R. Riches, in, The biology and control of *Alectra vogelii* Benth. (Scrophulariaceae) in Botswana. (Ph.D. Thesis, University of Reading, Reading, UK, 1989)
40. Riches, C. R., in, *Striga Research in Southern Africa and Strategies for Regionalized Control Options,* Ed., M.A. Mgonja, (International Crops Research Institute for the Semi-arid Tropics, Bulawayo, Zimbabwe, 2001), p. 89.
41. C. E. Mainjeni, in, The host range of *Alectra vogelii* Benth. from Malawi and resistance in common bean and cowpea (M.Sc. Thesis, University of Bath, UK, 1999).
42. B. B. Singh, A. M. Emechebe and I. D. K. Atokple, *Crop Sci* **33**, 70 (1993).
43. D. K. Atokple, B. B. Singh and A. M. Emechebe, *Crop Sci* **33**, 714 (1993).
44. J. T. Ouédraogo, B. S. Gowda, M. Jean, T. J. Close, J. D. Ehlers, A. E. Hall, A. G. Gillaspie, P. A. Roberts, A. M. Ismail, G. Bruening, P. Gepts, M. P. Timko and F. J. Belzile, *Genome* **45**, 175 (2002).
45. C. M. Menéndez, A. E. Hall, and P. Gepts, *Theor. Appl. Genet.* **95**, 1210 (1997).
46. J. T. Ouédraogo, J.-B. Tignegre, M. P. Timko, and F. J. Belzile, *Genome* **45**, 787 (2002).
47. A. Charcosset and L. Moreau, *Euphytica* **137**, 81 (2004).
48. H. C. Sharma, J. H. Crouch, K. K. Sharma, N. Seetharama, and C. T. Hash, *Plant Sci.* **163**, 381 (2002).
49. J. D. Kelly, P. Gepts, P. N. Miklas and D. P. Coyne, *Field Crops Res.* **82**, 135 (2003).
50. O. Boukar, L. Kong, B. B. Singh, L. Murdock and H. W. Ohm, *Crop Sci.* **44**, 1259 (2004).
51. J. T. Ouédraogo, M. Ouédraogo, B. S. Gowda and M. P. Timko, Manuscript in preparation (2007).
52. B. S. Gowda, J. L. Miller, S. S . Rubin, D. R. Sharma and M. P. Timko, *Euphytica* **126**, 365 (2002).
53. M. Touré, A. Olivier, B. R. Ntare, J. A. Lane and C. -A. St-Pierre, *Can. J. Plant Sci.* **78**, 477 (1998).
54. C. A. Fatokun, D. Danesh, D. Menancio-Hautea and N. D. Young, in, *A compilation of linkage and restriction maps of genetically studied organisms, Genetic maps 1992*, Ed. J. S. O'Brien (Cold Spring Harbor Laboratory Press, Cold Spring Harbor, NY, 1993), p. 6.256.
55. G. O. Myers, C.A. Fatokun and N.D. Young, *Euphytica* **91**, 181 (1996).

CHAPTER 10

INTROGRESSION OF GENES FOR *STRIGA* RESISTANCE INTO AFRICAN LANDRACES OF SORGHUM

Issoufou Kapran[a,*], Cecile Grenier[b] and Gebisa Ejeta[b]

[a]*Institut National de la Recherche Agronomique du Niger (INRAN), BP429 Niamey, Niger*

[b]*Dept of Agronomy, Purdue University, West Lafayette, IN 47907, U.S.A.*

[*]*E-mail: ikapran@yahoo.com*

Early maturing sorghum varieties are particularly suited to the low rainfall areas in Sahelian Niger. Landrace El Mota (EM) was selected by farmers for its adaptation to erratic rainfall, poor soil fertility and its high yield potential. However EM lacks resistance to *Striga*, which often results in major yield reduction. Natural resistance to *Striga* found in an African cultivar, SRN39 was introgressed into El Mota. A total of 103 BC_2F_3 introgression lines and two parental genotypes were tested in Niger for two consecutive seasons. Data on resistance to *Striga* and agronomic performance were recorded. The *Striga* count at 90 days after planting had homogeneous variance across years, heterogeneous variances among lines and the highest heritability. None of the introgression lines ranked as high in resistance as SRN39, but as many as 25 lines with the lowest *Striga* count had agronomic performances similar to El Mota. Laboratory assays revealed that the particular mechanism of resistance introgressed in the El Mota background was that of an incompatible reaction to *Striga*.

1. Introduction

Sources of *Striga* resistance have been identified[1] and characterized in sorghum germplasm.[2] Wide regional testing of these sources of resistance has resulted in formal releases in Sudan, Niger,[3] and Ethiopia (Chapter 15). Subsequent breeding has led to transfer of specific major genes for resistance to *Striga* from these sources into more productive sorghum germplasm backgrounds.

Nevertheless, there are several environments in Africa with unique agro-ecologies where improved cultivars do not perform as well as local landraces. Photoperiod sensitive late-maturing Guinea sorghums are cultivated in the high rainfall zone in Mali, maturing after the rains, which helps avoid grain deterioration caused by pests and diseases. Long grain-fill durra sorghums with cold tolerance are the most adapted in the highlands of Ethiopia. The so-called rice-type Guineas are also favored in eastern Tanzania for their unique adaptation to high rainfall. In Sahelian Niger and the low rainfall plains of the Sudan there is a preference for early maturity associated with rapid grain-fill, which makes El Motas and Feteritas, respectively, the cultivars of choice with farmers in the drylands. In these particular environments, the improved *Striga* resistant caudatums are less adopted because of problems of grain weathering and low food quality despite their higher yield potential.

The objective of this research is to transfer *Striga* resistance genes from known sources to well-adapted landrace cultivars. The goal is to deliver established cultivars with protection against *Striga* infestation afforded through a few genes as a stop-gap measure to local farmers through a phenotype-based, bioassay-mediated or marker-assisted introgression. This should ease the usual problems associated with transfer of new technology. It is believed that this approach may also enhance farmer adoption of subsequent new cultivars and associated technologies distributed by research and development agencies. We assessed the feasibility of selecting *Striga* introgression lines under field conditions in Africa.

2. Materials and Methods

2.1. *Development of Advanced Backcross Populations*

Gene introgression was based on a series of backcrosses performed to add *Striga* resistance from a donor parent to otherwise adapted local cultivars. In the introgression procedure, the landrace was used as a seed parent while pollen was obtained from the resistant parent. The resultant F_1 plants were then backcrossed (BC_1F_1) to the recurrent landrace. Additional backcrossing resulted in BC_2F_1 families that were then selfed for two successive generations to generate BC_2F_3 progenies used for phenotyping. El Mota (EM), a widely adapted local landrace cultivar in Niger was selected as the recurrent parent for backcrossing. This landrace has valuable agronomic characteristics, but lacks resistance to *Striga* although preliminary evaluations have shown that some landraces possess a certain level of tolerance. The donor parent we chose is SRN39, an African sorghum inbred line that has been extensively screened both in field and laboratory conditions and has an extremely good level of resistance to *Striga* (Chapter 12). At least two mechanisms of resistance were reported in SRN39.

2.2. *Field Trials*

An advanced backcross population made of a BC_2F_3 progeny between EM and SRN39 and containing 103 lines was evaluated for resistance to *Striga* at the Birni N'Konni location (13°82 N, 5°32 E) in Niger during 2002 and 2003. The two parents were included in the trial. Although the Birni N'Konni experimental station has a naturally infested *Striga* plot dedicated to such studies, artificial *Striga* infestation was added to the field in 2002 to increase the level and uniformity of *Striga* inoculum. No N fertilizer was used during the two years of experiment. The experiment was rainfed, with averages of 438 mm and 523 mm in 2002 and 2003, respectively.

The experiment was laid out in a randomized complete block design with three replications (blocks). Additionally, each parental genotype was repeated five times within each block. All entries were planted in single row plots, six and three meters long in 2002 and 2003, respectively. Row planting was done on hills spaced every 30 cm, providing 21 hills (2002) or 11 hills (2003) per row. Rows were 80 cm apart and separated by an empty row at both sides to increase chances for infestation and proper evaluation of each progeny. Plots were thinned to a single plant per hill.

Agronomic data on performance of host plants including a measure of maturity in number of days to half bloom and yield in grams per plot were recorded. *Striga* related traits were collected as the number of days to first *Striga* emergence, the number of *Striga* plants at 90 days after planting, and the ratio of the number of hills with flowered *Striga* over the total number of sorghum hills with *Striga*. Actual numbers of hills with sorghum plants were counted at 90 days after planting to parallel *Striga* counts.

2.3. *Laboratory Assays for Resistance in BC_2F_3 Lines*

The laboratory assays were carried out at Purdue University to screen for *Striga* resistance among a set of BC_2F_3 lines. These entries were selected based upon the field evaluation as the 14 most resistant and 14 most susceptible lines. We looked at mechanisms of resistance intervening before and after parasite attachment and that are reported in the SRN39 cultivar, notably the low germination stimulant production and the incompatible reaction (Chapter 7). Six seedlings were assayed for each line and the parental genotypes. The germination stimulant production was recorded as the distance between the host root and the furthest germinated *Striga* plants. For post-attachment resistance, *Striga* seeds were treated with GR24 at 3 days after infection to induce equal germination for all genotypes. At 7 days after infection, the number of attached *Striga* per sorghum seedling was recorded and the development stage of each parasite was subsequently observed at 14 and 21 days after infection. Three stages in *Striga* development are recognized including stage 1 characteristic of a newly attached *Striga* plant with the seed coat still intact; stage 2 whereby the seed coat is broken and a first pair of leaves appears; and stage 3 when several pairs of leaves have developed.

2.4. *Data Analysis*

Statistical analyses were performed on combined two years data using the GLM and MIXED procedures of SAS V8.2 (SAS Institute).[4] The following linear model was used:

$$Y_{ikjl} = \mu + Y_i + R_{(i)j} + G_k + YG_{ik} + \varepsilon_{ijkl}$$

where,
μ = overall population mean;
Y_i = effect of the ith year (random) $i = 1, 2$;
$R_{(i)j}$ = effect of the jth replication at the ith year (random) $j = 1, \ldots, 3$;
G_k = effect of the kth genotype (fixed) $k = 1, \ldots, 105$;
ε_{ijkl} = experimental error, assumed normally and independently distributed $(0, \sigma^2_\varepsilon)$.

Years, replications and backcross progeny (lines) were considered random while parental genotypes were considered fixed effects. Data were transformed, as described by Box *et al.*[5] Only days to half bloom, did not require transformation. Square root transformation was used for yield and the number of hills with flowered *Striga*, and log transformation was applied to number of hills with emerged sorghum plants and the number of *Striga* plants at 90 days after planting. Each source of variation in the model was tested with the corresponding error as defined by the expected mean squares in Table 1.

Table 1. Expected Mean Squares (EMS).

Source	df	Expected Mean Square
Year (Y)	y-1	$\sigma^2_\varepsilon + g\sigma^2_r + rg\sigma^2_y$
Rep/Y	y(r-1)	$\sigma^2_\varepsilon + g\sigma^2_r$
Genotypes (G)	g-1	$\sigma^2_\varepsilon + rg\sigma^2_{yg} + yrg\sigma^2_g$
Y × G	(y-1)(g-1)	$\sigma^2_\varepsilon + rg\sigma^2_{yg}$
Error	y(r-1)(g-1)	σ^2_ε

Entries were partitioned into two components: lines (progeny) and parents. The year x genotype interaction and the experimental error terms were subdivided accordingly. The partition of year x genotype were tested for homogeneity and pooled to offer a common error term for testing the following main effects: lines vs. parents (df = 1), among lines (df = 102), and between parents (df = 1). Similarly the experimental error was used as common error term for testing replications and years. Variation among the lines ($l = 1, \ldots, 103$) was used to calculate the genotypic variance component and broad-sense heritability (H_f) on a family-mean basis. The following formula was used:

$H_f = \hat{\sigma}_l^2/(MS_l/yr)$, where $\hat{\sigma}_l^2 = (MS_l - MS_{yl})/yr$ and $\hat{\sigma}_1^2$ is the genotypic variance.

Means were compared using the Least Square Difference (LSD) test. Pair-wise means comparisons were tested between two lines, between a line and a parent and between parents. Special interest was given to the performance of individual lines with extreme phenotypes, which were also compared to either the recurrent parent for sorghum traits or the resistant parent for *Striga*-related traits. The LSD was computed to test the significance of difference between the means of two observations with equal or unequal number of observations (n) as follows: $LSD_\alpha = t_{\alpha/2} \sqrt{2MS_{YG}/n}$ for equal n or $LSD = t_{\alpha/2} \sqrt{(MS_{YG} \times (1/n_1 + 1/n_2)}$ for unequal n, with MS_{YG} equal to the MS for the year interaction. In our case, n=y×r and thus $LSD_\alpha = t_{\alpha/2} \sqrt{2MS_{YG}/6}$ to compare two lines, $LSD_\alpha = t_{\alpha/2} \sqrt{2MS_{YG}/30}$ to compare the two parents and $LSD_\alpha = t_{\alpha/2} \sqrt{MS_{YG}(1/6 + 1/30)}$ to compare one line and one parent. The two-sided $t_{\alpha/2}$ value was obtained at $t_{(0.05;df)}$ with df=(y-1)(g-1).

The maximum germination distance was noted as the average of 3 measurements taken 3 days after infection in laboratory assay of pre attachment resistance for each genotype. The stages of *Striga* development and the general appearance of the parasite/host association were recorded to ascertain post-attachment resistance.

3. Results

3.1. *Analysis of Variance*

Results for sorghum and *Striga* related traits are shown in Table 2. Except for yield, year effect and year x genotype interaction were not significant among the 105 genotypes for all traits considered. Only for number of days to first *Striga* emergence genotypes effect was not significant. With genotypes segregated, there were highly significant differences between lines and parents for all traits measured. Similarly, there were highly significant differences between the two parents for all traits analyzed. Highly significant differences among the 103 lines were found for all sorghum traits, and only for the number of *Striga* plants at 90 days among the *Striga* related traits. This indicates more genetic variation for sorghum related traits than for *Striga* related traits in the lines.

3.2. *Heritability*

Estimates of broad sense heritability for all traits were calculated (Table 3). Heritability values were generally greater for sorghum related traits than for *Striga* resistance traits. This is probably due to the complexity of the mechanisms involved in *Striga* infestation and interaction with environments. Among the *Striga* related trait, the number of *Striga* plants at 90 days after planting showed the highest heritability (38%).

Table 2. Analysis of variance of sorghum and *Striga* related traits.

Source	df	Sorghum related trait		*Striga* related traits		
		FLO	YIE	SEME	S90D	SCAP
Year (Y)	1	393.35ns	13200.9**	0.0049ns	33.825ns	61.75ns
Rep / Y	4	84.94**	199.46**	0.0352**	4.795**	11.25**
Genotypes (G)	104	33.16**	67.53**	0.0081ns	0.409**	3.36*
Lines (L) vs. Parents (P)	1	90.14**	134.39**	0.0687**	6.577**	31.92**
Among L	102	30.88**	50.69**	0.0062ns	0.216**	2.73ns
Between P	1	209.07**	1718.52**	0.1352**	13.922**	38.94**
Y × G	104	4.17ns	19.12**	0.0077ns	0.131ns	2.56ns
Y × L vs. P	1	2.64ns	147.10**	0.0030ns	0.004ns	10.54ns
Y × L	102	3.75ns	14.91*	0.0076ns	0.133ns	2.41ns
Y × P	1	48.60**	321.13**	0.0199ns	0.011ns	9.89ns
E	464	3.26	11.01	0.0060	0.132	2.83
L error	4	3.63	3.61	0.0005	0.032	5.19
L vs. P error	408	3.37	10.83	0.0066	0.128	2.50
P error	4	1.33	11.74	0.0005	0.039	2.06
Pure error	48	2.50	13.09	0.0030	0.178	5.48

*, ** Significant at $p \leq 0.05$ and $p \leq 0.01$, respectively.

FLO: number of days to half bloom; YIE: yield in grams/plot; SEME: number of days to first *Striga* emergence; S90D: number of *Striga* plants at 90 days after planting; SCAP: ratio of the number of hills with flowered *Striga* over the total number of sorghum hills with *Striga*.

3.3. *Means Comparison*

Comparison of the two parents in Table 4 shows that parent EM was earlier maturing, and a higher yield than parent SRN39. SRN39 had fewer *Striga* plants than EM, as expected. *Striga* plants on SRN39 emerged later and fewer reached flowering stage than on EM.

The 103 individual lines were compared to each parent by LSD analysis (Table 4). Although some lines performed well for *Striga* related traits, no line was comparable to SRN39 for number of *Striga* plants at 90 days after planting. In terms of agronomic performance, 47 lines and 27 lines were similar to EM for number of days to half bloom, and yield, respectively.

Table 3. Estimates of broad sense heritability of sorghum related traits and *Striga* related traits.

Trait	H_f
Sorghum related traits	
Number of days to half bloom	0.88
Yield in grams/plot	0.71
Striga related traits	
Number of days to first *Striga* emergence	0.22
Number of *Striga* plants at 90 days after planting	0.38
Number of hills with flowering *Striga*	0.12

Lines scoring closest to SRN39 with the lowest number of *Striga* plants at 90 days after planting were compared to both parents for their agronomic performance. Out of the 41 selected lines, 17 showed performance comparable to EM for number of days to half bloom, and 21 for yield.

3.4. *Characterization of Resistance in BC_2F_3 Lines*

Among the 14 lines selected with the best level of field resistance to *Striga,* 13 entries had maximum germination distance similar to EM and all had significantly greater scores than SRN39 (results not shown). Based on this experiment it appears that the resistant progenies were high stimulant producing lines. This suggests that the field resistance found among these 14 lines is not linked to low germination stimulant production.

The post-attachment mechanism of resistance referred as incompatible reaction translates to retarded growth and development of *Striga* plants. Within the set of progeny we screened, only one was free of attached *Striga*, and 5 progenies had attached *Striga* that did not develop further than stage 1. The resistant SRN39 is typically characterized with either no *Striga* attachment or, occasionally, few attached *Striga* at growth stage 1. Thus the incompatible response of SRN39 was transmitted to a few of the BC_2F_3 lines through introgression.

As expected for a recessive trait, we may have lost the character of low germination stimulant production of the donor parent, but five progenies seem to express a resistance similar to an incompatible reaction, as *Striga* that attached did not develop.

Table 4. Comparison of 103 progenies to parental genotypes for sorghum and *Striga* related traits.

Parent	FLO	YIE	SEME	S90D	SCAP
SRN39	59.50a	10.43a	1.771a	0.559a	1.936a
EM	55.77b	21.13b	1.676b	1.523b	3.547b
LSD_{α}	0.93	1.71	0.040	0.188	0.870
No. of progenies similar to parent[a]					
SRN39 [b]			26	0	48
EM [c]	47	27			

[a] Similarity defined when progeny mean falls within the range of the parent means ±1LSD
[b] *Striga* related traits only
[c] sorghum related traits only
FLO: number of days to half bloom; YIE: yield in grams/plot; SEME: number of days to first *Striga* emergence; S90D: number of *Striga* plants at 90 days after planting; SCAP: ratio of the number of hills with flowered *Striga* over the total number of sorghum hills with *Striga*. Parental means with the same letter are not significantly different at the critical value $t_{\alpha/2}$: 2.002.

4. Discussion

No significant environmental variation was detected for the traits related to agronomic performance except for yield (Table 2). The significant difference between years for yield was in part due to midge attack, a serious unpredictable panicle insect pest at the Birni N'Konni location. Genetic differences in agronomic performance were detected among lines, between the two parents, and between parents and lines.

The largely non-significant environmental variations for all traits related to resistance to *Striga* in the analysis of variance (Table 2) indicate the predominance of genotypic effects. Variation among genotypes was highly significant for all traits except for number of days to first *Striga* emergence. Also variation among lines for number of days to first *Striga* emergence was not significant despite highly significant differences between parents. The mechanism(s) of resistance to *Striga* introgressed in the progeny may not affect the time for *Striga* to emerge above the ground. Highly significant differences revealed genetic differences between the two parents, and between the parents and lines for all *Striga* related traits. Only the number of *Striga* plants at 90 days after planting exhibited genetic differences among lines. These results show that important genetic variation for number of *Striga* plants at 90 days after planting was present in the parental lines and was largely transmitted to the progeny. *Striga* counts were one of the most effective field measurements of *Striga* resistance reported.[7] Our result is a clear indication that resistance to *Striga* has been successfully introgressed in this second generation backcross population.

In general, heritability was greater for agronomic traits than for *Striga* resistance traits (Table 3). This indicates that selection in this population is feasible but it would be more efficient for agronomic characters than for *Striga* resistance traits *per se*. The number of *Striga* plants at 90 days after planting could be a valuable trait to consider for selecting lines in an introgression program as it showed the highest heritability.

The mean performance of SRN39 in the Birni N'Konni environment shows that this cultivar is distinctly better for *Striga* resistance than the local variety EM but does not have the required agronomic traits (Table 4). EM has excellent adaptation and some degree of tolerance to *Striga*. A study of genotype x environment interaction for *Striga* resistance and grain yield stressed the need to combine both resistance and tolerance to *Striga*.[6] Hence in a sorghum improvement program where the goal is to retain agronomic performance of one cultivar and enhance it with *Striga* resistance, this research confirms the value of EM and SRN39 as parents.

The laboratory assay provides a very fast and inexpensive tool to screen for mechanisms of resistance (Chapter 7). As expression of resistance in the field may be confounded with other factors, laboratory assay allows more control of the environmental variance and therefore, gives a more reliable estimate of resistance.[8,9] The bioassay was useful to reveal the introgression of post-attachment resistance in the BC_2F_3 progeny.

5. Conclusions

Our goal is to develop sorghum lines that combine resistance to *Striga* introgressed from SRN39 with agronomic characteristics similar to those of EM. We then selected lines which approached the performance of SRN39 for number of *Striga* plants at 90 days after planting, and were close or superior to EM for number of days to half bloom and yield. A total of 17 to 25 *Striga* resistant progenies had a performance similar to EM for the two agronomic traits considered, including three lines that resembled EM for all agronomic traits simultaneously. After verifying the presence of resistance mechanisms among these introgressed lines, we can select best candidates for the next stage of backcrossing whereby *Striga* resistance will be further incorporated into EM background.

To date, our group has identified excellent sources of resistance and has widely tested them across a number of countries in Africa. The stability and adaptation of some of these *Striga* resistant sorghum genotypes have been well established.[1,3] Introgression breeding was successfully conducted with Niger landraces. Our data showed that there was a tendency to shift the resistance of progeny towards that of the resistant parent while retaining the phenotype of the EM landrace. As expected, the BC_2F_3 progenies expressed a level of resistance to *Striga* lower than the donor parent SRN39, and showed a phenotype more homogeneous and closer to the phenotype of the recurrent landrace. The use of laboratory assays to select lines with clear mechanisms of resistance should enhance the efficiency of the introgression approach for crop improvement.

Similar efforts are developed by other groups to introgress resistance to *Striga* into local crop varieties. Earlier, collaboration between CIMMYT and KARI was established to exploit resistance to *Striga* ssp. in *Tripsacum* species for introgression into local maize varieties.[10] A current collaborative initiative between the University of Hohenheim and ICRISAT aims to introgress QTL for resistance to *Striga hermonthica* into farmer-preferred sorghum varieties (Chapter 12).

Several approaches have been used often separately to fight *Striga* in farmers' fields. Whereas the introgression approach can improve breeding efficiency, the complexity of host-parasite interaction in the African context requires that genetic resistance itself be part of an integrated package to insure greater and durable crop protection against the parasite. As we are learning more about the biological phenomena of host-*Striga* interactions, we are discovering and understanding mechanisms for resistance. Soil fertility and water management of the plant environment has shown some success in reducing *Striga* damage. Biological control and emerging biotechnological tools are also given more consideration in view of their demonstrated or potential contribution. The integration of various research disciplines with local farmers' knowledge is convincingly emerging as the most rational and long lasting solution to the *Striga* scourge in Africa.

Acknowledgements

This work is supported by the Rockefeller Foundation under Project 2000FS024.

References

1. D. E. Hess and G. Ejeta, *Plant Breeding* **109**, 233 (1992).
2. G. Ejeta, A. Mohamed, P. J. Rich, A. Melake-Berhan, T. L. Housley, and D. E. Hess, in: Breeding for *Striga* Resistance in Cereals (eds. Haussmann, B.I.G. *et al.*) Margraf Verlag, Weikersheim. (2000), p. 29.
3. J. Clark, I. Kollo, I. Kapran, and O. Cherif, *Sorghum Newsletter* **32**, 44 (1991).
4. SAS Institute Inc., SAS/STAT software SAS language guide. Edition 8.03. SAS Institute Inc., Cary (2001).
5. G. E. P. Box, W. G. Hunter and J. S. Hunter, Eds., Statistics for experimenters. Wiley, New York (1978).

6. B. I. G. Haussmann, D.E Hess, B.V.S. Reddy, S.Z. Mukuru, M. Kayenao, H.G. Welz, and H.H Geiger, Euphytica, **122**, 297 (2001).
7. G. O. Omanya, B. I. G. Haussmann, D. E. Hess, B. V. S. Reddy, M. Kayenao, H. G. Welz, and H. H Geiger, *Fields Crops Res.*, **89**, 237 (2004).
8. C. S. Mutengwa, P. B. Tongoona, I. Sithole-Niang, *African Journal of Biotechnology* **4**, 1355 (2005).
9. J. Rodenburg, L. Bastiaans, E. Weltzien, D. E. Hess, *Field Crops Res.* **93**, 34 (2005).
10. D. Hoisington, M. Khairallah, T. Reeves, J.-M. Ribaut, B. Skovmand, S. Taba, and M. Warburton, *Proc. Natl. Acad. Sci.* **96**, 5937 (1999).

Section IV

Biotechnology: Opening New Frontiers

CHAPTER 11

SUCCESS WITH THE LOW BIOTECH OF SEED-COATED IMIDAZOLINONE-RESISTANT MAIZE

Fred Kanampiu[a,*], Alpha Diallo[a], Michael Burnet[b], Haron Karaya[a] and Jonathan Gressel[c]

[a]*CIMMYT, P.O. Box 1041, Village Market, Nairobi 00621, Kenya*
[b]*Hi-Cap Formulations, Paul Ehrlich Str 15, 72076 Tübingen, Germany*
[c]*Plant Sciences, Weizmann Institute of Science, Rehovot, Israel*
[*]*E-mail: f.kanampiu@cgiar.org*

Parasitic *Striga* spp. cannot be controlled underground by selective herbicides, except on crops with target-site resistance to systemic herbicides. Spraying herbicides is uneconomic in African conditions. Instead, targeted herbicide use via seed dressing of maize varieties bred with mutant ALS genes is being commercialized in Kenya. A multi-partnership is currently testing this technology with farmers to create awareness to facilitate its delivery to control *Striga* for eventual deployment in sub-Saharan Africa. This technology has tripled yields in heavily infested areas and provided season long control in short season maize. High rainfall can leach the imidazolinone herbicide in short season maize, and normal rainfall leaches it in longer season maize. Controlled release formulants (high capacity ion exchangers) are being developed for this seed treatment application to limit herbicide leaching.

1. Introduction

Striga spp. coupled with low soil fertility (due mainly to nitrogen deficiency), drought, and foliar diseases such as maize streak virus, have been the major reasons why maize yields in sub-Sahara Africa have not increased over the last two decades, and are hovering around 1.5 t/ha, well below the world average of 4.2 t/ha.[1] Although technologies discussed in other chapters can partially allay the problem, no single short term control measure has been developed that subsistence maize farmers find within their financial means, or that fit well into their traditional cropping systems with immediate effect. Moreover, many of these measures require several seasons of repeated use before they begin to produce yield benefits.[2] Thus, despite widespread extension efforts, they have not been widely adopted, as they are not what the farmers consider "appropriate" for their needs of providing sufficient food for their families on small, intensively cultivated holdings.

Striga can be selectively controlled by foliar applications of phenoxy herbicides after the *Striga* flower stalk has emerged, requiring spray equipment and high doses of herbicide. These treatments are too late to be reverse the impact of *Striga* on yield for the current season and are ineffective for coming seasons if there is a large seed bank. Spray applications of most herbicides would kill intercropped legumes, which are planted by many subsistence farmers in an effort to reduce risk and increase the dietary intake of protein that would otherwise come from maize alone. Thus there is an immediate need for cost-effective mechanisms meeting at least 3 criteria: controlling *Striga* itself, so that adequate crop yields can be achieved in same season; deplete the *Striga* seed bank in the soil; and allow legume intercropping. Such technologies are needed as a stopgap until crop varieties with adequate conventional genetic or transgenic resistance become available.

Subsistence farmers in Kenya and elsewhere cultivate maize with judiciously used, small inputs of fungicide and insecticide seed dressings, and weeks later, apply a few granules of insecticide into the whorl of maize leaves to control stem borers. We thought that small amounts of herbicide could control *Striga* while it is still underground, before the weed debilitates the crop.[3,4,5] Although economically feasible, such a strategy requires the adoption of new varieties and techniques and thus posed both technical and extension challenges. Doubling yields from 1 ton/ha would produce enough maize to provide two million Kenyans with their current average annual consumption on the over 250,000 ha of maize land severely infested with *Striga hermonthica*. Despite the apparent focus of the *Striga* problem in poorer areas, acceptance of a solution seemed likely because African subsistence farmers have adopted over the last two decades new maize varieties and technologies having perceived value, in places where *Striga* is not a major problem.

The agro-economic situation in the problem areas is, therefore, one that can respond to a new variety and implement in parallel micro-application of agrochemicals. Some transgenic[6] and mutant[3,4,7] herbicide-resistant crops with altered target enzymes[8] enable the early control of parasitic weeds before or during attachment to the host. The herbicides are exuded from crop roots and kill attached *Striga* as well as its nearby seeds in the soil, before germination.[9] These herbicide spray applications cannot, however, be used conventionally because of cost as well as the effect on intercropped legumes. We thus demonstrated that seed dressings of IR-(imidazolinone resistant) maize with small amounts of imazapyr or pyrithiobac could provide season long control of *Striga.*[9] The seed dressings allowed intercropping with legumes, as long as the legume seeds were more than 15 cm from the maize seeds.[9] Results are described below with the varieties approved by the regulatory authorities and being commercialized.

In Kenya, a multi-member partnership has been testing this technology with farmers to facilitate its delivery for eventual deployment in sub-Saharan Africa to control *Striga.* Although this strategy is effective, it has limits. When the soil is very dry during germination, a high local level of herbicide can cause a 2-3 day delay in germination of the IR-maize and in extreme cases result in reduced stands. Conversely, very high rainfall can wash the herbicide beyond the root zone, allowing establishment of late germinating *Striga.* It was clear that while the treatments were appropriate for Kenya with its 12-14 week maize, there might not be sufficient herbicide available in the longer (20-22 week) season maize, grown where there is only one rainy season per year. We are, therefore, developing the next generation of seed treatments based on high capacity ion exchangers, and report some results below showing that they facilitate *Striga* control under simulated high rainfall, as also described below.

2. Materials and Methods

The breeding of maize varieties that have been released is described in Ref. 10. The development of the seed dressing protocols being commercially used are described in Refs. 5, 9, 10. Micronized technical grade imazapyr acid is added to commercial fungicide/insecticide seed dressings and applied to the seeds.

The slow release formulations have imazapyr bound to anion exchangers such as Dowex-1 or DEAE cellulose, or other specially-synthesized high capacity anion exchangers, and then were mixed with the fungicide/insecticide seed dressings, or in some cases bound to the seed with polyvinyl pyrollidone, as described in Ref. 11.

3. Results and Discussion

The research and development described above began utilizing the initial temperate IR-maize developed for the temperate USA market. This material was very susceptible to turcicum leaf blight, leaf rust, gray leaf spot, and maize streak virus disease, as well as having a low yield potential in the tropics. A breeding program was therefore initiated to incorporate adaptations to the local environment. High yielding and disease resistant IR-maize inbred lines, hybrids and open pollinated varieties with increased yields were gradually achieved, while experiments were being performed with synthetic open-pollinated varieties with increasing levels of adaptation.

3.1. *First Generation of Released Technology*

This material was subjected to extensive multi-site testing in western Kenya (Figures 1 and 2). Multi-site field tests there and in seven countries demonstrated that herbicide seed-coating of herbicide-resistance maize controls both *Striga hermonthica* and *S. asiatica.*[10] The varieties adapted for western Kenya did not always outperform local varieties in yield, despite the *Striga* control, indicating a need to back-cross the recessive IR-gene conferring resistance into locally adapted material to control both *Striga* and improve yields.[10]

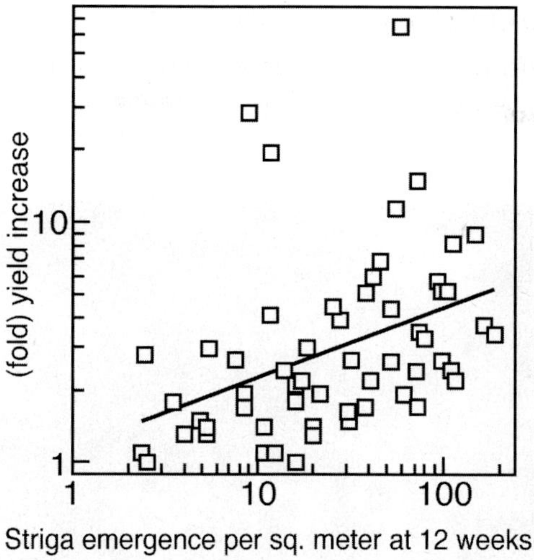

Figure 1. Incremental maize yield increase as a function of Striga *attachment on imazapyr seed treated IR-maize compared to the untreated local hybrid, in on-farm multi-site testing. In no case was the yield less than the local hybrid (ratio=1). Note that the scales are logarithmic.*

Following proof of concept in the field, imazapyr was registered as a seed treatment by BASF, trademarked as the "Strigaway" technology. The technology and hybrid varieties were tested and received their first regulatory approval in Kenya after finding excellent *Striga* control and high maize yields (Table 1). The four fully released hybrids have been allocated to three local seed companies and Kenya Agricultural Research Institute for commercialization. The technology was first commercially launched in Kenya in July 2005 after extensive pre-release demonstrations of the technology throughout western Kenya. Western Seed Company has produced over 100 tons of certified seed available for the March 2007 long rainy season. The first new commercialized maize hybrid is marketed under the common commercial name of *Ua Kayongo* (*Striga* killer) by the seed companies, with assistance in dissemination to the poorest farmers by a collection of NGOs and other international organizations that had conducted extensive on-farm demonstrations.

Based on the results of further large-scale, on farm testing of the most recently developed hybrids and open pollinated varieties (OPVs) across East Africa (Table 2), six early OPVs, five late OPVs and two hybrids have been allocated to seed companies and NARS to nominate them in the national performance trials for generation of information required by regulatory agencies for registration of the new varieties and their

Figure 2. No Striga hermonthica *emergence on imazapyr-resistant maize, seed coated with 30 gm imazapyr per hectare in a farmer's field heavily infested with* Striga. *Plants grown from coated seed are in the background and control plants in the foreground. (Photo courtesy of Dennis Friesen)*

Table 1. Grain yield increased and emerged *Striga* count reduced on IR-maize hybrids compared to local hybrids in 10 farmers' fields in western Kenya.

Germplasm	Yield (tons/ha)	*Striga* emergence (plants/m^2)	Status
Local susceptible hybrid			
H513	3.2	3.6	Check
	CIMMYT IR hybrids		
CKT036071-IR	7.2	0.45[a]	Released
CKT036069-IR	6.3	0.81[a]	Released
CKT026065-IR	6.1	0.83[a]	Released
CKT036067-IR	5.9	0.75[a]	Released
CKT026061-IR	5.6	0.76[a]	Not released

[a]The emerged *Striga* did not set seed.

subsequent commercialization. Wide-scale participatory field-testing, both on-station and in farmers' fields, of elite IR-maize material from breeding program is also being carried out by NARS and seed companies in several African countries. This exercise will result in selection of varieties nominated for registration in respective countries.

Table 2. IR-three way hybrids and OPVs outperform best local material both on non-infested, but especially *Striga* infested multiple sites in East Africa, 2004.

Variety		Maize yield Uninfested optimum (tons/ha)	Maize yield *Striga* infested (tons/ha)	Striga ($\#/m^2$)
Three-way hybrids	*Local H513 –HYBRID non-IR*	*2.3*	*1.7*	*117*
	KB03-0B43-11	4.1	2.1	7
	KB03-0B43-9	3.7	2.2	5
Mid-altitude late OPV	*KSTP 94 – LOCAL non-IR CHECK*	*3.3*	*1.5*	*50*
	ECA-STRIGOFF-VL-144	5.6	1.9	4
	ECA-STRIGOFF-VL-131	5.3	2.0	3
	ECA-STRIGOFF-VL-102	5.2	2.1	2
	ECA-STRIGOFF-VL-130	5.3	1.9	3
	ECA-STRIGOFF-VL-107	5.2	1.9	3
Mid-altitude early OPV	*KSTP 94 - LOCAL non-IR CHECK*	*2.5*	*0.9*	*24.8*
	ECA-STRIGOFF-VE-216	3.0	2.5	1.0
	ECA-STRIGOFF-VE-206	3.2	2.4	2.7
	ECA-STRIGOFF-VE-217	2.9	2.2	0.2
	ECA-STRIGOFF-VE-210	2.8	2.1	1.1
	ECA-STRIGOFF-VE-208	2.9	2.2	1.0
	ECA-STRIGOFF-VE-215	2.7	2.0	0.9

3.2. *Limitations of the First Generation Material*

It was clear from all the field tests (Figure 1, Table 1 and 2, and not shown) that the new hybrids and OPVs performed far better in Kenya than the local varieties, especially under heavy *Striga* infestations. The few *Striga* stalks that emerged did so late in the season, such that they failed to set seed, and thus did not replenish the seed bank. The higher yields at low infestation (Fig. 1) may be due to the superior disease resistance of the hybrids.

3.3. *Slow Release Formulations — The Second Generation*

Where late *Striga* emergence was observed, it was clear that the duration of *Striga* control was shortened either by herbicide leaching due to a longer season or higher rainfall. This small drawback on the use of IR-maize seed coating technology is more pronounced when planting is followed by very heavy rains, the herbicide gets leached and washed away making the technology less effective. Further, dry planting or a drought spell immediately after planting causes a reduction in germination due to the high concentration of herbicide around the germinating seed.

To reduce leaching and maintain control, the herbicide was combined with novel slow release seed dressings that were generated by binding imazapyr to high capacity ion exchangers (> 1 meq imazapyr bound/g exchanger). Previously generated formulations[12] using similar technologies had a more than ten fold lower exchange capacity, and would be far too bulky for seed dressings. The effectiveness in preventing leaching was demonstrated in a simulation experiment using large pails for cultivating maize with *Striga* (Fig. 3). The data show that 2-3-fold less herbicide was needed under all rainfall regimes for equivalent *Striga* control.

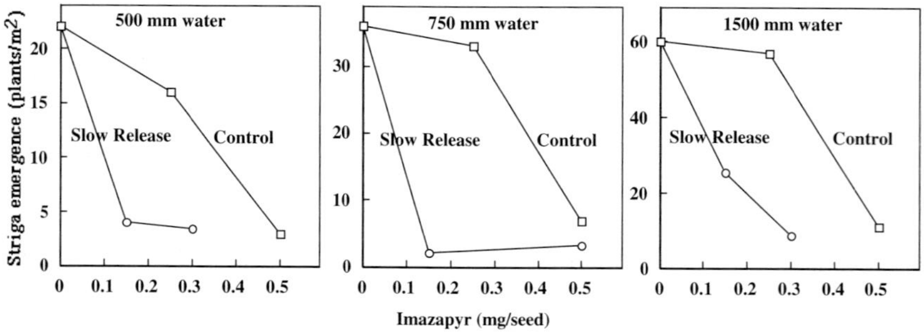

Figure 3. Enhanced control of Striga hermonthica *on IR-maize with slow release formulations of imazapyr (averaged) under different rainfall regimes.* Striga *emergence was measured 12 weeks after planting using either unformulated imazapyr or the formulated form. Natural rainfall was supplemented by sprinkler irrigation to achieve the desired regimes.*

Seed dressings with slow release herbicide are also needed to prevent crop phytotoxicity, early in dry seasons. It was seen in the field that the formulations also abolished the transient phytotoxicity observed with low rainfall. These results were obtained with prototype materials, and improved derivatives are in field testing. It appears likely that this approach to targeted pesticide application should be applicable with other pesticides, especially seed or soil applied compounds.

The slow release formulants tested are categorized as generally regarded as safe (GRAS) by the US FDA, generally bio-degradable, and the pesticides remain always in the parent form so that no "novel" pesticides are formed that would require registration of new molecules.

3.4. *Raising Striga Control Awareness*

In the last 2 years, a multi-partnership between farmers, seed companies, non-governmental organizations, extension agents and research organizations has conducted over 10,000 on-farm demonstrations to test this technology with farmers to facilitate its delivery for eventual deployment in sub-Saharan Africa. The focus was to create awareness and minimize *Striga* infestation using the herbicide seed-coating technology and other cultural methods, thereby improving maize yields, food security and well being among rural poor.[13-15]

Herbicide coated IR-maize from CIMMYT was bulked up by the Western Seed Company, and delivered to NGO's farm input supply facilities. They packaged the seeds into various quantities, which they provided to different cooperators for use in about 10,000 field tests, 140 on-farm experiments and 12 community demonstrations. There was continual monitoring and evaluation during the growing season and during field days. One of those independent large scale multi-site comparative tests of this technology was performed in comparison with legume rotations, standard intercropping with legumes and intercropping with perennial legumes.[15] It was demonstrated that this technology is the most immediately effective way to raise maize yields while reducing *Striga* infestation and seed-banks (see Chapter 16, Table 2), but the yields are even better after a few years of intercropping with *Desmodium* (Chapter 18), an effect that was not immediate.

Another avenue creating awareness was the printing of 2000 copies of an extension manual that was distributed to all cooperators. The seed companies marketing the IR-maize are training their stock-listers and creating awareness through their commercial channels. Awareness is also being created through exhibitions and the internet with *Striga* videos, extension booklets, project reports and publications. These exhibitions are attended by researchers, agro-based industries, farmers, policy makers and development workers. A special page dedicated to *Striga* has been established, *www.africancrops.net/striga* and continues to be updated to disseminate the outcomes of research. Seven articles on *Striga* eradication were disseminated through the African Crops News Service, a monthly newsletter on improving African crops including articles by partners working with *Striga* and those by newspaper reporters.

3.5. *The Technology Does Not Always Work*

At least one infested site was found where the technology was repeatedly ineffective in Kenya. We are trying to ascertain the reasons why, but so far have been stymied. Efforts are being to invested in trying to find out whether this may happen elsewhere as the technology becomes more widespread, so that it can be predicted where the technology will not be valuable.

4. The Long Term Sustainability of Herbicide Technologies

4.1. *Evolution of Resistance*

All technologies utilizing herbicide resistance are prone to having the weeds themselves evolve resistance to the technology, and resistance typically evolves very quickly to inhibitors of acetolactate synthase, the target of imazapyr.[16] It was initially predicted by modeling that resistance would rapidly evolve, and that there would be five resistant *Striga* plants establishing per hectare per year, based on previous weeds that evolved evolution to the same group of herbicides.[17] This would necessitate stringent field monitoring by farmers for early flowering *Striga* plants and rouging them before they set seed. Only then would the technology be sustainable for a lengthy period. No rare early flowering *Striga* plants have been seen where the technology has been effective, despite treatments of hundreds of hectares and many seasons of use, in all the field trials described above. Thus, it is clear that some assumption in the model must have been incorrect. The incorrect assumption became clear to us while back-crossing the resistance gene into elite African backgrounds. We had to use much less herbicide during backcross selection so as not to kill the heterozygous resistant individuals, while killing the susceptible ones. At the high localized concentrations normally used for the seed treatment in field, the maize had to be homozygous, as that dose used kills heterozygotes. All previous cases of evolution of weed resistance to this group of herbicides were to lower uniform doses, and it was a mistake, by a factor of a million, to use the hetereozygote mutation frequency in the models. When the more accurate homozygous frequency is inserted in the models, resistance comes out as being exceedingly rare, five resistant plants per million hectares per year.[18] Still, as the technology becomes widespread, it must be assumed that the inevitable will happen, and resistance will evolve somewhere, and pre-emptive, rapid reaction mechanisms must be in place to deal with this.

One must also consider the possibility of needing other resistances, as no herbicide resistance has been forever. Africa must shift from subsistence agriculture to production agriculture, as it must to feed its people as world grain prices will render grain aid unaffordable, with the shift to using grains for biofuels.[19] Thus *Striga* control will be even more necessary, along with general weed control. Other inexpensive herbicides such as glyphosate will surely be more widely used, and transgenic glyphosate-resistant maize has already been released in South Africa. Glyphosate is systemic and can be used to control the related parasite *Orobanche* on transgenic crops,[6] so there is no reason not to expect it to work on *Striga*. Glyphosate could be used as a general spray to kill all weeds, but could also be used as a seed treatment, specific for parasitic weeds,[20] if the seeds can withstand the local high concentrations. Pyramiding glyphosate resistance with IR may be an excellent long term strategy, with IR seed treatments, and mid season spray or topical applications of glyphosate, to delay resistance to both herbicides.

4.2. *The Herbicide Seed Treatment Technology for Other Crops*

It has been very easy to develop crops resistant to acetolactate synthase-inhibiting herbicides, both via tissue culture,[21] and even quicker by such techniques as pollen mutagenesis.[22] In retrospect, using such techniques would require a much shorter gestation time than it took from the first experiments over 15 years ago, to field commercialization of the present technology.

4.3. *Integration with Other Technologies*

This technology was successful in tests with intercropping with *Desmodium,* as described in Chapter 18. This technology could also be easily integrated with other methods of *Striga* control, especially biocontrol and with resistance breeding; subsequently further depleting the *Striga* seed bank in the long run. If transgenic herbicide resistant crops are generated, or if a gene for transmissible siRNA resistance is to be genetically engineered (Chapter 14), the two traits should be introduced simultaneously. Such integration of technologies would severely delay the evolution of *Striga* resistance to each technology, especially if any one technology is continually used separately.

The target area has poor soil fertility, and farmers apply little or no fertilizer. To fully realize the potential benefit of the technology, an effort should be made to combine it with appropriate soil fertility management.

5. Conclusions

IR-maize herbicide seed treatments provide affordable season long *Striga* control suitable for subsistence farmers, increasing yields two or three times and reduces the number of *Striga* seed and plants per unit area. Many farmers observed a general weed free zone around the maize plants, allowing later and easier hand weeding and less weed competition with the crop.

The success of the IR-maize seed coating technology will largely depend on the existence of a dynamic seed sector to process and market quality IR-maize coated seed. Finally, the target area has poor soil fertility, and farmers apply little or no fertilizer. To fully realize the potential benefit of the technology, an effort should be made to combine it with appropriate soil fertility management. The technology is relatively cheap, and marginal analysis indicates good returns to the investment. Further research is needed to address various drawbacks including slow-release formulations to address possibilities of leaching of herbicide under heavy rains, and scorching under dry spells. Other research needed is on resistance management and long term economic impact assessments.

Finally, whereas the deployment of IR-maize in Africa holds the potential for addressing *Striga* infestations on maize fields, it is worth noting that this will not be a panacea to *Striga* problems in the continent. The contribution of other *Striga* control measures is certainly noteworthy and an integration of all the existing control measures is therefore called for in the sustainable management of this weed.

Acknowledgements

The participation of Joel Ransom, George Odhiambo, Gordon Abayo, and Dennis Friesen in early stages of this research is greatly appreciated, as is the initial plant breeding by Kevin Pixley and David Jewel. The research was funded by the Rockefeller Foundation and the African Agricultural Technology Foundation; and herbicide was provided by BASF.

References

1. FAOStat, in *http://apps.fao.org/page/collections?subset=agriculture* (2003).
2. J. K. Ransom, *Crop Protect.* **19**, 759 (2000).
3. G. O. Abayo, J. K. Ransom, J. Gressel and G. D. Odhiambo, *in* M. T. Moreno, J. I. Cubero, D. K. Berner; D. M. Joel; L J. Musselman and C. Parker, Eds. *Advances in Parasitic Weed Research*; Cordoba, Spain: Junta de Andalucia, p. 762 (1996).

4. G. O. Abayo ,T. English, R. E. Eplee, F. K. Kanampiu, J. K. Ransom and J. Gressel, *Weed Sci.* **46**, 459 (1998).
5. F. K. Kanampiu, J. K. Ransom and J. Gressel, *Crop Protect.* **20**, 885 (2001).
6. D. M. Joel, Y. Kleifeld, D. Losner-Goshen, G. Herzlinger and J. Gressel, *Nature* **374**, 220 (1995).
7. D. K. Berner, F.O. Ikie and J.M. Green, *Weed Tech.* **11**, 704 (1997).
8. K. E. Newhouse, B. Singh, D. Shaner and M. Stidham, *Theor. Appl. Genet.* **83**, 65 (1991).
9. F. K. Kanampiu, J. K. Ransom, D. Friesen and J. Gressel, *Crop Protect.* **21**, 611 (2002).
10. F. K. Kanampiu, V. Kabambe, C. Massawe, L. Jasi, D. Friesen, J. K. Ransom and J. Gressel, *Crop Protect.* **22**, 697 (2003).
11. M. W. Burnet F. K. Kanampiu and J. Gressel, *US Patent Office* PCT, WO2004004453 (2004).
12. Y. G. Mishael, T. Undabeytia, O. Rabinovitz, B. Rubin and S. Nir, *J. Agric. Food Chem.* **50**, 2864 (2002).
13. P. L. Woomer and M. N. Omare, *in Ua Kayongo Hybrid Maize: The Striga Killer,* African Agricultural Technology Foundation, Nairobi, Kenya, p. 16 (2005).
14. P. L. Woomer, D. Okello, H. Otieno, J. Maina, B. Vanlauwe, L. Nyambega, G. Odhiambo, J. R. Okalebo, C. Savala and M. Omare, *Technical Report and Financial Statement,* FORMAT, Nairobi. p.48. (2005).
15. P. L. Woomer, J. Tungani, G. Odhiambo and F. M. Mwaura, in *African Crop Sci. Conf. Proc.* **7**, 479 (2005).
16. J. Gressel, *Molecular Biology of Weed Control.* Taylor and Francis, London, 2002.
17. J. Gressel, L. E. Segel and J. K. Ransom, *Intl. J. Pest Management* **42**,113 (1996).
18. J. Gressel, *Crop Protection,* **24**, 1007 (2005).
19. J. Gressel, African Crops News Service Newsletter, – March 2006.
20. J. Gressel and D. M. Joel, US Patent 6,096,686, (2000).
21. P. C. Anderson, and M. Georgeson, *Genome* **31**, 994 (1989).
22. J. A. Greaves, G. K. Rufener, M. T. Chang and P. H. Koehler, *Proc. 48th. Ann. Corn & Soybean Ind. Res. Conf.* (Amer. Seed Trade Assoc. Washington DC) 104 (1993).

CHAPTER 12

MARKER-ASSISTED SELECTION FOR *STRIGA* RESISTANCE IN SORGHUM

Cécile Grenier[a,*], Yahia Ibrahim[b], Bettina I. G. Haussmann[c], Daniel Kiambi[d] and Gebisa Ejeta[a]

[a]*Dept of Agronomy, Purdue University, West Lafayette, IN47907, U.S.A.*
[b]*ARC, P.O. Box 126, Wad Medani, Sudan*
[c]*ICRISAT, B.P. 12 404, Niamey, Niger*
[d]*ICRISAT, P.O. Box 39063, Nairobi, Kenya*
[*]*E-mail: grenier@purdue.edu*

Striga resistance breeding has progressed further in sorghum than any of the other crops. We summarize the contributions made in generating new genetic information as well as in the development of robust tools and methodologies for reliable implementation of marker-assisted selection (MAS) for *Striga* resistance in sorghum. Advances are reviewed in: developing proper field testing methodology to map QTL for field resistance; in the use of bioassays that dissect the complex mechanism of *Striga* resistance into simply inherited components; as well as in the genetic analysis and molecular mapping of these simple traits. Controlled introgression of *Striga* resistance into improved, local African landraces using the increased reliability and efficiency of MAS is underway with validated and robust molecular markers.

1. Introduction

Genetic variation for *Striga* resistance has been found in landraces of sorghum since 1933.[1] Empirical breeding through selection of progenies in populations of crosses grown in *Striga* infested fields has not resulted in sustainable progress. The reasons lie in insufficient knowledge about the inheritance of *Striga* resistance, the lack of appropriate laboratory assays based on the biology of host-parasite relationships, and the overall difficulty in producing reliable field data. Substantial progress has been made in the last two decades as a result of our and others' research. An array of useful information has been generated. The development of improved germplasm screening techniques has simplified the identification of new sources of *Striga* resistance and the characterization of more specific mechanisms of resistance (Chapter 7). Analyses of the genetic control, estimates of quantitative-genetic parameters, and the identification of genomic regions involved in the expression of resistance have also been facilitated by using laboratory techniques,[2-11] or by improved field screening methods.[4,9,12-14]

Identification of molecular markers associated with *Striga* resistance offers a significant advantage if the markers are robust and consistent across populations and environments. Marker-assisted selection can greatly accelerate breeding progress for *Striga* resistance, because screening for complex resistance under field conditions is difficult and sometimes unreliable. In addition, *Striga* is quarantined, confining tests to areas where *Striga* is endemic. Furthermore, some *Striga* resistance genes are recessive,[3,13,15] making selection during backcrossing schemes more difficult. This chapter summarizes the progress made in the genetic analyses of *Striga* resistance in sorghum. It deals with some of the challenges in the identification of molecular markers in populations and discusses the development of populations and techniques to enhance and facilitate marker-assisted selection for *Striga* resistance in sorghum.

2. Genetic Mapping of *Striga* Resistance

A number of crosses were made between *Striga* resistant and susceptible parents to generate an array of genetic populations including early generation populations as well as advanced recombinant inbred populations for genetic analyses. Genetic populations segregating for *Striga* resistance were evaluated either in crop fields infested with *Striga* or in the laboratory, using specifically developed assays. Some results obtained from analyses of these studies are reported below.

2.1. *Data from Phenotyping Based on Field Resistance to Striga*

A population of recombinant inbred lines generated from a cross between SRN39, a *Striga* resistant line and Shanqui Red (SQR) a susceptible Chinese sorghum cultivar was developed through the single seed descent method of plant breeding. Field testing was conducted for both *S. asiatica* and *S. hermonthica* in naturally infested field in North Carolina (USA) and Sudan. The level of resistance against the two *Striga* species was compared to map the genetic regions controlling the trait. Significant genetic variation for *Striga* infestation was detected among the recombinant inbred lines. *Striga* emergence was a week earlier and more parasites were counted per individual host plant on the susceptible than in the resistant plots.[6] Transgressive segregation was observed with some recombinant inbred lines more resistant than SRN39 and some others appearing more susceptible than SQR. Among the set of recombinant inbred lines tested, a significant and positive genetic correlation ($r=0.52$, $p<0.01$) was observed between counts of *S. hermonthica* and *S. asiatica*. This strong genetic correlation was supported by the interval mapping analysis where molecular markers for resistance to the two *Striga* species mapped to the same genomic regions. Six resistance QTL were identified and four were shared between the two species. These results suggested that selection for resistance to one species could result in resistance to the other, a phenomenon that has been corroborated in a wide range of field testing and deployment that we subsequently conducted. The low germination stimulant (*lgs*) gene mapped to one of the QTL regions, supporting the hypothesis that resistance to *Striga* is the product of one or a combination of several mechanisms that influences the development of parasitism.[16]

The predicted response to marker-based selection (where selection is solely based on genetic markers) was compared to marker-assisted selection (based on marker loci information plus phenotype) relative to a strictly phenotypic selection for resistance to *Striga*. The predicted responses were higher with marker-based selection (24% more efficient for selection for resistance to *S. hermonthica* and 37% for resistance to *S. asiatica*) and marker-assisted selection (41% for *S. hermonthica* and

43% for *S. asiatica*) than with phenotypic selection. The importance of molecular markers for enhancing efficiency of breeding for *Striga* resistance in sorghum via marker-based selection or marker-assisted selection was clearly demonstrated in this first stage of research.

We conducted a further experiment for mapping QTL for *Striga* resistance under field conditions. Two populations of recombinant inbred lines from crosses: IS9830 x E36-1 (RIP-1) and N13 x E36-1 (RIP-2) were evaluated. The partially resistant sorghum lines IS9830 and N13 are characterized by different mechanisms of resistance. Line IS9830 is a low inducer of *Striga* seed germination while line N13 stimulates abundant *Striga* seed germination but possesses a resistance mechanism yet to be adequately described. Both lines are attacked by *Striga* in highly infested fields, but to a much lesser extent than *Striga*-susceptible cultivars. The common parental line E36-1 is highly susceptible to *Striga* but possesses a certain degree of drought resistance through maintenance of green leaf area ("stay-green"). Each population was divided into two sets for the phenotyping. Set 1 of each population consisted of 116 $F_{3:5}$ lines tested with parents and checks in 1997; Set 2 comprised 110 $F_{3:5}$ lines tested with parents and checks.

Field trials were conducted at Samanko and Cinzana (Mali) during the rainy seasons of 1997 and 1998. In Kenya, where rainfall is bimodal, trials were conducted at Kibos and Alupe in the long rainy season, and at Alupe in the short rainy season of both years. Additional details of the protocols and results and field-testing methodology have been summarized elsewhere.[9] *Striga* seeds were added to the field at sowing via artificial infestation, and trials were planted in six replicates using a lattice design with each progeny planted in two-row plots and separated from each other by one blank row. Results were analyzed using novel resistance index; area under the *Striga* number progress curve (ASNPC), which was computed from four or five counts of emerged *Striga* plants performed at two-week intervals during the growing season. The ASNPC accounts for both intensity and speed of the epidemic. It was selected for QTL mapping because of its good differentiation at all sites and high heritability estimates in all four sets of material (0.66 and 0.74 in Sets 1 and 2 of RIP-1, and 0.81 and 0.82 in Sets 1 and 2 of RIP-2).[14]

Because of the low genetic polymorphism among the parental lines and initial low availability of SSR markers, it took several years to develop acceptable genetic maps of the two recombinant inbred populations with good genome coverage, a prerequisite for reliable QTL mapping.[14,17,18] The final maps revealed lengths of about 1550 cM in the two recombinant inbred populations, and contained relatively few gaps. In the RIP-1 (IS9830 x E36-1) population, 11 and 9 QTL were identified for ASNPC in Sets 1 and 2, together explaining 77 and 60% of the genetic variance, respectively.[14] Five of the QTL were common to both sets, i.e. they contributed to *Striga* resistance in all 10 test environments in both years and both genotypic samples (Sets 1 and 2). The most important QTL in this population mapped close to the *lgs* gene. The identification of additional QTL on other linkage groups in RIP-1 suggests that the parental line IS9830 may possess other resistance mechanisms besides low stimulation of *Striga* seed germination. It may also be that the newly derived resistant lines may possess genes that control different intensities of germination stimulants, or those that govern the synthesis of different germination stimulants.

Several QTL were identified in the two genotypic sets of RIP-2 (N13 × E36-1) and explained about 80% of the genetic variance for ASNPC. Again five QTL were common to both sets. A five-fold cross-validation of the results revealed a low genotype dependency of the QTL results for the N13 population. Because of the successful QTL validation across locations, years, and genotype samples, the five stable QTL identified in this population may serve as candidates for marker-assisted transfer into other cultivars via marker-assisted backcrossing (Table 1). These QTL analyses affirm that *Striga* resistance under field conditions is a quantitative trait affected by many genes. The results of these studies based on the two populations seem to suggest that several linkage groups may be involved in the expression of *Striga* resistance. Some loci probably have a stronger role in host-parasite interaction and may therefore be more stable across test locations and years.[14]

QTL-environment interactions were significant in both populations and resulted in variable QTL effects in individual test locations or years.[14] In the RIP-1 (IS9830 × E36-1) population, two of the five QTL that had been identified in both genotypic sets had high interactions with the test locations. These were manifested in a positive effect of a QTL in one location but a negative effect on resistance of the same QTL in another location. In the RIP-2 (N13 × E36-1) population, QTL x environment interactions were much less important, and all five QTL identified in both sets revealed a positive effect towards resistance in all test environments (Table 2). This suggests that parental line N13 may possess a more stable resistance than the parental line IS9830. These effects were validated across environments, years and independent samples of the same population. Our program has selected these QTL for marker-assisted selection using SSR markers that flank the QTLs for marker-assisted introgression.

Table 1. Percentage of phenotypic variance explained by a single QTL in position (LG-cM) for ASNPC in RIP-2.

Set	LG01-185	LG02-65	LG06-90	LG05-5	LG05-70
1	24	17	30	19	15
2	21	22	15	12	29

Linkage groups as defined by Kim,[20] and position on the linkage group in centiMorgans.

Table 2. QTL x environment interaction for ASNPC in RIP-2. Effect of QTL in position (LG-cM) at test location (Loc.).

Set	Loc.[a]	ASNPC Mean	LG01-185	LG02-65	LG06-90	LG05-5	LG05-70
1	Sko	10	1.3	0.9	2.0	0.6	1.1
	Cza	9	1.5	1.0	2.7	1.6	1.9
	Alul	22	6.6	5.5	3.8	2.8	5.0
	KibL	9	1.1	2.0	1.7	1.1	1.5
	AluS	22	3.2	3.1	1.7	1.4	2.9
2	Sko	6	0.5	0.8	1.0	0.4	0.8
	Cza	16	2.6	4.2	2.7	2.0	2.8
	Alul	21	4.4	3.9	1.1	0.9	2.4
	KibL	18	4.0	4.8	1.1	1.5	2.2
	AluS	6	1.4	2.1	0.5	0.6	1.3

Linkage groups as defined by Kim,[20] and position on the linkage group in centiMorgans.
[a] Sko: Samanko; Cza: Cinzana; Alul: Alupe long rain; KibL: Kibos long rain; AluS: Alupe short rain.

2.2. *Data From Phenotyping for Specific Mechanisms of Striga Resistance Based on Laboratory Assays*

As discussed in Chapter 7, the complex trait of *Striga* resistance has been dissected into simpler components using an array of laboratory assays. The low germination stimulant *lgs* gene was mapped using a recombinant inbred population generated from the cross between the low producer SRN39 and the susceptible Chinese sorghum SQR reported with high germination stimulant production. One hundred sixty four (SRN39 x SQR) F_7 progenies were phenotyped using *in vitro* bioassays and genotyped with molecular markers.[19] Genetic analysis conducted earlier had shown that *lgs* was inherited as a single recessive gene,[3] which was confirmed in the phenotypic analysis undertaken in conjunction with subsequent molecular mapping efforts. Using a sorghum consensus map, the *lgs* gene mapped to the sorghum linkage group LG07, flanked by a maize RFLP PIO200725 at 5.7 cM on one side and an ISSR allele ISSR617g at 7.9 cM on the other. Linkage groups were named according to the nomenclature used by Kim.[20]

The low haustoria initiation factor trait was only found in wild sorghum species PQ434.[21,22] Two mapping populations were generated from crosses with PQ434 as donor parent and two high stimulant lines (Shanqui Red from China and a line derived from a random mating population PP34) used as recipient to $F_{2:3}$ families. The two populations were evaluated for haustorium production using *in vitro* bioassays. The data suggest that *Lhf* is inherited as a single dominant nuclear gene.[21] One hundred twenty two families from the (PQ434 x SQR) $F_{2:3}$ population were genotyped using microsatellite markers and *Lhf* was subsequently mapped to 19.3 cM from the marker Xtxp358 on LG09.

Genetic analysis and mapping were also conducted for mechanisms of *Striga* resistance expressed after attachment of *Striga* to host roots. Strong expression of a hypersensitive response was found among a select group of sorghum cultivars (Framida, Dobbs, Serena) as well as a variant (P47121) found in a wild sorghum species, *S. arundinaceum.* Advanced backcross populations were developed from the cross between P47121 and two susceptible lines namely CK60 and KP9.[8] Two lines (CK32 and

K33) that possess a strong hypersensitive response were selected from these two populations. Each was further crossed into susceptible sorghum cultivars (TX430 and TX2737) to generate segregating populations for molecular mapping of this trait. Analysis of phenotypic data using laboratory assays determined that the hypersensitive response was controlled by two nuclear genes with dominant gene action; *HR1* and *HR2*. Genotyping was carried out with microsatellite markers on two populations of BC_3F_4 families derived from the P47121 parent. Two markers were associated with the hypersensitive response on the resulting genetic linkage map. *HR1* mapped at 7.5 cM from Xtxp96 on LG02 and *HR2* mapped at 12.5 cM from SbKAFGK1 on LG05.

The incompatible reaction form of *Striga* resistance was found in several lines including cultivar SRN39.[21] The recombinant inbred population (SRN39 x SQR) F_7 derived from this cultivar was characterized for the incompatible response. Data on this trait were recorded as the ratio of attached *Striga* that were developmentally suppressed, but without apparent necrosis at the site of attachment. The *in vitro* method available for phenotypic evaluation at the time was too cumbersome and inconclusive to rely on for detailed genetic analysis and mapping. A new bioassay has recently been developed (Chapter 7) that may be more amenable for large scale screening required in careful analysis and mapping of genetic populations.

3. Marker-Assisted Introgression

Both laboratory assays and molecular markers have been used to test for introgression of *Striga* resistance into selected genotypes. Recipient parents were either improved sorghum cultivars or landraces susceptible to *Striga*, but with otherwise desirable attributes. High-yielding *Striga*-resistant sorghum cultivars have been developed via bioassay mediated selection and released; a list of these lines and the local names ascribed to these selections in their respective countries of national release have been published.[10] We have also introgressed genes for *Striga* resistance into highly adapted sorghum. For example, El Mota is an early maturing sorghum preferred for its drought tolerance and wide adaptation in Niger. Crosses were made between the resistant line SRN39 and El Mota to generate a BC_2F_3 population. Progenies were found with both very good field resistance to *Striga* and valuable agronomic attributes similar to the original landrace parent (Chapter 10). Several other highly adapted African sorghums landraces were used in the *Striga* resistance introgression program, but have not yet been evaluated for either field resistance to *Striga* or closeness to the agronomic characteristics of their recurrent parent. Marker-assisted selection can be employed after validation of putative molecular markers to enhance introgression of genes for specific resistance components into highly adapted landraces. This can result in improved varieties that combine local adaptations and unique agronomic merit with badly needed genes for *Striga* resistance.

The introgression of genes for *Striga* resistance in parallel collaborative initiatives is currently underway with landraces from Kenya, Mali, Eritrea and Sudan.[23] Farmers together with scientists selected two *Striga*-susceptible farmer-preferred sorghum varieties as candidates for marker-assisted introgression. Initial crosses were made between N13 and the landrace selections. N13 was the *Striga* resistant parent in the RIP-2 population used to identified the 5 stable QTL associated to field resistance (Table 1) and for which none of the 5 QTL alleles showed strong interaction with the environment (Table 2). Backcrosses of the resultant F_1 have also been made to the local cultivars to produce BC_1F_1. Progenies have been advanced through both selfing and further backcrossing. A set of BC_2S_2 progenies has been generated to fix the desired QTLs. Using high throughput genetic fingerprinting, 712 backcrossed lines have so far been genotyped using 10 foreground SSR markers aimed at identifying backcross plants heterozygous for one up to three *Striga* resistance QTL. At least two markers flanking each side were selected for foreground screening of each QTL. Seventeen SSR markers were used for background screening with the aim of speeding up recovery of the recurrent parent. These represented 3 SSR markers on the other arm of the linkage group where the *Striga* resistance QTL were mapped and 14 SSR markers from the remaining 7 linkage groups (each with one SSR marker per chromosome arm). This genotyping, still in progress, has revealed that 256 plants from the second backcross generation (BC_2F_1) are heterozygous for 1 to 3 QTLs that represent four linkage groups. Selected BC_2S_2 plants with QTLs will be evaluated for *Striga* resistance in artificially infested fields through a farmer-participatory approach.

3.1. *Marker-Mediated Gene Pyramiding*

The stacking of several genes controlling each component of resistance into a single genotype is a preferred strategy to deliver more durable resistance. Gene pyramiding was successfully achieved in a variety developed from the cross between SRN39 and Framida.[10] This new variety combining *Striga* resistance with high yield and broad adaptation has been officially released in Ethiopia in 2002 under a local name "Brhan". This cultivar was selected from among progenies that were evaluated both in the field and with the *in vitro* assay after several cycles of selfing. Some progenies had both low germination stimulant production, hypersensitive response, and incompatible reaction.

Introgression and gene pyramiding was also conducted in another population using a widely adapted food grain sorghum selection SEPON82 (Ejeta, unpublished). Paired crosses were made between SEPON82 and each of the two *Striga* resistant parents SRN39 and PQ434. The resistant lines have been characterized for multiple mechanisms of resistance using our bioassays. The simultaneous introgression of genes associated with these resistance mechanisms was accomplished by generating a double cross between the two initial hybrids followed by backcrosses to the SEPON82 parent. Advanced backcross progenies selfed to homozygosity have recently been evaluated for agronomic performance under *Striga* free conditions. Field evaluation under *Striga* infestation will be conducted in Niger. Genotyping of progenies will first be done using markers identified in previous mapping populations. Selected progenies could be released as is, or enhanced through further crossing using marker-assisted selection to enhance recovery of the recurrent parent while pyramiding genes for durable resistance to *Striga*.

4. Conclusion

Progress has been made in controlled evaluation and selection for *Striga* resistance in sorghum through improved field evaluation techniques and development of an array of laboratory assays. Bioassay-mediated selection has resulted in the release of *Striga* resistant cultivars as well as a more effective characterization of specific resistance mechanisms. Loci controlling the inheritance of many of these individual components of *Striga* resistance have been mapped on the sorghum linkage map, though rather grossly for some, but at least they could be located in specific regions of the sorghum linkage groups.

Our work is part of a general effort toward the genetic mapping of genes or QTL associated with resistance to parasitic weeds. Genetic analyses conducted for several crops species have resulted in identification of molecular markers associated with resistance to root parasites. Four QTL were found to control the post-attachment resistance to *S. hermonthica* in rice.[24] Single dominant genes for resistance to race 1 and race 3 of *S. gesnerioides* were identified in cowpea.[25] Genetic maps were also used to identify and locate QTL associated with resistance to *O. crenata,* another parasitic weed that seriously attacks legume crops as well as wild legume species.[26-28] Two QTL were detected in peas[26,27] and three QTL were found in faba bean.[28] *Orobanche cumana* is specialized and parasitizes sunflower. Genetic maps have shown sunflower markers associated with QTL for race-specific resistance to race E and race F of *O. cumana.*[29]

Genetic linkage analyses have resulted in generating molecular markers as tools that allow easier selection and development of breeding material. The use of more accurate screening methods under controlled environmental conditions helped us locate QTL or single genes acting at different stages of the infestation process. Nevertheless, each gene or QTL controlling a particular mechanism of resistance has to be validated across environments and populations before marker-assisted selection is earnestly implemented. Specific QTL may need to get finely mapped to insure that the associated markers are in tight linkage disequilibrium with the gene or, ultimately, to identify the causative mutations that are responsible for the QTL. Allele-specific primers for the trait of interest would facilitate easier introgression of resistance into susceptible host plants. Several cowpea AFLP markers associated with race-specific resistance were converted into co-dominant markers, based on sequence information, which is promising for the use of marker-assisted selection for *Striga* resistance (Chapter 17). As map locations are ascertained and smaller map distances are obtained, opportunities arise to consider QTL-based cloning or search for candidate genes across related species. QTL analysis was integrated with transcriptomics to begin to identify candidate genes for resistance to *S. hermonthica* in rice (Chapter 13). Genetic markers closely associated with individual components of *Striga* resistance will foster the discovery and characterization of specific families of resistance genes for gene cloning and transfer into crop species that are devoid of as wide an array of natural sources of *Striga* resistance as sorghum.

Acknowledgments

Collaborative projects involving Purdue University and several African national agricultural research services (NARS) were funded by the Rockefeller Foundation. Projects spearheaded by ICRISAT and University of Hohenheim and conducted with African NARS were funded by the German Federal Ministry for Economic Development and Cooperation.

References

1. A. R. Saunders, in *South African Department of Agricultural Science*, Bulletin **128**, (1933).
2. D. E. Hess, G. Ejeta and L. G. Butler, *Phytochemistry* **31**, 493 (1992).
3. R. K. Vogler, G. Ejeta and L. G. Butler, *Crop Sci.* **36**, 1185 (1996).
4. B. I. G. Haussmann, D. E. Hess, H. G. Welz and H. H. Geiger, *Field Crop Res.* **66**, 195 (2000).
5. B. I. G. Haussmann, D. E. Hess, B. V. Reddy, H. G. Welz and H. H. Geiger, *Euphytica* **116**, 33 (2000).
6. Y. Ibrahim, *A sorghum linkage map and predicted response to phenotypic and markers selection for resistance to Striga in sorghum*, PhD Dissertation (Purdue University, 1999) [http://docs.lib.purdue.edu/dissertations/AAI9952104/].
7. B. I. G. Haussmann, D. E. Hess, G. O. Omanya, B. V. S. Reddy, H. G. Welz and H. H. Geiger, *Crop Sci.* **41**, 1507 (2001).
8. A. Mohamed, A. Ellicott, T. L. Housley and G. Ejeta, *Crop Sci.* **43**, 1320 (2003).

9. G. O. Omanya, B. I. G. Haussmann, D. E. Hess, B. V. S. Reddy, M. Kayentao, H. G. Welz and H. H. Geiger, *Field Crop Res.* **89**, 237 (2004).
10. G. Ejeta, in *In the wake of the double helix: From the green revolution to the gene revolution*, Eds. R. Tuberosa, R. L. Phillips and M. Gale (Edizioni Avenue media, 2005), p. 239.
11. C. S. Mutengwa, P. B. Tongoona and I. Sithole-Niang, *Afr. J. Biotech.* **4**, 1355 (2005).
12. B. I. G. Haussmann, D. E. Hess, B. V. Reddy, S. Z. Mukuru, M. Kayentao, H. G. Welz and H. H. Geiger, *Euphytica* **122**, 297 (2001).
13. B. I. G. Haussmann, D. E. Hess, B. V. S. Reddy, S. Z. Mukuru, M. Kayentao, H. G. Welz and H. H. Geiger, *Plant Breed.* **120**, 49 (2001).
14. B. I. G. Haussmann, D. E. Hess, G. O. Omanya, R. T. Folkertsma, B. V. Reddy, M. Kayentao, H. G. Welz and H. H. Geiger, *Theor. Appl. Genet.* **109**, 1005 (2004).
15. K. V. Ramaiah, V. L. Chidley and L. R. House, *Euphytica* **45**, 33 (1990).
16. G. Ejeta, L. G. Butler and A. G. Babiker, Eds., *New approaches to the control of Striga* (Agricultural Experiment Station, Purdue University, 1998).
17. B. I. G. Haussmann, V. Mahalakshmi, B. V. Reddy, N. Seetharama, C. T. Hash and H. H. Geiger, *Theor. Appl. Genet.* **106**, 133 (2002).
18. B. I. G. Haussmann, D. E. Hess, N. Seetharama, H. G. Welz and H. H. Geiger, *Theor. Appl. Genet.* **105**, 629 (2002).
19. A. H. Mohamed, P. J. Rich, T. L. Housley and G. Ejeta, in *7th International Parasitic Weed Symposium,* Eds. A. Fer, P. Thalouarn, D. M. Joel, L. J. Musselman, C. Parker and J. A. C. Verkleij, Conference Proceeding (Nantes, France, 2001), p. 96.
20. J.-S. Kim, P. E. Klein, R. R. Klein, H. J. Price, J. E. Mullet and D. M. Stelly, *Genetics* **169**, 1169 (2005).
21. A. H. Mohamed, *Identification and characterization of genetic variants in sorghum for specific mechanisms of Striga resistance*, PhD Dissertation (Purdue University, 2002) [http://docs.lib.purdue.edu/dissertations/AAI3099187/].
22. P. J. Rich, C. Grenier and G. Ejeta, *Crop Sci.* **44**, 2221 (2004).
23. R. T. Folkertsma, B. I. G. Haussman, H. K. Parzies, V. Hoffmann and H. H. Geiger, in *International research in food security, natural resource management and rural development,* Conference Proceeding (Stuttgart, Germany, 2005) [http://www.tropentag.de/2005/proceedings/node281.html].
24. A. L. Gurney, J. Slate, M. C. Press and J. D. Scholes, *New Phytol.* **169**, 199 (2006).
25. J. T. Ouedraogo, V. Maheshwari, D. K. Berner, C. A. St-Pierre and M. P. Timko, *Theor. Appl. Genet.* **102**, 1029 (2001).
26. M. R. Valderrama, B. Román, Z. Satovic, D. Rubiales, J. I. Cubero and A. M. Torres, *Weed Res.* **44**, 323 (2004).
27. D. Rubiales, C. Alcantara and J. C. Sillero, *Weed Res.* **44**, 27 (2004).
28. B. Román, A. M. Torres, D. Rubiales, J. I. Cubero and Z. Satovic, *Genome* **45**, 1057 (2002).
29. B. Pérez-Vich, B. Akhtouch, S. J. Knapp, A. J. Leon, L. Velasco, J. M. Fernàndez-Martínez and S. T. Berry, *Theor. Appl. Genet.* **109**, 92 (2004).

CHAPTER 13

THE MOLECULAR GENETIC BASIS OF HOST RESISTANCE TO *STRIGA* SPECIES: A WAY FORWARD

Julie D. Scholes*, Philip J. Swarbrick, Jon Slate and Malcolm C. Press

Department of Animal and Plant Sciences, Alfred Denny Building University, Western Bank, University of Sheffield, Sheffield S10 2TN, UK.
**E-mail: j.scholes@sheffield.ac.uk*

Understanding the molecular basis of host resistance to *Striga* species is a critical step in the identification of genes that can be used for improving crop productivity via biotechnology based approaches such as crop transformation, or via the development of molecular markers for use in marker assisted selection (MAS) programmes. An understanding of the mechanisms underlying both the detrimental effects of parasitic plants on susceptible hosts, and resistance to these parasites may benefit greatly from the application of genomic technologies. Here we describe how genomic technologies such as Quantitative Trait Loci (QTL) mapping and transcript profiling using oligonucleotide arrays can provide an insight into the molecular genetic basis of resistance to *Striga*.

1. Introduction

Eradication of *Striga,* so necessary in sub Saharan Africa and Asia[1] has proved difficult as the parasite life cycle is intimately linked to that of its host (Chapters 2, 4). Seed germination and haustorial development occur only in response to host-derived chemical cues.[2-5] In addition, the mechanisms underlying the negative impact of *Striga* on crop yield are complex. Firstly, the parasites have a profound effect on host plant growth and development within days of attachment, even a very small parasite biomass causes notable stunting of the plant.[6] Secondly, later in the infection cycle the parasites compete effectively for host carbon and nutrients further reducing growth and yield.[6,7]

An understanding of the mechanisms underlying both the detrimental effects of parasitic plants on susceptible hosts, and resistance to these parasites may benefit greatly from the application of genomic, proteomic and metabolomic technologies. It is clear that the use of *Striga*-resistant cultivars could represent a cost effective control measure, especially when used as part of an integrated control strategy, as their cultivation does not require costly inputs from farmers. However, the use of resistant cultivars is limited by a lack of resistant germplasm and information about the genetics of both host resistance to *Striga* and parasite diversity (race structure). Such information is critical for the identification of genes that can be used for improving crop productivity via biotechnology-based approaches and for the development of molecular markers for use in marker assisted selection (MAS) programmes. Here we review the current state of knowledge of host resistance to *Striga* and describe how the application of quantitative genetic and genomic technologies can contribute to our understanding of the molecular basis of host resistance with particular reference to the rice-*Striga* interaction as a model system.

2. Identification of Post Attachment Resistance in Cereals to *Striga* Species

There is a need to identify crop genotypes that show post attachment resistance to *S. hermonthica*, *S. asiatica* and *S. aspera* in order to exploit modern molecular genetic techniques. While some tolerant cultivars (defined as yield improvement in the presence of the parasite) of both maize and sorghum have been identified,[8-11] there appear to be remarkably few sources of good post attachment resistance among the very large numbers of sorghum and maize genotypes screened to date.[10] There are however, a limited number of sorghum cultivars and wild relatives of sorghum that show different types of post attachment resistance to *Striga*, e.g. in two sorghum cultivars, Framida and Dobbs and a wild accession P47121.[12] The resistance was characterized by a hypersensitive response where tissue surrounding the sites of parasite attachment rapidly became necrotic thus preventing the parasite from invading further into the root cortex. Such a reaction is also seen when the cowpea cultivar B301 is parasitized by *S. gesnerioides* race SG3 from Nigeria.[13] The hypersensitive response is characteristic of race specific or "gene for gene" resistance in which specific resistance (R) genes in the host confer resistance to specific genotypes or races of the parasite (which carry complementary avirulence (*avr*) genes).

There have been few attempts to elucidate the genetic structure of *S. asiatica* and *S. hermonthica* populations,[14,15] and these have not been linked directly with studies of resistance in host germplasm. This situation contrasts markedly with our knowledge of the *S. gesnerioides*-cowpea association in West Africa (Chapter 9), where at least five distinct races of the parasite have been identified and their interaction with different host cultivars described.[16-18] A relatively new, high throughput genotyping technology, Diversity Arrays Technology (DArT), would allow the genetic diversity of parasite populations to be rapidly determined,[19,20] and should be a priority in *Striga* research. The identification and molecular characterization of R genes conferring resistance to *S. hermonthica* and *S. asiatica*, together with a knowledge of parasite diversity is crucial as it would allow the 'pyramiding' (by plant transformation or conventional breeding techniques) of appropriate resistance genes in cultivars suitable for use in different regions of Africa.

In contrast to situations where there is an active resistance gene recognizing a parasite avirulence gene, some *Striga*-host interactions exhibit different forms of resistance that are probably under the control of many genes (broad spectrum resistance). Many of the *Striga*-resistance phenotypes described in cereals are probably polygenic. For example, when the sorghum cultivar N-13, Nandyal Local, was infected with *S. asiatica*, the parasite germinated and penetrated the root cortex but did not penetrate the endodermis and pericycle, both of which accumulated thickening materials.[21] A wild relative of maize, *Tripsacum dactyloides*, is also resistant to *S. hermonthica*, but in this case the parasite attaches, penetrates the root cortex and establishes vascular continuity with the host.[22] However, the haustorium fails to differentiate and the parasite then dies.

Recently we have shown that a cultivar of rice, Nipponbare, has almost complete post-attachment resistance to one population of *S. hermonthica* seed, while several other cultivars showed varying degrees of resistance to this same population of the parasite.[23] These differences are quite clear in the root systems of IAC 165 and Nipponbare inoculated with *S. hermonthica* (Figure 1A and B). IAC 165 is a very susceptible cultivar and supports many large parasites. In contrast, most parasites that attach to Nipponbare die after a few days although one or two parasites per plant have been observed to develop slowly. We attribute this to the fact that *S. hermonthica* is an obligate out-crossing species and the seed is therefore likely to be genetically diverse. The early stages of *Striga* development were similar on IAC 165 and Nipponbare; *Striga* attached to the host root system within 24 h of inoculation and by 72 h the parasitic endophyte had successfully penetrated the host root cortex. This demonstrates that host-specific factors were present in both cultivars that are necessary for early haustorial formation and successful penetration of the cortex. At this early stage, the tissue surrounding attached parasites on Nipponbare was slightly discolored, indicative of the early stages of necrosis. This was not a typical hypersensitive response as observed in the sorghum cultivar Framida, following infection by *S. asiatica.*[12] By 21 days after inoculation, the parasites attached to the susceptible cultivar had well developed haustoria, extensive parasite-host xylem-xylem connections and the parasites had developed between 2-5 leaf pairs (Figure 1 C and E). In contrast, parasites attached to Nipponbare rarely breached the root endodermis and were therefore unable to form xylem continuity with the host (Figure 1 D and F). Lack of access to host nutrients and factors required for parasite differentiation probably resulted in the death of the parasite. The reason for the inability of the parasite to penetrate the endodermis and pericycle is unclear as there was no obvious structural difference between the endodermis of Nipponbare and that of susceptible cultivars. The endodermis did not appear to be more heavily lignified or thickened either before or following infection by *Striga*. This contrasts with the phenotype observed in the sorghum cultivar N-13,[21] where thickening of

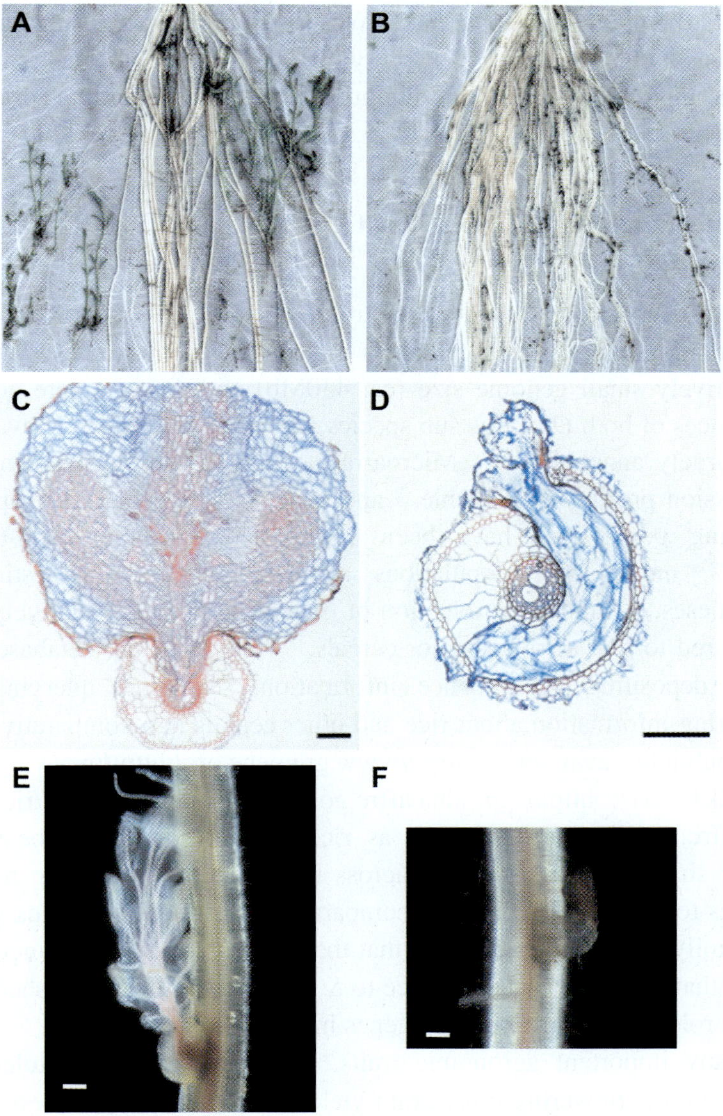

Figure 1. During a resistant interaction between S. hermonthica *and the rice cultivar Nipponbare, the parasite penetrates the root cortex but fails to form vascular connections with the host and dies. (A and B) The root systems of susceptible (IAC 165) and resistant (Nipponbare) rice cultivars. (C and D). Transverse sections through embedded tissue of IAC 165 and Nipponbare illustrating the extent of parasite development and (E and F) whole sections of tissue stained with phloroglucinol to show areas of lignification (red colour). The scale bar represents 0.1 mm. Adapted from Ref. 23*

these structures was apparent. We are currently investigating the hypothesis that the lack of ability of the parasite to penetrate the endodermis may reflect an alteration in auxin signalling or auxin sensitivity.

3. Understanding the Molecular Genetic Basis of Host Resistance to *Striga*: The Use of Genomic Technologies

The discovery of resistance in rice to *Striga* is of great significance as it is currently the best model cereal for molecular genetic studies. Rice has a relatively small genome size (ca. 430Mb) and the complete genome sequences of both *O. sativa* sub species *japonica* and *indica* are available and largely annotated.[24,25] Microarray technology for studying mRNA expression profiles is available,[26] and high resolution linkage maps and mapping populations have been constructed.[27-29] Transposon-tagged (*Tos17*) mutant rice populations are available for the testing of hypotheses,[30-32] and the production of transgenic plants is relatively easy compared to that of other major cereals.[33-34] In addition, databases that allow depositing of sequence information, searching, querying and analyzing information about rice and other cereals in a comparative way are publicly available (http://www.gramene.org/;http://www.tigr.org/tdb/e2k1/osa1/; http://rgp. dna.affrc.go.jp/E/index.html; http://ricegaas.dna.affrc.go.jp/). Cereals such as rice, maize, sorghum, wheat, and barley share extensive synteny across their genomes, allowing for one species to serve as the base for comparative functional genomics within the family.[35] Thus it is possible that the identification of the function of genes that confer resistance in rice to *S. hermonthica* will also shed light on the role of the corresponding genes in other cereal hosts.

Many important agronomic traits, for example drought tolerance, heading date, flowering time, grain yield and broad spectrum resistance, are each controlled by many Quantitative Trait Loci (QTL) genes. Because the inheritance of such traits is complex, their identification and hence use in breeding programmes has proved difficult. However with the availability of genome sequences it is possible to design molecular markers based on genome information. Mapping populations such as Recombinant Inbred Lines (RILs) and Backcross Inbred Lines (BILs) are required to carry out a QTL analysis together with detailed linkage maps based on molecular markers. Each member of the mapping population is scored for the trait of interest. A statistical calculation of linkage is carried out using the linkage maps to localize QTLs underlying the trait on the genome. This information can then be utilized in Marker Assisted Selection programmes. Using such techniques progress has been made in identifying QTL underlying field tolerance/resistance to *Striga* using mapping populations of sorghum,[36-37] and this information is being used in MAS Programmes in Africa (Chapters 7, 10).

We have carried out a QTL analysis using a mapping population (Nipponbare/Kasalath//Nipponbare) of Backcross Inbred Lines (BILs) to begin the identification of the molecular genetic basis of post-attachment resistance to *Striga* in rice.[23] The mapping population consisted of 391 plants (mean *c*. 4.0 replicate plants per BIL). Each of the BIL plants was established in the presence of *S. hermonthica.*[23]. Plants were scored for *S. hermonthica* resistance 21 days after inoculation, with host resistance being defined as the proportion of *S. hermonthica* parasites that were attached to the roots but had not developed further together with those attachments that were clearly dead. Resistance QTL were mapped by composite interval mapping.[23] QTL explaining a large proportion of resistance were discovered on five chromosomes; 4 alleles providing resistance from Nipponbare and 1 allele from Kasalath (Table 1). Each of these QTL was statistically significant at the stringent genome-wide $P < 0.001$ threshold. Allelic substitutions at each QTL altered the phenotype by at least 0.5 of a phenotypic standard deviation (SD) relative to the parental lines (Table 1). This suggests that, although the resistance trait is polygenic, it is likely to be due to a few genes of major effect.[23]

Although QTL mapping allows regions of a chromosome associated with a particular phenotypic trait to be identified, the regions are often large and contain thousands of potential genes. Fine mapping of the genes is required to narrow down candidate genes. However, this alone is not sufficient to identify a small enough number of genes for proof of function analysis. A number of recent studies have combined QTL mapping with gene expression profiling using microarrays to identify potential candidate genes. This novel approach allowed the successful identification of 34 candidate genes for ovariole number, a quantitative trait, in *Drosophila melongaster.*[38] Similarly, positional candidate genes conferring resistance to Marek's disease (a herpes virus-induced T cell cancer in chicken) were identified by integrating DNA microarrays and genetic mapping.[39] These analyses of changes in gene expression were performed on the parental lines of the mapping populations, or on a small number of lines exhibiting contrasting phenotypes.

Table 1. QTL explaining a large proportion of resistance to *S. hermonthica* were discovered on five chromosomes; 4 alleles providing resistance from Nipponbare and 1 allele from Kasalath.

Chromosome	Position (cM)	LRT	PVE	Allelic substitution (SD)	Number of differentially regulated genes*
4	79	66.13	7.6	-0.064 (-0.8)	-
5	77	19.65	1.9	0.039 (0.49)	17
6	97	45.04	4.2	0.051 (0.64)	5
8	32	21.59	2.1	0.038 (0.48)	5
12	41	63.28	7.4	0.075 (0.94)	9

LRT = likelihood ratio test statistic where the null hypothesis is no QTL; PVE = percentage phenotypic variance in the mapping population explained by the QTL; the additive effect on mean resistance (arc sine transformed) of an allelic substitution from a Kasalath allele to a Nipponbare allele (effect size is also measured in standard deviations, where the phenotypic standard deviation in the parental races is 0.08). An additive effect with a positive coefficient means that the Nipponbare derived allele confers increased resistance. * Number of genes within each QTL region that were significantly up or down regulated $p < 0.05$. Adapted from Gurney *et al.*[23]

We have profiled changes in gene expression in *Striga*-infected roots using the Affymetrix whole genome rice oligonucleotide array to begin to dissect the Nipponbare resistance QTL into their underlying genetic determinants and to compare differences in gene expression between a resistant and a susceptible interaction. IAC 165 and Nipponbare plants were grown in rhizotrons for 3 weeks in a controlled environment room and roots inoculated with pre-germinated *S. hermonthica* seed to ensure synchronous attachment and parasite development. Root samples were harvested 2, 4 and 11 days after inoculation and unattached *Striga* seeds, the external part of the haustorium and, in the susceptible interaction, the *Striga* shoot were carefully removed from the roots. Total RNA was prepared, labeled and hybridized to the arrays and data analyzed following standard protocols.

A small number of genes have been identified so far within the QTL regions of the Nipponbare genome that are significantly up or down regulated during the resistance response (Table 1). The expression of these genes is not altered in roots of the susceptible cultivar IAC 165. These genes are potential candidates for *Striga* resistance and currently proof of function studies are being carried out using reverse genetic approaches. In addition to linking transcriptomic studies to QTL analysis, comparison of changes in gene expression in the resistant versus susceptible cultivar is revealing important information about pathways and processes that may be important in resistance and susceptibility.

One of the most striking differences between the susceptible and resistant interaction is the extent of the down-regulation of gene expression that takes place as *Striga* develops on roots of the susceptible cultivar; of the 2588 genes that are differentially regulated, 553 are up regulated whereas over 2000 are down regulated (Table 2). The down regulated genes include those purportedly involved in metabolism, cell cycle and DNA processing, transcription, protein synthesis and fate, cellular communication and signal transduction (Table 2). Such changes in gene expression are consistent with the reduction in host growth that occurs shortly after infection by *Striga*. In contrast, in the resistant interaction, similar numbers of gene are up and down regulated (Table 2).

Interestingly, many of the up-regulated genes in the resistant cultivar are those classically associated with defence responses to fungi and bacteria. They include pathogenesis related (PR) genes, genes encoding defence response proteins, genes containing leucine rich repeat (LRR) motifs that are characteristic of resistance genes, cytochrome P450s and transcription factors, such as those of the WRKY family. The latter are particularly interesting, as many transcription factors are thought to act as "master-switches" controlling the expression of several genes in a single pathway. Therefore, it may be possible to produce large changes in a single trait by manipulating such genes, for example resistance to parasitic plants. Although it is often very difficult to determine the function of individual transcription factors there has been some success; over expression of three WRKY genes in *Arabidopsis thaliana* resulted in enhanced resistance to the bacterial pathogen *P. syringae.*[40]

Table 2. Genes significantly up and down regulated in a susceptible cultivar (IAC165) and a resistant (Nipponbare) cultivar following infection by *Striga hermonthica.* ($p < 0.05$; following the application of the Benjamini Hochberg correction)

Functional category	Number of genes significantly up regulated		Number of genes significantly down regulated	
	Nipponbare	IAC 165	Nipponbare	IAC 165
Metabolism	59	77	95	316
Energy	21	24	16	30
Cell cycle and DNA processing	6	3	22	86
Transcription	63	46	49	167
Protein synthesis and fate	35	39	32	111
Cellular transport	33	43	82	201
Cellular communication/signal transduction mechanism	45	38	69	209
Cell rescue, defence and virulence	92	81	71	141
Interaction with the environment	18	16	28	49
Development	16	11	28	75
Biogenesis of cellular components	13	12	49	99
Unknown proteins	150	163	237	551
TOTAL	551	553	778	2035

We are currently investigating the importance of some of the genes that are up regulated in the rice-*Striga* resistance response by carrying out more detailed studies of spatial and temporal changes in gene expression using quantitative RT-PCR and *in situ* localization of mRNA. In addition, we are using DNA insertion mutants to examine the effect on *Striga* development and transcript fingerprints, of 'knocking out' specific genes where appropriate.

4. Conclusions

Integrating genomic strategies such as QTL mapping and transcript profiling will certainly increase our ability to identify genes, suites of genes, and pathways that are causally linked to resistance phenotypes. In addition, coupling these techniques with high throughput proteomic and metabolomic analyses in the future will provide a more comprehensive view of the complex interactions between parasitic angiosperms and their hosts, and pave the way for the design of novel control strategies.

Acknowledgements

We would like to thank the Rockefeller Foundation and the Syngenta Foundation for Sustainable Agriculture for funding the rice-*Striga* work.

References

1. C. Parker, C. R. Riches, *Parasitic Weeds of the World: Biology and Control* (CAB International 1993).
2. W. J. Keyes, J. V. Taylor, R. P. Apkarian, and D. G. Lynn, *Plant Physiol.* **127**, 1508 (2001).
3. J. I. Yoder, *Curr. Opin. Plant Biol.* **4**, 359 (2001).
4. A. G. Palmer, R. Gao, J. Maresh, W. Kaya Erbil, and D. G. Lynn *Ann. Rev. Phytopath.* **42**, 439 (2004).
5. A. J. Humphrey and M. H. Beale, *Phytochemistry* **67**, 636 (2006).
6. A. L. Gurney, M.C. Press and J.D. Scholes, *New Phytol.* 143, 573 (1999).
7. D. L. Frost, A. L. Gurney, M. C. Press and J. D. Scholes, *Plant, Cell Environ.* **20**, 4873 (1997).
8. A. Van Ast and L. Bastiaans *Weed Res.* **46**, 264 (2006).
9. A. Menkir, *Plant Breeding* **125**, 131 (2006).
10. A. Oswald J. K. Ransom, *Crop Protect.* **23**, 89 (2004).

11. A. L. Gurney, A. Taylor, A. Mbwaga, J. D. Scholes and M. C. Press, *Weed Res.* **42**, 299 (2002).
12. A. Mohamed, A. Ellicott, T. L. Housley, G. Ejeta, *Crop Sci.* **43**, 1320 (2003).
13. C. J. Botanga and M. P. Timko, *Phytopathology* **95**, 1166 (2005).
14. J. G. Gethi, M. E. Smith, S. E. Mitchell and S. Kresovich, *Weed Res.* **45**, 64 (2005).
15. C. S. Mutengwa, P. B. Tongoona, I. Sithole-Niang, *African J. Biotech.* **4**, 1355 (2005).
16. O. Boukar, L. Kong, B. B. Singh, L. Murdock and H. W. Ohm, *Crop Sci.* **44**, 1259 (2004).
17. J. T. Ouedraogo, J. B. Tignegre, M. P. Timko and F. J. Belzile, *Genome* **45**, 787 (2002).
18. C. J. Botanga, J. G. Kling, D. K. Berner and M. P. Timko, *Euphytica* **128**, 375 (2002).
19. D. Jaccoud, K. Peng, D. Feinstein and A. Kilian, *Nucleic Acids Res.* **29**, 25 (2001).
20. P. Wenzl, J. Carling' D. Kudrna, D. Jaccoud, E. Huttner, A. Kleinhofs, and A. Kilian, *Proc. Natl. Acad. Sci. USA* **101**, 9915 (2004).
21. R. K. Maiti, K. V. Ramaiah, S. S. Bisen and V. L. Chidley, *Ann. Botany* **54**, 447 (1984).
22. A. L. Gurney, D. Grimanelli, F. Kanampiu, D. Hoisington, J. D. Scholes and M. C. Press, *New Phytol.* **160**, 557 (2003).
23. A. L. Gurney, J. Slate, M. C. Press and J. D. Scholes JD, *New Phytol.* **169**, 199 (2006).
24. S. A. Goff *et al.*, *Science* **296**, 92 (2002).
25. J. Yu *et al.*, *Science* **296**, 79 (2002).
26. W. A. Rensink and C. R: Buell, *Trends Plant Sci.* **10**, 603 (2005).
27. M. Yano and T. Sasaki, *Plant Mol. Biol.* **35**, 145 (1997).
28. N. Kurata *et al.*, *Nat. Genet.* **8**, 365 (1994).
29. Y. Harushima *et al.*, *Genetics* **148**, 479 (1998).
30. D. Sallaud, C. Gay, P. Larmande, M. Bès, P. Piffanelli, B. Piégu,G. Droc, F. Regad, E. Bourgeois, D. Meynard, C. Périn, X. Sabau, A. Ghesquière, J. C. Glaszmann, M. Delseny and E. Guiderdoni, *Plant J.* **39**, 450 (2004).
31. H. Hirochika, E. Guiderdoni, G. An,Y. I. Hsing, M. Y. Eun, C. D. Han, N. Upadhyaya, S. Ramachandran, Q. Zhang, A. Pereira, V. Sundaresan and H. Leung, *Plant Mol. Biol.* **54**, 325 (2004).
32. T. Kolesnik, I. Szeverenyi, D. Bachmann, C. S. Kumar, S. Jiang, R. Ramamoorthy, M. Cai, Z. G. Ma, V. Sundaresan, and S. Ramachandran, *Plant J.* **37**, 301 (2004).
33. Y. Hiei, S. Ohta, T. Komari, and T. Kumashiro, *Plant J.* **6**, 271 (1994).
34. T. Komari, Y. Hiei, Y. Ishida, T. Kumashiro, and T. Kubo, *Curr. Opin. Plant Biol.* **1**, 161 (1998).
35. K. M. Devos, *Curr. Opin. Plant Biol.* **8**, 155 (2005).
36. B. I. G. Haussmann, D. E. Hess, G. O. Omanya, R. T. Folkertsma, B. V. S. Reddy, M. Kayentao, H. G. Welz and H. H. Geiger, *TAG* **109**, 1005 (2004).
37. B. I. G. Haussmann, D. E. Hess, B. V. S. Reddy, S. Z. Mukuru, M. Kayentao, H. G. Welz and H. H. Geiger, *Plant Breeding* **120**, 49 (2001).
38. M. L. Wayne and L. M. McIntyre, *Proc. Natl. Acad. Sci. USA* **99**, 14903 (2002).
39. H. C. Liu, H.H. Cheng, V. Tirunagaru, L. Sofer and J. Burnside, *Animal Genet.* **32**, 351 (2001).
40. C. Chen and Z. Chen, *Plant Physiol.* **129**, 706 (2002).

CHAPTER 14

EFFECTS ON *STRIGA* PARASITISM OF TRANSGENIC MAIZE ARMED WITH RNAi CONSTRUCTS TARGETING ESSENTIAL *S. ASIATICA* GENES

Anic de Framond[1], Patrick J. Rich[2,*], John McMillan[1] and Gebisa Ejeta[2]

[1]*Syngenta Biotechnology Inc., 3054 Cornwallis Road, Research Triangle Park, NC 27709-2257, USA*

[2]*Purdue University, Department of Agronomy, 915 W. State St., West Lafayette, IN 47907-2054, USA*

[*]*E-mail: pjrich@purdue.edu*

We are attempting to engineer transgenic maize for resistance to the parasitic weed *Striga,* based on RNA interference silencing technology (RNAi). In this approach, the transgenic maize produces double-stranded RNA (dsRNA) molecules targeted against genes essential for *Striga* survival. As *Striga* establishes on the maize roots, the silencing agent could systemically spread to its cells, shutting down the targeted essential genes and thereby killing the parasite. Five *Striga* genes were chosen as targets in either of 13 dsRNA interference constructs and 55 transgenic maize lines containing an RNAi construct were selected for testing. We report the results of an initial screening of these materials with *S. asiatica*. Although some events still need to be assayed, none of the transgenic maize in 11 events tested is obviously resistant to *Striga* parasitism within 4-5 weeks of infestation. Some *Striga* plants were able to develop and survive on all transgenic materials tested. There are indications that *Striga* grows slower when attached to maize with an RNAi construct in at least half the transformation events tested relative to non-transgenic segregants of those events. Further testing is needed to confirm these results, and ascertain their field relevance.

1. Introduction

RNA interference silencing technology was first discovered in *C. elegans*.[1] It is a very general phenomenon in organisms as diverse as protozoa, animals, plants, and fungi. The natural function of RNAi in plants is believed to be a defense response to silence invasive nucleic acids from viruses and transposable elements. The technology of using RNAi to target genes for silencing is only a few years old and, it is not yet completely understood, and much has been learned about its mechanism. Double-stranded RNA (dsRNA) molecules can induce the degradation of homologous RNA transcripts, resulting in post-transcriptional gene silencing. The dsRNA is generally processed by plant cells into small interfering RNA (siRNA) molecules of 20-26 nucleotides that are believed to be intermediates in the silencing effects of dsRNA. RNAi allows for efficient and specific gene silencing. RNAi is extremely potent and requires only a few dsRNA molecules. It is 1000 times more efficient in gene silencing than antisense.[2] Silencing of reporter genes as well as endogenous plant genes has been shown using RNAi.[3-5] Furthermore, and more importantly for our project, RNAi can spread systemically within a plant via some uncharacterized signal, and can also be transmitted from a transformed plant through a graft union to untransformed plants.[6,7] The RNAi signal was able to move from transgenic *Arabidopsis* to silence the targeted virulence gene in root-knot nematodes resulting in resistance to parasitism.[8] Evidence suggests that the systemic signal travels via the phloem.[9]

The possibility of using RNAi in plants to control parasitic weeds has not been reported. In such an approach, a transgenic plant would express dsRNA molecules targeted against genes essential for parasitic weed survival. As the parasite establishes on the host root, it will presumably take up dsRNA molecules, siRNA molecules or some unknown systemic signal molecule, which in turn will trigger silencing of its essential genes. Whether this approach fails or succeeds will depend on the efficient uptake of the RNAi systemic signal by the parasite. Although no direct phloem connections have been observed in *Striga hermonthica* with its maize and sorghum hosts,[10] there is a complex movement of solutes and carbon assimilates through what appears to be the xylem sap in such associations.[11,12] Specialized transfer cells within the *Striga* haustorium may also be involved in drawing vital factors from the host.[13] There is also evidence of an exchange of developmental signals between *Striga* and a cereal host affecting the growth of both.[14,15] Direct phloem connections have been observed in the parasitic association of *Orobanche crenata* with a legume host.[16] Given the ability of the dicotyledonous C_3 *Striga* to obtain a diversity of compounds from its monocotyledonous C_4 maize host, it is conceivable that the RNAi systemic signal could traverse the largely uncharacterized connective tissues of the haustorium.

We wanted to test this idea in maize to target *Striga asiatica*. The *Striga* genes chosen as targets for our dsRNA constructs are known essential genes in plants, such as herbicide targets (EPSP synthase, target of glyphosate), as well as genes shown to be essential in plants, such as AdSS (adenylosuccinate synthetase, the first enzyme in AMP biosynthesis) or VCL1 (Vacuoleless1), a gene required for vacuole formation and morphogenesis. The *Striga* gene sequences used in the constructs were chosen from regions of less than 80% homology to their counterparts in maize. Therefore, the dsRNA resulting from transcription of these transgenes are not toxic to maize.

2. Materials and Methods

2.1. *Striga Genes Chosen as RNAi Targets and Their Source*

Striga asiatica leaves, roots, and haustoria were collected from an infested maize field in Horry County, South Carolina, with the help of USDA-APHIS and an import permit from the North Carolina Department of Agriculture. Total RNA was isolated from underground white tissue of *Striga* that contained roots and haustoria. Essential genes or gene fragments were then cloned from total RNA by RT-PCR. The following targets were chosen:

1- **EPSP synthase** (5-enoylpyruvylshikimate 3-phosphate synthase) is required for the synthesis of aromatic amino acids in plants and is the target of the herbicide glyphosate.[17]

2- **CTase** (**άCTase**) is part of a 4-protein ACCase (acetyl-CoA carboxylase) complex. This activity is needed for the initiation of fatty

acid biosynthesis in the plastid. Maize does not have this particular gene, but uses a different type of ACCase for this activity.[18]
3- **ENR** (enoyl-ACP reductase) is also involved in fatty acid biosynthesis and was also shown to be essential in plants.[19]
4- ***VCL*1** (an ortholog of *S. cerevisiae Vps*16) is an essential gene required for vacuole formation and morphogenesis. *Arabidopsis VCL*1 is expressed throughout development, but especially in growing organs.[20]
5- **AdSS** (adenylo-succinate synthase) is a key step in adenosine monophosphate (AMP) synthesis.[21]

2.2. *Vectors and RNAi Design*

The binary backbone vector is pNOV2117. The intron used as the spacer fragment for the loop of the dsRNA is an intron from maize *Adh1*. The promoter driving the dsRNA is CMPS from Cestrum Yellow Leaf Curling virus. The promoter region includes a TATA box and enhancer factors. The plant selectable marker is phosphomannose isomerase (PMI) driven by ZmUbiInt. In addition to the dsRNA constructs built using each of the target genes described above, we also made a chimeric construct that contains a fragment of each of the five target genes.

2.3. *Transgenic Maize Lines*

All constructs passed quality control and were transformed into maize. Single-copy events were obtained and confirmed via genomic Southern analysis. Functionality of the CMPS promoter sequence used in the constructs was confirmed in another construct in which CMPS was linked to a GUS reporter gene. Two to ten lines (events) per construct were chosen for testing on *Striga*. There are two constructs for each target gene; one has the sense strand-spacer-antisense strand, the other antisense strand-spacer-sense strand. Three constructs for EPSP synthase were used containing various parts of the gene due to the high homology between maize and *Striga* EPSP synthase genes.

Thirteen constructs were made and depending on the construct, between two and ten independent transformation events were selected after plant analysis and quality control. The selected events were then either selfed, or when the primary event was not self-fertile, crossed back to wild-type maize. Therefore, the seeds available for testing represent a segregating population. Fifty-five transformation events were tested with *Striga asiatica* in the Purdue parasitic weed containment facility.

2.4. *Laboratory Testing of Maize Transformation Events with* **Striga asiatica**

As *Striga* establishes vascular connection with its host within a few days after attaching to maize roots, and the RNAi constructs target essential *Striga* genes, it was assumed that the any silencing effect would be manifest in parasites on transgenic maize early in the association. Screening therefore focused on the early stages of parasite establishment, which are illustrated in Figure 1.

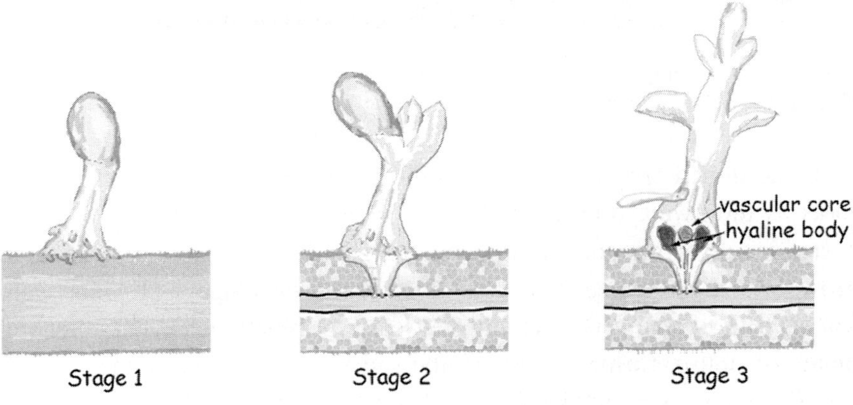

Figure 1. Early post-attachment Striga *growth stages:* Stage 1 – Visible attachment. *The* Striga *haustorium adheres to the host root and begins penetrating the various root cell layers (epidermis, cortex, endodermis). Growth is confined to the endophytic haustorium and no apparent shoot development occurs.* Stage 2 – First leaf pair emergence. *Appearance of the first leaf primordia from the* Striga *seed coat is believed to coincide with penetration of the haustorium to the host root stele.* Stage 3 – Shoot development beyond the first leaf pair. *New scale leaves appear in pairs with alternating orientations from the shoot apex. Internally, the haustorium development continues and vascular connection to individual host xylem elements are established and fortified.*

The materials were screened by co-culture with the Carolina strain of *S. asiatica* in the transparent cup system (Figure 2). In this method, five-day-old maize seedlings were transplanted between the sides of transparent plastic cups and a glass fiber cone into which sand is placed.

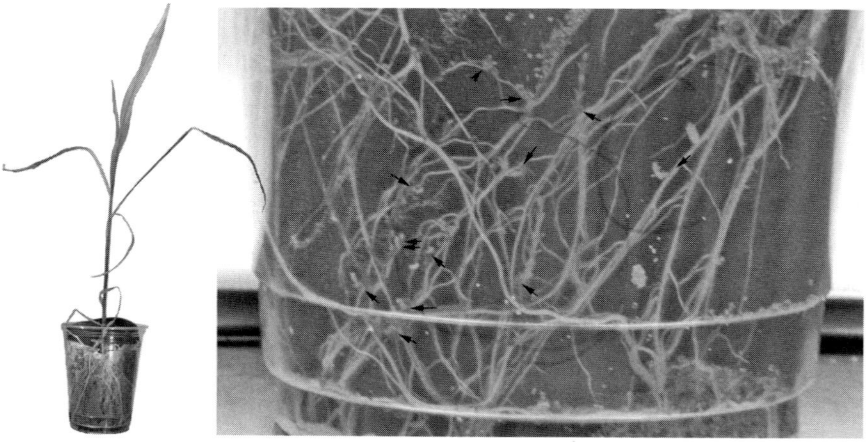

Figure 2. The transparent cup method for growing Striga *on maize. Arrows show developing parasites on maize roots.*

Conditioned *Striga* seeds were applied with a paintbrush to the maize roots as they were transplanted to the cups. Each transparent cup was nested inside an opaque cup to exclude light. This method allowed us to grow *Striga* on maize roots for five weeks or longer. Events were screened in twelve sets, including at least six plants per event. Several plants of untransformed maize in the same genetic background used in transformation were included in each set as controls. The infection rate of each batch of *Striga* used was determined by its ability to form attachments on these untransformed maize plants. Maize roots were scanned five weeks after infesting, and *Striga* attachments reaching stage 3 (two or more scale leaf pairs) were counted and size (number of leaf pairs and shoot length) of the most developed parasites was recorded. The presence of a transgene in infected plants was determined by testing a ground fresh leaf sample with an immunostrip (Strategic Diagnostics, Inc.) specific for the phosphomannose isomerase (PMI) selectable marker present on each construct. The screen sought to eliminate those events on which *Striga* grew equally well on transgenic and non-transgenic segregants.

3. Results and Discussion

Some difficulties were encountered during the transparent cup screening of this transgenic material. We attempted to test representatives from all 55 transgenic events. Seedlings from 49 events survived to five weeks. Three of the twelve sets in which the materials were screened were discounted due to low attachment rates on non-transformed control plants. The events of those sets are being retested. Additional difficulties in the screening were due to limited amounts of maize seed, poor germination and low seedling vigor for some events. Maize from a few events did not germinate, others did not form a shoot or died a week or two after transplanting. A total of 373 maize plants were infested with *S. asiatica* in the transparent cup system and 78% of these (294) survived the five weeks until measurements were made. Of these, only some carried a transgene. Oddly, progeny from some events contained no transgenic individuals, most notably those from events transformed with dsENR constructs. Of the 49 events for which there was testable material, that is seedlings which survived to five weeks, only 40 included some transgenic (PMI positive) segregants. Of these, only 11 had sufficient numbers of attached *Striga* ($\geq$ 25) to compare parasite growth on transgenics with their corresponding non-transgenic segregant controls. A summary of *Striga* growth on these 11 events is presented in Table 1. Further testing is underway.

All transgenic plants tested supported stage 3 *Striga* with at least three leaf pairs within the 35-day co-culture period. Typical root scans of infected transgenic and control plants are presented in Figure 3. Out of all the materials screened, the percentage of transgenics supporting at least one *Striga* plant with four leaf pairs was 97% and 91% supported some parasites to at least five leaf pairs. The average maximum number of leaf pairs present on *Striga* attached to the non-transgenic segregants of all events was nine. Only about a third (31%) of the largest *Striga* on the screened events with transgenic segregants reached the 9-leaf pair stage. It generally appears that *Striga* growth was less on some transgenics relative to controls.

Table 1. Transgenic maize tested in the transparent cup system with ≥ 25 *Striga asiatica* attachments. Data presented are from those events for which at least two transgenic plants were available to compare with non-transgenic segregants from the same event. Chimeric constructs contain portions of all five targeted *Striga* sequences (EPSP, CTase, ENR, VCL1 and AdSS).

Event	Construct	Target	Avg. No. Parasites Reaching Stage 3 per Maize Plant		No. Leaf Pairs on Largest Parasite	
			Transgenic	Non-Transgenic	Transgenic	Non-Transgenic
5627	pSTR39	CTase	12	6	12	7
5630	pSTR39	CTase	3	30	7	11
5636	11057	VCL1	8	1	12	4
5638	11058	VCL1	12	15	9	13
5651	11238	AdSS	20	14	8	10
5653	11238	AdSS	8	24	6	4
5656	11239	AdSS	10	16	7	9
5659	11240	Chimeric	32	42	7	9
5664	11240	Chimeric	3	13	7	7
5677	11241	Chimeric	9	12	9	9
5678	11241	Chimeric	8	14	8	6

Of the eleven events reported in Table 1, 72% had fewer *Striga* attachments reaching stage 3 relative to non-transgenic segregants from the same transformation event. Considering only the largest parasites growing on the maize roots, slightly less than half (45%) of the transgenic segregants had smaller parasites than their non-transgenic counterparts. These largest parasites were smaller than those on controls both in terms of the number of leaf pairs and shoot length. All the tested plants containing chimeric constructs supported fewer *Striga* than their respective controls. Only plants from one of these, however, had smaller parasites than the controls. In one event each of CTase, VCL1, and AdSS, plants containing constructs had both fewer and smaller parasites with respect to the non-transgenic segregants of those events.

Figure 3. Striga asiatica *growing on the roots of non-transgenic (top) and transgenic (bottom) segregants from transformation event 5659. Arrows show developing parasites on maize roots. Red bar = 5 mm.*

This initial screen was not set up as an experiment to test for subtle differences between *Striga* growth on transgenics and controls. Rather, the screen was to look for immediate and obvious effects of RNAi on *Striga* attaching to these materials. If the RNAi signal got into the parasite, and effectively silenced genes required for normal growth and development, one would expect all attached parasites to deteriorate quickly after penetrating transgenic roots. None of the attached *Striga* observed on any of the transgenics actually died within the 35 days of infestation. So depending on where criteria are set to define resistance to *Striga* at this early stage, most constructs appear to have no effect on *Striga*, as ten of eleven (Table 1) tested events supported *Striga* with up to the five leaf pairs. Alternatively, if the criteria are set at nine leaf pairs, which was the average size of the largest parasites on non-transgenic segregants, then most (seven out of eleven) events tested appear to have gained some resistance from the RNAi constructs.

Whatever the case, further testing is required. There is no swift and obvious effect of these RNAi constructs on *Striga* attaching to transgenic maize. In all cases, no *Striga* died within the 4-5 weeks after attaching to transgenic maize roots, except when the particular root branch where they attached died, but this also occurred on control plants. Given the supposed essential nature of the targeted *Striga* genes, it is hard to imagine how the parasite could survive if the silencing signal passed to them from their transgenic hosts.

Some possible reasons for the lack of the anticipated dramatic *Striga* resistance from these materials include:

1. The constructs are faulty in that the transcription products in maize do not form the kind of double-stranded molecule that is required to cause the silencing effect of targeted parasite genes.
2. The RNAi systemic silencing signal is not passing from the transgenic maize into the *Striga* parasitizing its roots.
3. Any siRNAs passing from transgenic hosts to their parasites are not sufficient to completely and effectively silence essential *Striga* genes.
4. If targeted *Striga* gene transcription levels are negatively affected, the parasite can compensate for the loss, either by functions of other gene products in its own genome, or obtain vital metabolites from its host.

5. The effect occurs later in the parasitic association of *Striga* with maize, beyond the period observed in our assays.

It is definitely too soon to conclude that the technology represented by these unique materials is ineffective against *Striga*. On the contrary, the limited comparison of transgenics with controls to date suggests that *Striga* growth on most transgenics is subtly limited. Further testing of these materials includes conducting replicated experiments with appropriate numbers of transgenic plants to test for subtle effects of selected constructs on early *Striga* growth. This would allow mean comparisons between transgenic and non-transgenic segregants within selected events to determine the statistical significance of any differences. There is a possibility that *Striga* with impaired EPSP synthase might compensate by obtaining the aromatic amino acids they need from the maize. If this is happening, it should be evident in their free amino acid profiles. The bulk of the transgenic material that remains to be tested targets EPSP synthase, so we can compare free amino acids from *Striga* attached to transgenics in these tests with those growing on controls. mRNA levels of targeted genes should be ascertained by quantitative RT-PCR to see if these messages are reduced in parasites on transgenics relative to controls.

There is promise that RNAi could work in protecting host crops from parasitic plants. Lettuce transformed with a ds*GUS* construct was able to silence *GUS* expression in transgenic *Triphysaria* that attached to its roots (Chapter 3). Steve Runo *et al.*, in a poster at this conference reported mRNA movement from tomato and alfalfa to stem parasitic dodder (*Cuscuta pentagona*). This group is developing transgenic sorghum and tobacco with RNAi vectors targeting *KNOX1* genes required for meristem maintenance in parasitic plants.

4. Constraints and Integration

The RNAi approach could be used to augment genetic resistance in *Striga* hosts, particularly in maize where native resistance genes may be limited. Because of its transgenic nature, certain regulatory issues will complicate deployment of the technology. As with all control measures, it should be combined with other technologies to avoid possible virulence that could quickly develop in *Striga* populations growing on host plants with only a single defense mechanism.

Acknowledgements

This work is supported by a grant from the Rockefeller Foundation (grant # 2004-FS-088).

References

1. M.K Montgomery, X. SiQun and A. Fire, *Proc. Natl. Acad. Sci. USA* **95,** 15,502 (1998).
2. A. Thakur, *Electronic J. Biotech.* **6**, 39 (2003).
3. P. M. Waterhouse, M. W. Graham and M.B. Wang, *Proc. Natl. Acad. Sci. USA* **95**, 13959 (1998).
4. C. F. Chuang, and E. M. Meyerowitz, *Proc. Natl. Acad. Sci. USA* **97**, 4985 (2000).
5. Z. L. Levin, A. J. de Framond, A. Tuttle, M. W. Bauer and P. B. Heifetz, *Plant Mol. Biol.* **44**, 759 (2000).
6. R. A. Jorgensen, R. G. Atkinson, R. L. S. Forster, and W. J. Lucas, *Science* **279**, 1486 (1998).
7. H. Vaucheret, C. Beclin, T. Elmayan, F. Feuerbach, C. Godon, J.-B. Morel, P. Mourrain, J.-C. Palauqui and S. Vernhettes, *Plant J.* **16**, 651 (1998).
8. G. Z. Huang, R. Allen, E. L. Davis, T. J. Baum, and R. S. Hussey, *Proc. Natl. Acad. Sci. USA* **103**, 14302 (2006).
9. B. Tournier, M. Tabler, and K. Kalantidis, *Plant J.* **47**, 383 (2006).
10. I. Dörr, *Ann. Bot.* **79**, 463 (1997).
11. M. C. Press, in *Parasitic Plants*, Ed. M. C. Press and J. D. Graves (Chapman and Hall, London, 1995), p. 103.
12. K. Pageau, P. Simier, B. Le Bizec, R. J. Robins and A. Fer, *J. Exp. Bot.* **54**, 789 (2003).
13. J. M. Hibberd and W. D. Jeschke, *J. Exp. Bot.* **52**, 2043 (2001).
14. J. D. Graves, in *Parasitic Plants*, Ed. M. C. Press and J. D. Graves (Chapman and Hall, London, 1995), p. 206.
15. A. L. Gurney, D. Grimanelli, F. Kanampiu, D. Hoisington, J. D. Scholes and M. C. Press, *New Phytologist* **160**, 557 (2003).
16. I. Dörr and R. Kollmann, *Bot. Acta* 108:47 (1995).
17. M. R. Boocock and J. R. Coggins, *FEBS Lett.* 154, 127 (1983).
18. T. Konishi, K. Shinohara, K. Yamada, and Y. Sasaki, *Plant Cell Physiol.* **37**, 117 (1996).
19. Z. Mou, Y. He, Y. Dai, X. Liu and J. Li, *Plant Cell.* **12**, 405 (2000).
20. E. Rojo, C. S. Gillmor, V. Kovaleva, C. R. Somerville and N. V. Raikhel, *Develop. Cell* **1**, 303 (2001).
21. L. Prade, S. W. Cowan-Jacob, P. Chemla, S. Potter, E. Ward and R. Fonne-Pfister, *J. Mol. Biol.* **296**, 569 (2000).

Section V

Agronomic Options: The First and Essential Line of Control and Policy Considerations

CHAPTER 15

AN INTEGRATED *STRIGA* MANAGEMENT OPTION OFFERS EFFECTIVE CONTROL OF *STRIGA* IN ETHIOPIA

Tesfaye Tesso[a,*], Zenbaba Gutema[a], Aberra Deressa[a] and Gebisa Ejeta[b]

[a]Ethiopian Institute of Agricultural Research, P.O. Box 2003, Addis Ababa, Ethiopia; [b]Purdue University, Department of Agronomy, West Lafayette, IN 47907-2054 USA

[*]E-mail: tesso1970@yahoo.com

Three *Striga* resistant cultivars, P-9401, P-9403 and PSL85061 selected from among a series of resistant varieties developed were recommended for commercial production in *Striga*-infested regions of Ethiopia. The varieties 'packaged' along with a soil moisture conservation and a soil fertility management scheme were demonstrated on farmers' fields in four *Striga* endemic regions of Ethiopia over three cropping seasons. The objective was to expand the use of integrated *Striga* management (ISM) package through participatory evaluation and demonstration of the technology, and to facilitate the establishment of an informal community-based seed multiplication and distribution system. The package effectively suppressed *Striga* and dramatically increased yields. *Striga* count from the ISM package plots was ten to fifteen times lower while sorghum yield was two to three times higher than plots planted to local varieties. Farmers in all regions overwhelmingly positively evaluated the efficacy of the ISM package. NGOs and local farmers cooperatives responded to the growing demand for the ISM package by engaging in production and distribution of seeds of the resistant varieties.

1. Introduction

The limited selection of alternative crops to sorghum that can be grown in marginal soil fertility conditions and *Striga* infested dry lands[4,5] preclude the use of traditional control practices of fallowing and crop rotation. The use of resistant cultivars is a most robust and effective approach to control parasitic weeds. We assess the on-farm performance of *Striga* resistant varieties under *Striga* infested conditions in Ethiopia and discuss the role of integrating *Striga* resistant varieties with improved agronomic options in reducing *Striga* emergence and increasing yield. We also report the success in disseminating these varieties through informal seed multiplication and distribution systems.

2. The Release of *Striga* Resistant Varieties in Ethiopia

The steps in the breeding of the *Striga* resistant varieties are outlined in Chapter 7. *Striga* resistant varieties were tested in Ethiopia on *Striga* infested experimental plots beginning in 1995. The field trials were coordinated by the national sorghum research program in collaboration with regional research centers. The tests were carried out in several locations across the country in both *S. hermonthica* and *S. asiatica* infested fields. P-9401 and P-9403 were consistently superior to the other 6 varieties tested as well as the standard and the local farmers' variety for four consecutive seasons. Besides their resistance to *Striga* as shown by the low number of *Striga* plants supported, the varieties also had excellent grain quality, drought tolerance and good agronomic adaptability in all test environments. The varieties were thus officially released for commercial production in *Striga* infested regions of Ethiopia under the local names Gobiye (P-9401) and Abshir (P-9403) and registered by the national Seed Industry Agency.[14] A third variety, PSL85061, tested in the next batch of resistant varieties was released under the local name Berhan. The vernacular names given to the varieties come from either the name of the area where the varieties were first tested and attracted public attention (Gobiye and Abshir) or from the superior performance of the variety under heavy *Striga* infestation. Only the first two varieties are discussed in this chapter.

3. The Approaches

3.1. *Packaging of Technology Options*

A pilot project was launched to conduct on-farm demonstration of an integrated *Striga* management package. The resistant varieties were integrated with selected agronomic options, soil moisture conservation using tied-ridges, and fertility amendment using locally recommended rates of nitrogen and phosphate fertilizers. The package was then tested in four *Striga* infested regions of Ethiopia: Amhara, Oromia, the South and Tigray. Seeds of resistant varieties were multiplied at Melkassa Agricultural Research Center and distributed to different regions along with chemical fertilizers. Tie-ridgers used for building dykes for soil-moisture conservation were fabricated at the local sheet metal industry from a prototype developed by Melkassa Agricultural Research Center.

3.2. *On-farm Testing of the Package*

The project was organized around three different sets of activities: participatory-evaluation of the ISM technology ("demonstration"); testing of the *Striga* resistant sorghum cultivars by interested farmers ("popularization"), and training of carefully selected farmers in the multiplication and redistribution of seeds of resistant varieties ("seed production"). The primary focus was to allow farmers to evaluate the benefit of combining host plant resistance with improved agronomic practices (soil fertility management and soil moisture conservation) . Progressive farmers with infested fields in each region were selected and provided with seeds of the resistant varieties, fertilizers, and a tie-ridger for soil moisture conservation (Table 1). They were given onsite training on the test protocols and on management of the demonstration and seed multiplication plots. Agricultural development agents and representatives of various peasant associations were also given formal training on the biology of *Striga*, crop management systems and guidelines for implementation of this project. The package (a *Striga* resistant variety, 50 kg/ha urea and 100kg/ha di-ammonium phosphate, and tied ridges) was laid out on 0.25 ha infested plots of each participating farm. The farmers' own landraces were planted at each test site with local management practice as a check next to the ISM test plot. The ISM plots were planted in rows after the onset of the main rainy season using a seed rate of 10 kg/ha, while the check plots were planted 45 to 60 days before the ISM package using the traditional planting method (broadcasting) and seed rates. *Striga* counts were recorded at flowering and physiological maturity in 1m square quadrants from both ISM and the check plots. Counts from 10 quadrants per plot were averaged to represent *Striga* count per plot. Grain yields were recorded from both the ISM and local check plots.

Farmers who desired to participate in the ISM demonstration activity, but whose fields were inaccessible for routine supervision by development agents were given seeds of a *Striga* resistant variety to test the genetic component of the package with or without additional inputs. The seed production activity was designed to promote organized production and distribution of seeds of resistant varieties and encourage a local seed business. There is little private or government effort in production and marketing of seeds of improved sorghum varieties in Ethiopia. Seed farms were selected based on superior soil type and fertility as well as isolation from other fields with sorghum or its wild and weedy relatives. Farmers interested in engaging in seed production were requested to commit *Striga*-free plots of at least 0.5 ha for seed multiplication. Weeds in seed production fields were removed by hand, and care was taken during harvest, threshing, and processing of seed to avoid contamination by *Striga*. Farmers were advised to adopt the use of improved agronomic practices to ensure quality seed. Basic seed and chemical fertilizers were only provided free to seed producers during the first year. Ministry of Agriculture technicians and researchers from the implementing agencies regularly inspected seed production fields.

4. Farmer-Participatory Evaluation of the ISM Technology

The efficacy of the ISM package was evaluated by farmers, development agents, and research technicians on test plots conducted in farmers' fields. Interest in the ISM technology increased as farmers expressed overwhelming approval on its efficacy. The number of farmers and the area in the project (Table 1) significantly increased from year to year. A great deal of the increase in activities, was made in the distribution of resistant cultivars (popularization). Data shown are only for farms that received input directly from project staff. The total number of farmers and acreage planted to resistant cultivars through secondary and tertiary redistribution of seed is not known, but is estimated to be much higher.

Table 1. Farmer participation in the extension of integrated *Striga* management (ISM).

	Number of farmers			
Region	Demonstration	Seed production	Popularization	Total
2002				
Amhara	36	13	71	120
Oromia	29	27	94	150
South	10	21	29	60
Tigray	70	22	50	142
Sub-Total	145	83	244	472
2003				
Amhara	32	30	321	383
Oromia	30	38	415	483
South	63	11	92	166
Tigray	10	33	265	308
Sub-Total	135	112	1093	1340
2004				
Amhara	76	30	820	936
Oromia	141	129	1250	1520
South	63	38	220	321
Tigray	107	108	1335	1550
Mean	387	305	3625	4327
Grand mean	**667**	**500**	**4962**	**6139**

4.1. *Demonstration*

Over 600 farmers took part as lead demonstrators of the ISM package during the three years activity (Table 1). Fields planted to the ISM package supported remarkably fewer *Striga* and gave significantly higher yields than plots planted to the local landrace under traditional farmer practice (Table 2). Across regions, mean *Striga* count recorded from the ISM package over the three years period was 50 times less than in plots planted with farmer practice. Similarly, mean grain yield from the ISM package plots over the three year period was more than three times greater than the local practice. This result was fairly consistent across regions and seasons in both *S. hermonthica* and *S. asiatica* infested fields (Table 2). ISM plot average yields of as high as 3.4 tons ha^{-1} were recorded compared to a maximum of only 1.6 tons ha^{-1} from plots planted to local landraces under local practices. Some of the demonstrations were conducted in plots that had previously been abandoned due to severe infestation by *Striga*.

Table 2. *Striga* count and grain yield recorded on ISM and control plots in different *Striga* endemic regions of Ethiopia.

	Grain yield (t ha^{-1})		*Striga* count at crop maturity (m^{-2})	
Region	ISM package	Local practice	ISM Package	Local practice
2002				
Amhara	3.40	0.80	6	2052
Oromia	1.12	0.12	32	1110
Mean	2.26	0.46	19	1585
2003				
Amhara	2.67	1.33	5	95
Oromia	2.02	0.29	7	104
South	0.53	0.00	4	128
Mean	1.74	0.38	6	108
2004				
Amhara	2.61	1.55	12	158
Oromia	1.02	0.25	12	122
South	0.13	0.00	0	0
Tigray	2.13	1.37	24	163
Mean	1.51	0.79	12	123
Grand mean	1.84	0.54	12	605

Differences in drought tolerance between cultivars and inherent yield potential of the varieties may have contributed to some of the variation. However, much of the disparity in grain yield can be attributed to differences in the level of *Striga* control between the two practices. In fact, some of the cultivars used in the local practice are improved varieties that under *Striga* free condition could give comparable or better yields than the resistant varieties. However, these varieties could not stand heavy *Striga* pressure and thus gave very low yields and in some plots produced no grain. While some other cultivars used are tolerant landraces that despite *Striga* pressure give reasonable yield. Yield differences were more pronounced in dry years where the problem of *Striga* infestation was compounded with severe moisture stress. Farmers' cultivars in many areas failed totally while the resistant varieties produced grain despite the severe stress (Table 2).

When compared with similar packages tested in sorghum and other cereals, this package appears to be extremely effective both in reducing *Striga* infestation and increasing yield as well as in its ease of application. An integrated approach that involved short fallow period and crop rotation marginally increased yield and reduced *Striga* emergence in pearl millet.[15] An integrated package tested for maize[7,15,16] in west Africa that also included crop rotation, resistant variety and trap crops and intercropping options reduced *Striga* infestation by 35-46% and increased yield by 76-100%. The difference, however, disappeared after two seasons of trap cropping and two seasons of crop rotation.

Striga control with a mycoherbicide (*Fusarium oxysporum*) coated seeds and host plant resistance reduced *Striga* emergence by 95% and increased sorghum yield by 50%[17]. Inoculation with arbuscular mycorrizal fungi also reduced damage by *S. hermonthica* in both *Striga* tolerant and susceptible cultivars.[18,19] While these approaches had remarkable effect both in reducing *Striga* emergence and improving yield, the ISM package under current test has been much more effective and convenient to use. This may be either due to proper compatibility of the component options included in the package or due to the specific strength of individual components, especially the stability of the resistant varieties. Some of the components tested elsewhere, such as the mycoherbicide coating of seeds may be included in this package as a fourth option to further improve the efficacy of the package.

4.2. *Popularization*

Over the first three project years, more than 20 tons of seeds were distributed as popularization to nearly 5,000 farms covering over 1,400 hectares of *Striga* infested land (Table 1). Grains produced by these farmers were shared as seed with other farmers through the informal seed market, but no record could be kept of this activity. Although no *Striga* counts were made on popularization plots, yield estimates provided by farmers indicate that resistant varieties yielded much more than the local varieties when planted in infested fields. In contrast to the demonstration plots where integration of multiple *Striga* control options had a synergistic effect on enhancing grain yield, performance of resistant varieties in the unfertilized popularization plots gave excellent control of *Striga* though the yields were not as high. Hence, even without chemical fertilizers and tied-ridges, resistant varieties provided effective control of *Striga*. We separately compared the efficacy of different control options, and verified that resistant varieties effectively reduced *Striga* emergence with and without other options, indicating that host plant resistance alone can be used in situations where integration of all options is impossible. Integration of soil moisture conservation and fertility management practices with susceptible varieties also contributed to reduced *Striga* emergence and increased yields in both susceptible and resistant varieties (Table 3).

Table 3. The relative effectiveness of components in reducing *Striga* infestation and increasing sorghum yield.

Treatments	Yield (t ha^{-1})	*Striga* count at crop maturity (m^{-2})
LV x F_0 x M_0	0.73e	216c
LV x F_0 x M_1	1.02d	680a
LV x F_1 x M_0	1.14cd	250c
LV x F_1 x M_1	1.46b	527b
SR x F_0 x M_0	0.80e	16d
SR x F_0 x M_1	1.22c	26d
SR x F_1 x M_0	1.15cd	11d
SR x F_1 x M_1	1.68a	15d
Mean	1.16	227
LSD	0.2	110

Means in a column followed by same letter are not significantly different; LV=Local variety (Jigurte), SR= *Striga* resistant variety (P-9401), F_1 and F_0= with and without chemical fertilizer, respectively, M_1 and M_0= with and without soil moisture conservation, respectively

4.3. *Seed Production*

Because of tremendous success in adoption and diffusion of improved crop cultivars and agronomic practices in Ethiopia, demand for seed far exceeds supply. Private seed entrepreneurship does not exist in Ethiopia and government seed production efforts have not kept pace with demand. Consequently, despite the enormous potential of *Striga* resistant crop varieties in minimizing yield loss associated with *Striga* infestation, lack of a mechanism for sufficient supply of quality seeds limits wider adoption. The seed production component of the ISM project was thus included to encourage and promote organized seed production and distribution. Just over half of the 500 farmers participating in organized seed production (Table 1) satisfied the minimum seed production standard. They harvested 119 tons of acceptable quality seed that was redistributed to local farmers through various channels (Table 4). Each participating grower opened a new distribution network in subsequent years.

An informal survey conducted in the project area indicated that many farmers who had not directly participated in the pilot project were growing *Striga* resistant varieties acquired from neighbors and friends. This was particularly evident in the Hararghe and Humera zones of Oromia and Tigray regions, respectively, where farm communities are known to actively seek out new and improved technologies. In 2004, a total of 17 tons of seed was produced in Oromia, all of which was purchased by local NGOs at a premium price and redistributed to farmers in remote villages. Similarly, in Tigray, 37 tons of seed of *Striga* resistant sorghum were produced in Humera and Shiraro zones and redistributed in the region.

In addition to maintaining breeder seed of these varieties, Melkassa Agricultural Research Center was also engaged in multiplication of certified seeds of the resistant varieties for wide distribution. In the last three years, the center produced over 100 tons of seeds of these varieties. Similarly, Sirinka Research Center produced and distributed over 40 tons of seed. Most of this seed was purchased by local and international NGOs as well as the Ethiopian Seed Enterprise and distributed to farmers in *Striga* infested regions. It is estimated that 25,000-30,000 new farmers have received seed of resistant varieties produced through formal channels and sold to various organizations. In addition, several thousand farmers in each of the four regions have accessed seed of resistant sorghum varieties through informal farmer to farmer distribution networks. Coupled with seed distribution efforts made by NGOs, the ISM project, and the informal seed exchange network, estimate are that over 100,000 farmers are presently growing *Striga* resistant sorghum varieties in Ethiopia. Unfortunately, however, this number represents only a small fraction of sorghum farmers in *Striga* endemic areas of the country. With infestation rapidly expanding,[20] a more coordinated production and distribution of *Striga* resistant crop varieties will be needed.

Table 4. Farmer participation in production and distribution of seeds of *Striga* resistant varieties in Ethiopia.

Region	Number of farmers	Area (ha)	Seed produced (tons)
2002			
Amhara	13	2.1	4.1
Oromia	27	3.4	8.4
South	21	10.5	7.5
Tigray	22	14.5	6.5
Total	**83**	**75.5**	**26.5**
2003			
Oromia	26	7.32	8.2
South	11	7.5	2.4
Tigray	31	20.9	24.1
Total	**68**	**35.7**	**34.7**
2004			
Amhara	19	3.5	3.7
Oromia	60	19.1	17.4
Tigray	43	19.7	37.4
Total	**122**	**42.3**	**58.5**

5. Adoption and Diffusion of the ISM Technology in Ethiopia

Field days were routinely organized each season to inform and educate farmers and development agents on efficacy of the ISM technology and its application. Local government representatives and extension agents were often present at field days. The farmers' reactions to the technology have been consistently positive (Table 1). Besides the dramatic effect of the package in reducing damage by the parasite (Fig. 1), farmers were impressed by the resistant varieties for their drought tolerance, early maturity, excellent grain quality and processing attributes. The stalks of the resistant varieties are a preferred source of animal feed. According to the farmers, the texture of injera made from P-9401 was better than that of local landraces and stays fresh for a longer period of time. Formal studies conducted on utilization aspect of the varieties indicated that P-9401 produced best quality injera close to 76T1#123, a commercial sorghum variety known for its excellent injera quality. Adding with up to 20-30% wheat flour yielded to sorghum flour of the varieties gave normal quality cookies and breads.

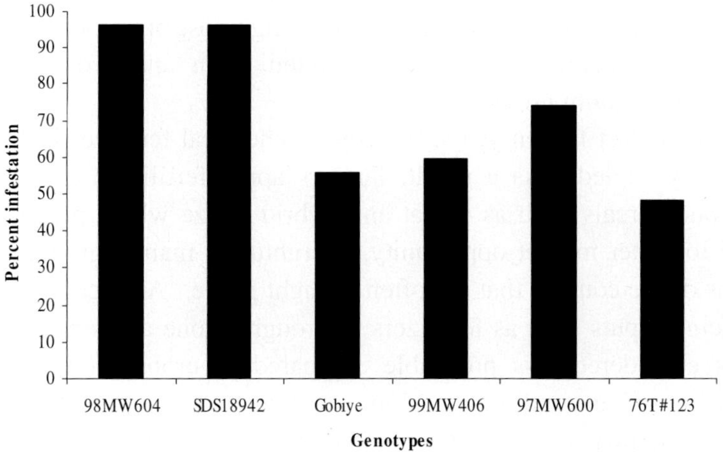

Figure 1. Comparison of reaction of Striga *resistant variety, P-9401 (Gobiye), to* C. partheles *as tested in hot spot area (central Rift Valley of Ethiopia) for cereal stem borer.*

The main impact of this activity is the creation of high level of demand for the resistant varieties. Farmers in all *Striga* infested regions including areas where the project was not implemented have been made very aware that these varieties offer effective control. Demand for seeds of resistant varieties dramatically increased. In Hararghe region, prices of seeds of the resistant varieties were always 15 to 20 percent higher than local sorghum seeds. In Tigray, the varieties are exchanged 1:1 for tef grain, where tef normally fetches a 60-100% higher price than sorghum grain.

6. Factors Affecting Further Diffusion of the ISM Technology

Although over 100,000 farmers at present are estimated to be growing the *Striga* resistant sorghum varieties in Ethiopia, several factors hamper greater diffusion of the technology. Demand for seed of *Striga* resistant sorghum cultivars is high and could not be met from current supply. Participatory evaluation of the technology has convinced farmers of the robustness of the genetic technology and the synergy expressed in the ISM package. However, there may be other factors negatively affecting diffusion of the ISM technology and use of *Striga* resistant sorghum cultivars. Some of these are lack of effective seed production and distribution mechanism, the unreasonably high cost of associated inputs, primarily fertilizers, and the very limited grain and product market outlets for sorghum growers.

Over the last fifteen years, the cost of chemical fertilizer in Ethiopia has nearly tripled. As a result, farmers apply fertilizers only to cash crops and cereals such as wheat and hybrid maize where production is linked to better market opportunity. Sorghum is mainly grown in drier regions of the country that are often drought prone. Application of yield enhancing inputs such as fertilizers in drought-prone areas carries a risk and is considered less profitable compared to crops produced under optimal growing conditions. Moreover, much of the sorghum crop is produced in areas far away from market centers and thus grain prices are always low and seed and fertilizer prices tend to be high because of added transportation costs.

Lack of a functional seed production and marketing mechanism is another constraint. Of the 35 million tons of seeds of all crops required each year, only 2.5 million tons, representing less than 10% of the total annual seed demand, are made available through organized private, governmental, and parastatal organizations. The balance is planted with landrace cultivars with excellent adaptation but very low yields and limited response to good growing conditions. Seed production is particularly a bottleneck for sorghum as there is no government or private agency engaged in sorghum seed production and distribution. The Ethiopian Seed Enterprise is focused mainly on production of more profitable wheat and maize seeds.

7. Conclusion

The ISM technology is a reliable *Striga* control package available to subsistent farmers in Ethiopia. To reach more farmers, the whole ISM package will have to be continually available to farmers. Even more significantly, rising fertilizer prices need to be checked. Systems should be devised for private and government enterprises to produce and distribute quality seeds or to empower farmers to fill the gap. With approximately 600,000 hectares of sorghum fields highly infested by *Striga* resulting in annual loss of over 640,000 tons of sorghum grain, control of parasitic weeds needs to be given serious consideration on the national agenda. The results of this study show that a proven technology that minimizes loss from *Striga* is now available, if only input components can be delivered at reasonable prices. Sorghum varieties are available that meet the multiple needs of farmers of *Striga* control, tolerance to drought, and good grain quality to make traditional food products. Soil moisture and fertility management techniques have been developed that can be further modified to fit prevailing situations of farm communities. Policy interventions that encourage delivery of essential inputs, create markets, and encourage profitable farm enterprises for sorghum growers in Ethiopia are urgently needed.

Acknowledgements

The authors wish to thank INTSORMIL-CRSP for financial support of the release of the varieties and demonstration of the ISM package and the Institute of Agricultural Research for facilitating its implementation.

References

1. National Research Council (NRC). 1996. *Lost Crops of Africa: Volume 1: Grains.* (National Academy Press, Washington, D.C, 1996).

2. Central Statistical Authority. Agricultural Sample Survey: Result on area, production and yield of major crops by sector and seasons (CSA, Ethiopia, 2003).
3. S. T. Lagoke, V. Parkinson and R. M. Augunbiade, in *Combating Striga in Africa,* Ed Kim (IITA, ICRISAT and IDRC, Ibadan Nigeria, 1991).
4. D. K. Berner, J. G. Kling and B. B. Singh, *Plant Dis.* **79**:652-660(9) (1995).
5. A. Oswald, J. Ransom and J. Kroschel, International Plant Protection Congress, Plant protection toward the third millennium – where chemistry meets ecology, (IPPC Jerusalem Israel, 1999).
6. G. Ejeta, L.G. Butler and A.G. Babiker, *Purdue University Agricultural Experiment Station* (Research Bulletin RB-991, 1991).
7. I. Kureh, A. Y. Kamara and B. D. Tarfa, *J. Agri. Rural Dev. Tropics Subtrop.* **107**, **41** (2006).
8. O. Samake, T. J. Stomph, M. J. Kropff and E. M. A. Samling, *Plant Soi.l* **286**, 245 (2006).
9. P. S. Marley, D. A. Aba, J. A. Y. Sebayan, R. Musa and A. Sanni, *Weed Res.* **44**, 157 (2004).
10. D. Musambasi, O. A. Chivinge and I. K. Mariga, *Afri. Crop Sci. J.* **10**, 163 (2002).
11. M. A. Osman, P. S. Raju, and J. M. Peacock, *Plant Soil* **131**, 265 (1991).
12. F. A. Showemimo, C. A. Kimbeng, and S. O. Albi, *Crop Protect.* **21**, 867 (2002).
13. G. Ejeta, in In the Wake of the Double Helix. From the Green Revolution to the Gene Revolution, Eds. Tuberosa *et al.* (Bologna, Italy, 2005).
14. G. Ejeta, A. Mohamed, P. Rich, A. Melake-Berhan, T. L. Housley and D.E. Hess, in *Breeding for Striga resistance in cereals*, Eds. B. I. G. Haussmann, D. E. Hess, M. L. Koyama, L. Grivet, H. F. W. Ratunde and H. H. Geiger (Margraf Verlag, Weikersheim, Germany, 2000).
15. National Seed Industry Agency, *Crop variety Registration. No.4*, NSIA, Addis Ababa, Ethiopia, 2001).
16. J. Ellis-Jones, S. Schulz, B. Douthwaite, M. A. Hussaini, B. D. Oyewole, A. S. Olanrewaju and R. White, *Exp. Agric.* **40**, 353 (2004).
17. A. C. Franke, J. Ellis-Jones, G. Tarawali, S. Schulz, M. A. Hussaini, I. Kureh, R. White, D. Chikoye, B. Douthwaite, B. D. Oyewole and A. S. Olanrewaju, *Crop Protect.* **25** 868-878(11) (2006).
18. V. W. Lendzemo and T. W. Kuyper, *Agri. Ecosys. Environ.* **87**, 29 (2001).
19. V. W. Lendzemo, T.W. Kuyper, M. J. Kropff, A. van Ast, *Field Crops Res.* **91**, 51 (2005).
20. A. O. Esilaba, T. Mulatu, F. Reda, J. K Ransom, G. Woldewahid, A. Tesfaye, I. Fitwy and G. Abate, in *The 16th East African Biennial Weed Science Conference* (1997).

CHAPTER 16

INTEGRATING CROP MANAGEMENT PRACTICES FOR *STRIGA* CONTROL

Joel K. Ransom[a,*], Abdel Gabar Babiker[b] and George D. Odhiambo[c]

[a]North Dakota State University, Fargo, ND, USA,
[b]Agriculture Research Corporation (ARC), Wad Medani, Sudan
[c]Maseno University, Kisumu, Kenya
[*]E-mail: joel.ransom@ndsu.nodak.edu

Striga spp. are pernicious pests of food crops in Africa for which an integrated control program is needed. An effective integrated program must have components that are adaptable to the biophysical and socio-economic environment where implemented. It must be proactive and farmers should understand the response time of and the effect of environment on the proposed interventions. It should also combine tactics that protect or enhance yield, with those that reduce seed production and/or reduce seed banks. Research in sorghum and maize systems in eastern Africa demonstrates that integrated approaches can be effective in controlling *Striga* and improving farmer profitability. Significant resources are needed to implement an integrated *Striga* control program that requires improved understanding by farmers and extension educators of the *Striga* problem and the recommended interventions. Soil fertility enhancement should be a component of all integrated *Striga* control programs.

1. Introduction

Numerous *Striga* control practices have been identified and the most prominent of these have recently been reviewed.[1,2] Few *Striga*-control practices have yet been widely adopted and *Striga*-related yield losses continue to impact food security in Africa. Adoption of *Striga*-control practices is hindered by limited knowledge of the problem, *Striga* biology and of potential control practices, lack of resources to invest in control practices, and the lack of immediate returns from the proposed control practices.[1] Most currently available *Striga* control practices are only partially effective and may require several years of continued application before their effects are noticeable. For these reasons, a *Striga*/crop-management approach that integrates two or more control practices is recommended in almost every recently published paper dealing with *Striga* control. *Striga asiatica* control and eradication was achieved in the USA through a program that integrated a wide-range of actions that included surveys to identify infested areas, policies that contained the spread and the integration of an array of practices that reduced seed numbers in the soil and eliminated the production of new seeds.[3] Eradication of a pest is only practical for relatively small and confined infestations, but the integrated approaches used in the witchweed eradication program have application to the containment and control of *Striga* in general.

Integrated pest management (IPM) is a concept of pest control developed decades ago to reduce dependence on pesticides and delay the evolution of pesticide resistance in insects and diseases. The concepts of IPM are continually evolving and have been used to reduce the reliance of farmers on a single intervention in a broad range of pest and crops situations, thereby reducing risk, and in many cases improving control and profitability. Though the concepts that comprise integrated pest management were initially developed for use with insects and diseases that have the capacity to reproduce and build-up to damaging levels within a growing season, many of these concepts have value in the development of an integrated *Striga* control approach.

This chapter discusses the concepts and application of integrating crop management practices for the control of *Striga*. Our focus will primarily be on integrating agronomic practices that have been developed and tested at the field level, as the breeding approaches are well covered in other chapters.

2. Components of an Integrated *Striga* Control Program

2.1. *Adaptable and Appropriate*

An effective integrated approach should include interventions that are selected in response to the level of infestation and the local resources that are available to deal with the problem. Prescribing a single cultivar of a rotation crop, for example, may offer the best option for *Striga* control in a given environment, but if seed is unavailable or if there is no market for the crop at harvest, then the chance for real impact at the farm level will be minimal. Moreover, interventions for differing levels of infestations are needed. Hand weeding, for example, may be the best option to prevent the build up of *Striga* in fields with very low *Striga* levels or as a supplementary treatment with resistant varieties, but is impractical as well as useless for heavy infestations. The challenge of availing a wide range of *Striga* control options that can then be adapted by farmers to match their needs, circumstances and interests is indeed daunting.

Extension services in Africa generally lack the expertise and resources to effectively extend the array of information needed and the useful information to extend is often far too limited. Only 48% of Kenyan farmers interviewed had received information about *Striga* control from the extension service.[1] Nearly 40% of the farmers interviewed had not received information on *Striga* control from any source!

Many "long-term" interventions are not profitable in the short-term and need to be combined with practices that do provide a short-term return for there to be adequate incentive for them to be adopted.[4] Social-economic factors such as the amount of labor available, the importance of the cereal crop, etc. must also be considered.

2.2. *Proactive Rather Than Reactive*

An effective integrated *Striga* control program develops a plan for interventions based on an understanding of the problem and the solutions that are available. Plans should include the big picture view, which incorporates the effect of time, environment and socio-economics into the equation.

Farmers and those who advise/teach farmers are better able to choose interventions and be motivated to implement them, when they understand how *Striga* develops and reproduces. Only 11% of the farmers surveyed in Kenya knew that *Striga* is spread by seed.[1] There is obviously much to be done with regards to helping farmers understand basic *Striga* biology.

Knowing the level of infestation is also critical to choosing appropriate interventions. Many interventions are simply a waste of resources when *Striga* seed levels are extremely high. Incipient infestations should be easiest to tackle as practices such as hand weeding are doable and the returns to other management practices such as the use of fertilizer are more likely. The challenge with developing a threshold for *Striga* is that the number of *Striga* seeds in the soil is not always a good predictor of the level of *Striga* emergence. Correlations between seed numbers in the soil and emerged *Striga* density were significant in only 3 of 10 seasons in a long-term trial in Kenya.[5] Poor correlations can result from seed densities that exceed the number of *Striga* plants that can be supported by the host and/or because of environmental effects on *Striga* attachment (see data in Table 1 as an example).

Table 1. Seed numbers in the soil are poor predictors of *Striga* emergence in a long-term trial in Kenya where hand-weeding was compared to no hand-weeding of *Striga*.[a]

	Kibos				Homa Bay			
	No weeding		Weeded		No weeding		Weeded	
Season	Seeds	Plants	Seeds	Plants	Seeds	Plants	Seeds	Plants
1991	584	14.5	451	17.6	134	18.3	130	16.1
1992	514	20.7	204	37.9	239	20.0	80	19.2
1993	66	6.6	72	5.3	375	14.7	37	9.1
1994	144	8.3	73	5.5	204	22.3	65	14.1
1995	87	1.8	13	1.7	150	8.3	55	4.4
1996	125	7.0	42	6.3	174	0.9	98	0.9

[a]Plants m^{-2} or seeds kg^{-1} of soil. Data are for the long rainy season. Treatments were also applied during the short rainy season, though data are not shown. Correlations (r^2) between seed numbers and emerged *Striga* were $0.29^{n.s.}$ and $0.09^{n.s.}$ for Kibos and Homa Bay, respectively.

2.2.1. *Putting Response Time in Perspective*

The amount of time can vary significantly before crop management practices have a visible impact on *Striga* control. Many currently recommended control practices have failed to be widely adopted because they require several seasons of implementation before they have a noticeable impact. Conversely, incipient infestations can explode to damaging levels after a single season, if left unchecked. Controlling incipient infestations is much easier than controlling heavy infestations. Unfortunately, farmers with new infestations may not intervene, as they do not understand the potential dangers of a few plants left to seed.

When attempting to reduce seed banks (with the exception of ethylene on *Striga* in the USA) most interventions require several seasons to be effective. This is because the seeds are generally plentiful throughout the soil profile, and may be dormant and unresponsive to germination stimuli.[6] Though a single season rotation out of a cereal can be beneficial in reducing *Striga* numbers and improving cereal yield in the following season in some environments,[7] in other environments *Striga* remained at damaging levels even after four seasons of continuous cultivation of a trap crop.[8] In Kenya, hand weeding failed to eliminate damaging levels of *Striga* seeds even after 10 seasons (Table 1).

2.2.2. *Understanding the Impact of Environment*

Environment can dramatically affect the impact of *Striga* on a susceptible crop. Typically, the most important environmental variable impacting the *Striga*-crop interaction is soil moisture, and indirectly the amount and distribution of rainfall. Soil moisture influences how crop roots develops, the rate of soil biological activity, the conditioning of *Striga* seeds and the interaction of these factors. Good soil moisture in the surface layers of the soil, where most of the *Striga* seeds are located, favors more extensive root development in these regions and enables greater parasitism. These conditions also favor microbiological activity that can hasten the breakdown of organic matter, including *Striga* seeds. Low rates of *Striga* seed degradation in the soil may be one reason that *Striga* is most problematic in the drier cropping zones of Africa. *Striga* suppressive soils, soils where *Striga* seed banks decline even in the absence of any germination, have been reported. The rapid decline in seed numbers in the Kibos location where more rainfall is received than in Homa Bay, is thought to be due to the development of *Striga* suppressive soils (Table 1). Ethylene-producing microbes that induce suicidal germination may also contribute to *Striga*-suppressiveness in soils.[9]

The variable response in multi-location testing of imazapyr treated seed (Chapter 11), illustrates the potential interaction that control practices have with environment.[10] The fact that the environment can interact so significantly with *Striga* control practices means that integrated approaches need to be tested and verified as to their effectiveness in each major and microenvironment. Furthermore, as environment can play a dominant role in the level of *Striga* parasitism, the concept of a seed-bank threshold becomes a moving target. The need for testing and technology development at multiple environments is indeed a challenge given the paucity of human resources and operating funds for most research and extension systems in Africa.

2.3. *Combining Complementary Tactics*

An effective integrated *Striga* control program combines control practices appropriate to the level of infestation and to the socio-economic and environmental circumstances of the farmer that complement one another. *Striga* control tactics can be broadly categorized into those that protect and/or enhance yield, those that reduce the production of new seed, and those that decrease the level of infestation in the soil. An ideal integrated program combines components of all three tactics.

2.3.1. *Practices that Protect and/or Enhance Yield*

Most farmers plagued with *Striga* are subsistence farmers, so integrating control practices that protect and/or enhance the yield potential of the crop is vital. Yield protecting/enhancing interventions are those that impact the productivity of the crop the year that they are applied. Currently there are a number of interventions available that offer some level of protection against *Striga*-related yield losses. Of these, resistant genotypes probably have the greatest chance of having wide-scale adoption and long-term impact. Still, providing resistance in adapted genotypes with traits preferred by farmers in the many environments where *Striga* is problematic will be challenging. Host plant resistance, when available, is often the cornerstone upon which an integrated pest management program is built. Progress in the breeding of adapted *Striga*-resistant varieties has been relatively slow. More progress has been achieved with sorghum than with maize. Furthermore, the current sources of resistance are quantitative in nature and yield losses can still be significant at high *Striga* levels. Genetic resistance needs to be verified in each environment, as it is probable that resistance may be *Striga* biotype specific. Cultivars with improved resistance relative to those commonly grown have failed to be adopted due to low yield potential[11] or they lack other traits valued by farmers such as grain color or plant height.[12] Marker assisted selection offers hope for improving the level of resistance and hastening the process of incorporating that resistance into farmer-preferred cultivars.[2]

Until recently, herbicides failed to provide an acceptable means of protecting crops from *Striga*. Dicamba can provide some protection when applied after *Striga* attachment and before its emergence, but timing of the application is critical both in terms of crop safety and *Striga* control.[13] The most promising currently available chemical intervention for maize is imazapyr applied to seeds of genotypes with resistance to ALS-inhibiting herbicides. This system of chemical control has been widely tested and can increase yields by three to four fold (Chapter 11). This system is very effective in reducing *Striga*-related yield losses in maize and could be used as the base upon which an integrated management program could be established in those environments for which adapted herbicide tolerant genotypes are available. Furthermore, introgressing a single herbicide resistance gene into adapted material in the short-term is simpler than the incorporation of polygenic host plant resistance. The fact that the herbicide is applied to the seed prior to planting makes this technology especially attractive to farmers that have little experience in applying herbicides to their fields. As with most chemical interventions, it should be integrated with other tactics to reduce the risk of the evolution of resistance by *Striga* to the herbicide,[14] a risk now considered lower than originally predicted.[15]

The application of selective strains of *Fusarium oxysporum* that are pathogenic to *Striga* significantly reduced the emergence of

S. *hermonthica* and increase crop yields (Chapter 21). A current challenge to the use of this technology is its delivery to the farming community, along with the methodologies for the production and application of these biocontrol agents. Restrictions on the movement of biocontrol agents from one country to another may also limit the availability of this technology.

Applying nitrogen containing fertilizers and organic materials can in some environments reduce the amount of *Striga* parasitism. Even though the level of control of *Striga* with nitrogen containing inputs can be minimal and/or erratic,[16] improving the fertility of the soil is often as critical as controlling *Striga* to maintaining and enhancing yield. Because the *Striga* problem is tightly linked to the decline in soil fertility in Africa, soil fertility improvement should be addressed concurrently with all *Striga* control extension programs.[4]

2.3.2. *Practices that Reduce Seed Production*

Striga has the capacity to produce thousands of dust-like seeds. At a density of 20 plants m^{-2} the amount of seeds produced could be in the millions. As few as two or three flowering *Striga* plants m^{-2} may be sufficient to maintain seed numbers at a damaging level.[17] Current levels of genetic resistance generally do not reduce *Striga* emergence below this threshold. Aside from hand weeding, there are limited options for controlling seed production once the plant has emerged. Hand-weeding is often impractical due to the numbers of plants involved. Furthermore, in many environments there is little or no incentive to hand weed as it may take several years, if at all, before seed banks are depleted to the point that *Striga* emergence is reduced (e.g. see Table 1). Nevertheless, in an integrated program where another component limits the number of plants that reach the point of seed production (i.e. seed dressing with imazapyr or highly resistant varieties), hand weeding may be doable and may produce important payoffs in the midterm. It may also be the key to prolonging the effectiveness of a whole range of control practices for which the evolution of resistance can occur.

2.3.3. *Practices that Reduce* Striga *Seed Banks*

Depleting *Striga* seed banks once that have built up to damaging levels can be a formidable task. Even after 10 years of cropping at Homa Bay, Kenya in the absence of new seed, seed numbers remained high enough to significantly reduce maize yield (Table 1). Ethylene injected into the soil after *Striga asiatica* seeds were conditioned in the spring, was used for reducing seed numbers in the soil as part of the witchweed eradication program in the USA. A single application was usually sufficient to eliminate nearly all seeds. Ethylene was not as effective in reducing S. *hermonthica* seed numbers in western Kenya[17], however, possibly due to seed dormancy.[6] The logistics of transporting and applying a gaseous chemical in Africa also presents a challenge and the expense can hardly be justified.

Many crops can be grown in rotation with susceptible cereals to reduce seed banks and to improve yields.[19-21] Crop rotation has been proposed as the central focus of an integrated program.[22] Even when they do not impact *Striga* seed numbers, crop rotations make good biological sense and can improve system productivity. However, in most cereal-based subsistent cropping systems, farmers have been reluctant to adopt crop rotation. Factors such as reduced cereal production, land pressure due to rising populations, limited markets, lack of experience in managing the rotation crop and lack of seed are reasons that negatively impact on the adoption of crop rotation as a *Striga* control option. Developing robust markets for non-cereal rotation crops and training farmers in their production could greatly facilitate the increased use of rotation as a *Striga* control tactic. Rotations are particularly attractive as a component of an integrated program when *Striga* seed pressure is high and when a nitrogen fixing crop is used in an area with depleted soil fertility.

Soils that have a high rate of *Striga* seed mortality are not uncommon in Africa.[4,23,24] *Striga* seed numbers declined dramatically after 3 years in a *Striga* suppressive soil at Kibos, even in the treatment where *Striga* was allowed to produce seed (Table 1). The causes of *Striga* suppressiveness in soils are not well understood, but improved soil fertility and organic matter content can be facilitating factors.[4,24]

Combining fresh organic matter and adequate levels of N to encourage the decomposition of the organic matter induced *Striga* suppression at one site in Kenya (Table 1). Inducing *Striga* suppressiveness in soils through the use of organic and inorganic inputs that improve the rate of biological activity in the soil could be an important tool for reducing seed banks in certain environments. Practices that improve soil organic matter and the nitrogen status of soils should therefore be a component of all integrated programs. Any practice that improves yield has the potential to improve organic matter even if the stover is removed as there is greater root biomass associated with a more productive crop.

Recent research in western Kenya shows the effect of a range of practices on *Striga* seed banks (Table 2). Selected intercropping treatments as well as imazapyr applied to a herbicide resistant maize variety reduced seed banks relative to the cultivation of a susceptible hybrid. Intercropping with legumes can reduce *Striga* emergence and in some cases reduces *Striga* seed numbers but does not always ensure greater cereal yield.[26] *Desmodium* did not establish well in the experiment summarized below, but *Desmodium* intercropping holds promise for controlling *Striga* and in improving cereal yield in those environments where it is adapted (Chapter 18).

Table 2. Management interventions can reduce *Striga* seed banks. Data from eight farms in Bondo district, western Kenya before and after eight different treatments in 2004.[25]

Striga Management Options	Initial	Final
	--(Seed # kg^{-1}	soil)--
H513 (a *Striga* susceptible hybrid)	309	544
WS 909 (a *Striga* tolerant hybrid)	189	393
Maize/*Desmodium* intercrop	326	383
KSTP 94 (a tolerant OPV)	304	334
Maize/bean/*Desmodium* intercrop with 100 cm maize spacing	287	262
Maize/soybean/groundnut intercrop	268	195
Imazapyr-resistant OPV with applied Imazapyr	289	194
$LSD_{0.05}$	n.s.	185

3. Integrating Crop Management Control Practices — Examples from Eastern Africa

3.1. *Sorghum*

Integrating crop management practices for *Striga* control based on resistant varieties, increased soil fertility and herbicides with the objectives of increasing yield, curtailing replenishment of seed reserves and depleting seed reserves in soils was adopted in sorghum. Experiments at the Gezira Research Station, Sudan showed clearly that emergence of the parasite was more intense and earlier on Gadam Elhamam, a *Striga* tolerant variety, than on SRN39, a *Striga* resistant variety (Tables 3 and 4). Urea at 190 kg ha^{-1} had an inconsistent effect on the tolerant variety, however, but was consistently suppressive to emergence of the parasite on the resistant variety. Dicamba, alone and when applied subsequent to urea suppressed *Striga* emergence on both varieties. Chlorsulfuron, an ALS inhibitor, alone and in a tank mixture with dicamba irrespective of the preceding urea treatment, effectively suppressed emergence of the parasite on both varieties. Unrestricted *Striga* parasitism reduced grain yield of both varieties. However, the grain yield obtained from the resistant variety was about twofold that attained by the tolerant cultivar. Urea, alone, increased grain yield significantly in one out of two seasons. Dicamba, when applied subsequent to urea increase yield of the *Striga*-tolerant cultivar,

Table 3. Urea and herbicides suppress *S. hermonthica* emergence on sorghum.

	Sorghum Variety			
	1991		1992	
Treatments	G/H	SRN39	G/H	SRN39
	Striga plants/m^2 at 60 days after sowing			
Untreated control	50	34	38	13
Urea	52	6	26	5
Dicamba	21	6	3	3
Dicamba + urea	35	3	3	1
Chlorsulfuron	5	0	3	3
Chlorsulfuron + urea	2	0	4	1
Chlorsulfuron + dicamba	2	0	5	0
Chlorsulfuron + urea + dicamba	2	0	3	0

Urea was applied at 190 kg ha^{-1}, dicamba 300 g ha^{-1}, chlorsulfuron 2.4 g ha^{-1}, G/H = Gadam Elhamam.

significantly. The increment in yield of the *Striga* resistant cultivar, was not significant. Chlorsulfuron, alone and in a tank mix with dicamba, irrespective of the preceding urea treatment, increased yield of the tolerant variety significantly. However, increments in yield of the *Striga* resistant variety were often not significant. The effectiveness of chlorsulfuron at a low rate (2.4 g a.i. ha^{-1}) in suppressing *Striga* infestation and increasing sorghum growth and yield was confirmed in several arrays of environments with different crop varieties and entries including land races.

Table 4. Effects of urea and herbicides on sorghum grain yield (t ha^{-1}) under *Striga* infestation.

	Sorghum variety			
	1991		1992	
Treatments	G/H	SRN39	G/H	SRN39
Untreated control	0.21	0.40	0.87	1.88
Urea	1.15	1.38	1.42	2.50
Dicamba	0.92	0.86	1.52	1.81
Dicamba + Urea	2.02	1.31	3.47	2.23
Chlorsulfuron	2.05	1.73	3.48	1.70
Chlorsulfuron + Urea	3.96	2.11	4.59	2.80
Chlorsulfuron + Dicamba	1.99	1.45	3.08	3.07
Chlorsulfuron + Urea + Dicamba	3.89	1.94	3.39	3.17
S.E.±	0.325		0.383	

Urea was applied at 190 kg ha^{-1}, dicamba 300 g ha^{-1}, chlorsulfuron 2.4 g ha^{-1}, G/H = Gadam Elhamam.

The adoption of chlorsulfuron and its tank mixture with 2,4-D for control of *Striga* in the rainfed area is progressively increasing in Sudan. The treated area increased from 8,000 hectares last season to over 40,000 hectares the current season because of its low cost (herbicide plus application is $6 for chlorsulfuron and $8.5 for its tank mix with 2,4-D). The tank mix with 2,4-D controls other broad-leaved weeds in addition to *Striga*. Chlorsulfuron does not influence *Striga* seed germination when applied late in season. Induction of *Striga* seed germination by sorghum root exudates coupled with reduced emergence of the parasite enhances depletion of *Striga* seed bank in soils. The effects of the treatment on the seed bank may be further accentuated by hand weeding of *Striga* plants escaping the treatments. The timing of application of the herbicides is, however, critical with respect to toxicity and effectiveness. The herbicides have to be applied as a soil directed spray 3-4 weeks after sorghum emergence. Emerging *Striga* plants can be killed by 2,4-D or dicamba applied in a tank mix with chlorsulfuron.

3.2. *Maize*

Data from a long-term trial conducted in western Kenya illustrates the application of many of the principles of integrated *Striga* management previously discussed. This trial was established in 1992 in two locations, Kibos and Homa Bay, and was carried out for 10 growing seasons (there are two seasons each year in western Kenya). A factorial combination of stover management (incorporated or removed), fertilizer application (80 kg ha^{-2} N at planting, or no applied N), and *Striga* removal before seed set (all *Striga* removed or *Striga* left to produce seed) was applied to the same plot each season. The data (Table 5) are averages of 10 growing seasons. Additional information from this experiment is available from Odhiambo.[5]

Table 5. Stover management, fertilizer, and hand weeding of *Striga* influence maize yield and *Striga* emergence. Data are an average of 10 seasons (1991-1997) of continuous treatment.

	Yield (t ha^{-1})		Emerged *Striga* (# m^{-2})	
Management practices	Kibos	Homa Bay	Kibos	Homa Bay
- Stover, - fertilizer, - *Striga* removal	1.03	0.83	6.3	13.0
- Stover, - fertilizer, + *Striga* removal	1.05	1.68	8.5	14.6
- Stover, + fertilizer, - *Striga* removal	1.01	1.57	8.3	12.1
- Stover, + fertilizer, + *Striga* removal	1.00	2.50	6.0	10.2
+ Stover, - fertilizer, - *Striga* removal	0.98	1.70	10.5	8.0
+ Stover, - fertilizer, + *Striga* removal	1.30	1.70	8.5	6.2
+ Stover, + fertilizer, - *Striga* removal	1.55	2.40	6.4	7.0
+ Stover, + fertilizer, + *Striga* removal	1.80	2.35	7.5	5.1
LSD 0.05	0.41	0.31	1.2	1.2

Yields were relatively low in both locations, due in part to *Striga* pressure, but also due to many seasons of drought. The two environments differed significantly. *Striga* numbers at Kibos were relatively similar regardless of the treatment, in part due to the fact that the soil at Kibos became suppressive to *Striga* in the third year of the experiment (seed number data are summarized in Table 1). *Striga* emergence at Homa Bay was reduced in treatments where stover was retained and incorporated. Hand weeding did not consistently reduce *Striga* emergence at Kibos, but did in Homa Bay when combined with stover and or fertilizer incorporation. Yields were highest at both locations when stover retention and fertilizer application were combined and when fertilizer and *Striga* removal were combined at Homa Bay. Hand weeding resulted in yield increases at Homa Bay when combined with all other factors except the incorporation of stover without the addition of fertilizer.

These data illustrate the potential for treatment by environment interactions in an integrated control program and the need for adaptive/on-farm type trials to identify interventions that may have an impact in a given environment. They also show how complementary interventions can be additive in controlling *Striga* and in increasing maize yields. Moreover, the significant interaction between environments and treatments and the lack of obvious correlation between *Striga* numbers and maize yield underscore the complexities of segregating the effect of *Striga* on yield from the other effects, such as drought and inadequate fertility. The interaction between *Striga* and other constraints on yield strengthens the argument that educational programs directed towards integrated *Striga* control need to address the management of other factors that concomitantly affect yield, especially soil fertility.

4. Conclusions

An integrated *Striga* control program is the key to success in controlling *Striga*. Components of an integrated approach need to be adaptable to the environment and circumstance of the farmers. Furthermore, farmers or those that advise and educate farmers need a high level of knowledge about *Striga* and the control options that are available so that control strategies can be proactive rather than reactive. This means that additional resources are needed in Africa for training and for on-farm demonstrations and research. An effective integrated program should combine tactics that are complimentary and should include a component that protects or enhances yield. Host plant resistance, intercropping with *Desmodium* spp., and imazapyr applied to herbicide resistant maize are examples of currently available technologies that protect yield potential. Rotation is a practice that should be encouraged, even in the absence of high levels of *Striga*. Resources are needed to identify productive and profitable rotation crops, develop markets for them, and educate farmers in how to grow them. Interventions that stop the reproduction of *Striga*, such as hand weeding, become important when combined with practices that drastically reduce the number of emerged *Striga,* as they can delay the evolution of resistance to that practice. Finally, the issue of declining soil fertility in Africa must be addressed. Soil fertility enhancement should always be an important component of an integrated *Striga* control program.

References

1. A. Oswald, *Crop Prot.*, **24**, 333 (2005).
2. D. M. Joel, G. Ejeta, P. J. Rich, J. K. Ransom, J. Sauerborn, and D. Rubiales, *Hort. Rev.* **33**, 267 (2007).
3. P. F. Sand, and J. D. Manley, in *Witchweed Research and Control in the United States*, Eds. P. F. Sand, R. E. Eplee, and R. G. Westbrooks, (WSSA 1990) p. 141.
4. J. K. Ransom, *Crop Prot.*, **19**, 759 (2000).
5. G. D. Odhiambo, *University of Reading*, 198 (1998).
6. G. Gbehounou, A. H. Pieterse, and J. A. C. Verkleij, *Weed Sci.* **48**, 561 (2000).
7. J. Ellis-Jones, S. Schulz, B. Douthwaite, M. A. Hussaini, B. D. Oyewole, A. S. Olanrewaju, and R. White, *Expl. Agric.* **40**, 353 (2004).
8. G. D. Odhiambo, and J. K. Ransom, (1996). in *Advances in Parasitic Plant Research. Proceedings of the Sixth International parasitic Weed Symposium*, Eds. M. T. Moreno, J. I. Cubero, D. Berner, D. Joel, L. H. Musselman, and C. Parker, (Cordoba, Spain) p. 834.
9. D. K. Berner, N. W. Schaad, and B. Volksch, *Bio. Control* **15**, 274 (1999).
10. F. K. Kanampiu, V. Kabambe, C. Massawe, L. Jasi, D. Friesen, J. K. Ransom, and J. Gressel, *Crop Prot.* **22**, 697 (2003).
11. A. Oswald, and J. K. Ransom, *Crop Protect.* **23**, 89 (2004).
12. E. Ezeaku, and S. D. Gupta, *African J. Biotechnol.* **3**, 324 (2004).
13. G. D. Odhiambo, and J. K. Ransom, *African Crop Sci. J.* **1**, 105 (1993).
14. J. Gressel, L. Segel, and J. K. Ransom, *Intl. J. of Pest Manag.* **42**, 113 (1996).
15. J. Gressel, *Crop Protect.* **24,** 1007 (2005).
16. H. Pieterse, and Verkleij, in *Proceedings of the 5th International Symposium of parasitic weeds* Eds. J. K. Ransom, L. J. Musselman, A. D. Worsham, and C. Parker, (CIMMYT, Nairobi, Kenya 1991). p. 457.
17. G. Van Delft, Ph.D. thesis, Univ. York, York, U.K. (1997).
18. J. K. Ransom, and J. Njoroge, in *Proceedings of the 5th International Symposium of parasitic weeds* Eds. J. K. Ransom, L. J. Musselman, A. D. Worsham, and C. Parker, (CIMMYT, Naibrobi, Kenya 1991) p. 391.

19. R. J. Carsky, D. K. Berner, J. G. Kling, A. Melake-Berhan, and S. Schulz, *Inter. J. Pest. Manag*. **46**, 115 (2000).
20. A. Oswald, and J. K. Ransom, *Crop Prot*. **29**, 113 (2001).
21. S. Schulz, M. A. Hussaini, J. G. Kling, D. K. Berner, and F. O. Ikie, *Expt. Agr*. **39**, 99 (2003).
22. D. K. Berner, J. G. Kling, and B. B. Singh, *Plant Dis*. **79**, 652 (1995).
23. G. Gbehounou, A. H. Pieterse, and J.A.C. Verkleij, *Experientia* **52**, 264 (1996).
24. J. Sauerborn, B. Kranz, and H. Mercer-Quarshie, *Appl. Soil Ecol*. **23**, 181 (2003).
25. G. D. Odhiambo, and P. L. Woomer. *African Crop Sci. Conf. Proc.*, **7**, 473 (2005).
26. I. Kureh, A. Y. Kamara, B. D. Tarfa. *J. Agric. Rural Dev. Tropics Subtropics* **107**:41 (2006).

CHAPTER 17

CULTURAL AND CROPPING SYSTEMS APPROACH FOR *STRIGA* MANAGEMENT — A LOW COST ALTERNATIVE OPTION IN SUBSISTENCE FARMING

Fasil Reda[a,*] and J. A. C Verkleij[b]

[a]*Melkassa Agricultural Research Center, P.O. Box – 436, Nazreth, Ethiopia*
[b]*Department of Ecology and Physiology of Plants, De Boelelaan 1087, 1081 HV Amsterdam, The Netherlands*
[*]*E-mail: maku1987@yahoo.com*

Striga is a major biotic constraint in the subsistence agriculture regions of Ethiopia. Thus, emphasis is placed on low-cost integrated systems. The benefits of intercropping and relay cropping were investigated for crop yield improvement and *Striga* control. Intercropping with cowpea produced the highest supplemental yield of grain and biomass. Relay cropping of sorghum with perennial legume shrubs (*Sesbania sesban* and *Cajanus cajan*) significantly improved yield at a site with relatively better weather and soil conditions. The legume shrubs resulted in significantly lower sorghum yield in a dryland location. The effect of the improved cropping systems on *Striga* infestation was not consistent. However, the systems ensured improved land and crop productivity, providing a viable option to farmers in the *Striga* prone areas, which are characterized by accelerated decline in natural resource base. A five-year rotation trial revealed that yearly alternate cropping of sorghum with legumes could sustain productivity of crops in dryland environments. Continued sorghum production (local practice) led to sharp decline in yield over the years. Integrated use of resistant varieties (row planted), fertilizer, 2,4-D or hand pulling significantly improved yield through effective control of *Striga*.

1. Introduction

The *Striga hermonthica* problem in Ethiopia is aggravated by the inherent low soil fertility, recurrent drought and overall natural resource degradation. Previous efforts to alleviate the scourge through available technologies, which were mostly high input demanding, have met with little success. A potentially viable technology should be low-cost and within reach to the small-scale farming community, and address at least the two highly interrelated problems of low soil fertility and *Striga* infestation. Cropping system approaches such as inter and relay cropping, and crop rotation could satisfy those two important concerns. Researchers have reported the multiple benefits of cereal/legume association in *S. hermonthica* affected areas.[1,2] Relay cropping and improved fallow systems, which involve the use of perennial legume shrubs, are receiving increased research attention as a promising method for resource poor farming communities. Improved fallow requires interruption of cereal production, which may not be favorably accepted by subsistence farmers. Relay cropping could be an attractive option in areas where population density is high, fallow periods are decreasing, and additional land is unavailable.[3] Similarly, intercropping has shown promise as a low-cost method of controlling *Striga*. Experience from elsewhere showed that the density of emerged *S. hermonthica* plants was reduced when sorghum was intercropped with groundnut[4] and *Dolichos lablab*.[5] Legume intercrops can induce seed germination of different *Striga* species without supporting further growth eventually leading to seed bank depletion.[6] Other recent evidence suggests that legumes could positively influence soil microbial ecology[7] and the microclimate,[8] possibly making the environment less favorable for the parasite.

Rotation of infested land into non-susceptible crops or into fallow is theoretically the simplest of all solutions, but hardly practical at present because of increased population pressure and shortage of land. At least 4-5 years of rotation are likely to be needed, emphasizing the practical limitations of this technique. Few farmers will be prepared to give up growing their preferred cereal for a long period, and in most infested areas, the choice of alternative crops is extremely limited. Nevertheless, rotation with crops that are not attacked by *Striga* is important and should be considered whenever possible. Integrated use of compatible and effective control methods holds promise for the management of parasitic weeds. This chapter reviews research in cultural and cropping systems for the control of *Striga hermonthica*, and improvement of land and crop productivity in Ethiopia.

2. *Striga* Management Research

2.1. *Improved Cropping Systems — Intercropping*

Intercropping is a potentially viable, low-cost technology that would enable addressing two important and interrelated problems of low soil fertility and *Striga* infestation. An important prerequisite is to identify the optimal spatial and temporal arrangements, and select effective, compatible and adapted legume crops, depending on the environmental conditions and existing populations of *Striga.* At Sirinka site in Amara region (North Ethiopia), one row of legume (cowpea or haricot bean) for every two rows of sorghum was an optimum arrangement both in terms of reduction in parasitic weed incidence and increase in cereal yield (Table 1). At Adibakel, a dry highland location in Tigray region, the same planting arrangement of sorghum and cowpea was superior in terms of crop productivity and *Striga* control (Table 2). Intercropping had a rather detrimental effect on sorghum yield and had no obvious suppressive effect on *Striga*, under non-fertilized conditions at Sheraro (Table 3). Fertilizer use was required, and inorganic fertilizer alone improved crop performance and created a non-conducive environment for *Striga* at this site located in the lowland plains in northwestern Ethiopia. In another environment, in Tigray, alternate row planting of sorghum and legumes, with staggered planting of the crops (sowing of legume intercrops 3 to 4 weeks after the cereal), was more productive and led to overall reduction in infestation, over two seasons.[9] Two cowpea varieties – cv. TVU 1977 OD and cv. blackeye pea were productive and most compatible with sorghum. Groundnut produced the highest biomass of 1.5 t ha^{-1}, which could be a valuable bonus and an important source of fodder and green manure in the dryland environments.

Table 1. Intercropping cowpea or haricot bean with sorghum best increased sorghum yield and reduced *Striga* at Sirinka.

Treatment	Striga count	Yield (kg/ha)	
	(Shoots/plant)	Sorghum	Legume
Intercropping (I)			
Sole sorghum	1391	2984	-
Sorghum/soybean	811	2164	354
Sorghum/cowpea	32	1326	1543
Sorghum/haricot bean	290	1509	1622
Planting arrangement (A)			
Within row	458	1641	1265
Alternate row	400	2066	1010
Broadcasting	275	1292	1244
LSD (0.05) (I)	458	537	303
LSD (0.05) (A)	NS	465	NS
LSD (0.05) (I X A)	NS	NS	NS
CV (%)	134	33	27

Table 2. Intercropping cowpea with sorghum best increased sorghum yield and reduced *Striga* at Adibakel.

Treatment	*Striga* count (Shoots/plant)	Yield (kg/ha) Sorghum	Biomass
Intercropping (I)			
Sole sorghum + fertilizer	97	321	5066
Sole sorghum, no fertilizer	95	444	5067
Sorghum/soybean	63	360	4867
Sorghum/cowpea	41	443	5517
Sorghum/haricot bean	77	466	5783
LSD (0.05) (I)	29	NS	845
Planting arrangement (A)			
BC/30 DAS	72	402	5600
BC/0 DAS	79	474	5889
AR/30 DAS	45	383	4800
EOR/0 DAS	44	435	5267
LSD (0.05) (A)	34	NS	975
LSD (0.05) (I X A)	59	NS	NS
CV (%)	45	24	14

Note: BC – Broadcast planting, AR – alternate row planting, EOR – legume intercrop planted every other row, DAS – days after sorghum sowing.

Table 3. Effect of intercropping on *Striga* control and sorghum yield at Sheraro.

Treatment	*Striga* count (Shoots/plant)	Yield (kg/ha) Sorghum	Biomass
Intercropping (I)			
Sole sorghum + fertilizer	12	2476	13000
Sole sorghum, no fertilizer	57	1076	7000
Sorghum/soybean	47	1020	6167
Sorghum/cowpea	62	1309	7417
Sorghum/haricot bean	53	1296	7083
LSD (0.05) (I)	10	NS	1689
Planting arrangement (A)			
BC/30 DAS	55	1195	7000
BC/0 DAS	55	1116	5889
AR/30 DAS	53	1276	7222
EOR/0 DAS	53	1246	7444
LSD (0.05) (A)	NS	NS	NS
LSD (0.05) (I X A)	20	NS	3291
CV (%)	17	28	22

2.2. *Improved Cropping Systems – Relay Cropping*

Relay cropping and improved fallow systems that involve the use of perennial legume shrubs are receiving a growing research attention as a promising method for resource-poor farming communities.[10] Experience with *Sesbania sesban* and *Cajanus cajan* in Adibakel and Sheraro, in Tigray, showed that the outcome from such an intervention could depend on environmental factors such as rainfall and inherent soil fertility.[9] Transplanting of the legume shrubs into sorghum fields, one month later led to consistent increase in cereal yield and decline in parasitic weed incidence at Sheraro, the site endowed with conducive weather and edaphic conditions (Table 4). This system sometimes resulted in significantly lower sorghum yields, under moisture stress and non-fertilized conditions, at the dry highland Adibakel site (Table 5). Inorganic fertilizer helped to maximize yields, particularly whenever there was a response to the input in good years.

2.3. *Crop Rotation*

Rotation of infested land into non-susceptible crops or into fallow is theoretically the simplest of all solutions, but impractical. At least 4-5 years of rotation are probably needed, further emphasizing the practical

Table 4. Relay cropping of sorghum and legume shrubs sustained crop yield but did not reduce *Striga* at Sheraro.

Treatment	1998 Grain yield (kg/ha)	1998 Striga count (n/plot)	1999 Grain yield (kg/ha)	1999 Striga count (n/plot)	2000 Grain yield (kg/ha)	2000 Striga count (n/plot)
Shrubs						
Control (no-tree)	343	4530	547	162	1330	556
Sesbania	394	4190	584	92	1920	261
Cajanus	330	4380	558	110	1760	330
P>0.05	NS	NS	NS	NS	NS	*
Fertilizer						
Control (no-fertilizer)	106	3940	166	110	907	503
20.5 N/23 P_2O_5 kg.ha^{-1}	287	4730	618	87	1760	360
41 N/46 P_2O_5 kg.ha^{-1}	674	4430	904	52	2450	284
P>0.05	**	NS	**	NS	**	NS

Table 5. Relay cropping of sorghum and legume shrubs for crop yield improvement and *Striga* control at Adibakel.

Treatment	1998 Grain yield (kg/ha)	1998 Striga count (n/plot)	1999 Grain yield (kg/ha)	1999 Striga count (n/plot)	2000 Grain yield (kg/ha)	2000 Striga count (n/plot)
Shrubs						
Control (no-shrub)	148	278	639	206	693	148
Sesbania	86	316	396	379	453	158
Cajanus	131	417	444	319	533	152
P>0.05	NS	NS	*	NS	NS	NS
Fertilizer						
Control (no-fertilizer)	91	284	352	134	464	123
20.5 N/23 P_2O_5 kg.ha^{-1}	145	385	535	332	640	186
41 N/46 P_2O_5 kg.ha^{-1}	131	343	593	439	587	151
P>0.05	NS	NS	*	*	NS	NS

limitations of this technique. Nevertheless, sustainable agronomic practices need to be widely adopted to curb the unabated decline in soil fertility. Thus, a five-year rotation experiment was conducted to compare alternate cropping of sorghum and annual legumes with the existing system of cereal monoculture, under *Striga* infested conditions.

Alternating sorghum and legume cultivation was a significantly advantageous system compared to the traditional cereal mono-cropping practice. Fertilizer input led to enhanced growth enabling sorghum to mature early, a critical attribute in those areas, which are frequently affected by terminal drought. However, increased *Striga* infestation was noted following chemical fertilizer use. Others have also reported that fertilizer could often lead to increased *Striga* emergence on infertile and highly degraded soils.[11] The most interesting observation in the initial year was that intercropped sorghum had a significantly higher grain yield than monoculture, which was comparable to that of the fertilized sole-sorghum treatment. Therefore, intercropping showed promise, from the outset, especially considering the additional gains that could be obtained from the companion food legume crops without compromising the main cereal yield. In the second season, the highest grain yield of 2130 kg/ha and biomass yield of 23 T/ha was obtained from sorghum grown after cowpea (data not shown). Similarly, fertilized sorghum, sorghum grown after haricot bean and continuous sorghum/cowpea intercropping resulted in improved overall sorghum performance. This was not followed by concomitant reduction in *Striga* infestation, except the trends for low parasite incidence on plots that were under legume crops the previous season. The control, continuous sorghum without fertilizer, had stunted growth and significantly lower yield.

The low yielding, short cycle, local sorghum variety – Jigurti was used in the final season because of late onset of the rainy season. Yet, results confirmed once again the superior performance of sorghum in one-year rotation with legumes compared to the traditional practice. Almost three-fold increase in grain and over two-fold increase in biomass yield was registered using the improved practices (Table 6). Furthermore, the cereal crop showed vigorous and relatively more accelerated growth. Rotation with food legumes, particularly haricot bean, produced up to 3700 kg ha^{-1} sorghum yield, in the intervening seasons (data not shown), which could serve as an additional incentive to farmers because of the considerably high current market prices fetched by the crop. The cowpea intercrop was compatible with sorghum. It was planted 3-4 weeks later and matured early without significantly affecting the performance of the main cereal crop. Inorganic fertilizer significantly lowered early *Striga* shoot counts, but infestation has increased and differences

Table 6. Crop yield improvement and *Striga* reduction after 5 years of sorghum rotation with food legumes at Sirinka (1999).

Treatment	Grain yield (kg/ha)	Biomass yield (t/ha)	Days to heading	Days to maturity	*Striga* count (n/plot)
Continuous sorghum -F	518	3.6	66	149	3150
Continuous sorghum +F	1450	7.2	76	143	2110
Sorghum/cowpea alternate cropping-F	1300	7.1	76	143	4140
Sorghum/haricot bean alternate cropping-F	1390	7.6	76	144	3130
Continuous sorghum/cowpea intercropping -F	1440	7.4	77	144	3210
CV (%)	18.4	12.9	1.1	1.1	29.7
P>0.05	**	**	**	**	NS

Note: ±F - with and without fertilizer.

were not significant later in the season. Therefore, improved crop growth conditions have not been matched by diminishing *Striga* incidence, a typical demonstration of the controversial effect of enhanced fertility on the pest in dry land environments. Nevertheless, the experiment clearly demonstrated that subsistence farmers could make their system more sustainable, in terms of increased yield and possibly improved soil fertility, by incorporating legumes as rotation crops.

2.4. *Integrated Control*

Integrated use of compatible and effective control methods holds great promise for the management of parasitic weeds. Our results demonstrated that the integrated use of weed control and crop management practices could enhance productivity and suppress *Striga.*[12] At Sirinka, a treatment consisting of row planting, mineral fertilizer (42 kg N/ha) and 2,4-D herbicide (0.6 kg a.i./ha, sprayed four weeks after *Striga* emergence) led to a 40% increase in cereal yield and appreciable reduction in *Striga* infestation, compared to the control (broadcast planting, no fertilizer and early weeding) (Table 7).

Table 7. Effect of integrated management practices on *Striga* infestation and sorghum yield at Sirinka.

Treatment	*Striga* control (shoots/plot)	Grain yield (kg/ha)
Variety (V)		
Degalit (local)	1229	4557
SRN-39	157	1541
LSD (0.05) (V)	700	463
Management (M)		
BC-F+HP	1440	2242
RP+F+HP	426	2210
RP+F+2,4-D	702	3142
LSD (0.05) (M)	NS	463
LSD (0.05) (V x M)	NS	NS
CV (%)	104	23

Note: BC – broadcast planting, RP – row planting, ±F - with and without fertilizer (41/46 N/P2O5 kg ha^{-1}), HP – hand pulling.

The combined use of row planting, fertilizers and hand pulling during flowering increased grain yield by half and halved *Striga* shoot counts compared to farmers' practices at Adibakel in the Tigray region (Table 8).

3. Conclusions

The acceptability of *Striga* control technologies could significantly improve if they are integrated and capable of simultaneously addressing constraints such as drought and low soil fertility. Thus, the research program was oriented to focus on cropping systems approach i.e., integration of annual and perennial legumes; crop rotation and integrated methods for the improvement of land and crop productivity, and *Striga* control. Encouraging results were achieved, but more needs to be done to understand these various systems, which involve complex interactions within and between plant species, and plants and the environment.

Table 8. Integrated fertilizer and herbicide reduced *Striga* infestation and increased sorghum yield at Adibakel.

Treatment	*Striga* control (shoots/plot)	Grain yield (kg/ha)
Variety (V)		
Local	262	307
ICSV-1006	42	621
LSD (0.05) (V)	105	162
Management (M)		
BC-F+HP	198	381
RP+F+HP	92	564
RP+F+2,4-D	73	541
LSD (0.05) (M)	117	181
LSD (0.05) (V x M)	235	362
CV (%)	80	35

Attempts will have to be made to unravel the various mechanisms involved for their manipulation to maximize benefits. Efforts have to be made to ensure improved access of farmers to fertilizers. Whenever there is no access to this input, farmers have to be advised to use all possible means of restoring the fertility of *Striga* infested soils through the introduction of soil improving legumes in relay- and inter-cropping arrangement and practice less cereal mono-culture.

We believe that the general tendency of viewing *Striga* as an ordinary biological problem is erroneous, simplistic and unhelpful. *Striga* is a natural resource problem, a biological problem and a socio-economic problem combined in one. Implementing a holistic approach, developed to suit the delicate socio-economic conditions of subsistence farmers is the only way forward in the battle against the diverse and formidable problem of *Striga* in developing countries such as Ethiopia.

References

1. G. D. Odhiambo, and J. K. Ransom, - *Proc. 4th East and Southern African Maize Workshop*. Pp. 250-254. CIMMYT. Harare, Zimbabwe (1995).
2. C. Parker and C. R. Riches, *Parasitic Weeds of the World: Biology and Control*. CAB International. Wallingford (1993).

3. S. Franzel,. *Agroforestry Syst.* 47, 1(1999).
4. A. G. Carson, *Trop. Pest Mgmt.* **35**,130 (1989).
5. A. G. T. Babiker, N. E. Ahmed, A. H. Mohamed, M. E. El Mana and S. M. El Tayeb, *Proc.Brighton Crop Protect. Conf. – Weeds.* (1993).
6. G. Gbehounou,. PhD Thesis, Vrije Universiteit, Amsterdam, The Netherlands (1998).
7. R. S. Gallagher, E. C. M. Fernandes and E. L. McCallie, *Agrofor. Syst.* 47, 197 (1999).
8. R. B. Matthews, S. N. Azam-ali, R.A. Saffell, J. M. Peacock and J.H . Williams, *Agric.For. Meteorol.* **53**, 285 (1991).
9. F. Reda, *Striga hermonthica* in Tigray (northern Ethiopia) –. PhD Thesis. Free University, Amsterdam, The Netherlands (2002).
10. ICRAF, Annual Report. Pp. 78-85. International Center for Research in Agro-forestry, Nairobi, Kenya (1992).
11. H. Doggett, *Sorghum* (2nd Edition), pp. 368-404. Longman Scientific and Technical, Harlow, UK (1988).
12. F. Reda, G. Woldewahid, B. Zemichael and W. Bayou, *Pest Manage. J. Ethiopia* **1**, 43- (1997).

CHAPTER 18

FIELD DEVELOPMENTS ON *STRIGA* CONTROL BY *DESMODIUM* INTERCROPS IN A "PUSH-PULL" STRATEGY

Zeyaur R. Khan[a,*], Charles A. O. Midega[a], Ahmed Hassanali[a] and John A. Pickett[b]

[a]*Plant Health Division, International Centre of Insect Physiology and Ecology (ICIPE), P.O. Box 30772, Nairobi 00100, Kenya.*
[b]*Biological Chemistry Div., Rothamsted Research, Hertfordshire AL5 2JQ, UK*
[*]*E-mail: zkhan@mbita.mimcom.net*

During investigations into the control of insect damage to maize crops in Kenya, which involved intercropping with repellent plants, the fodder legumes silverleaf (*Desmodium uncinatum*) and greenleaf (*D. intortum*) reduced *S. hermonthica* infestations of maize. This effect was significantly better than other food legumes. Although soil shading and additional nitrogen contributed to reduced levels of *S. hermonthica* infestation, an allelopathic mechanism associated with *D. uncinatum* was a major factor, as seen in pot elution experiments. Root exudates of *D. uncinatum* contain isoflavanones that stimulate germination of *S. hermonthica* and related constituents that inhibit its lateral root growth. Other *Desmodium* spp., have similar effects, indicating comparable phytochemical and physiological attributes. *Desmodium*-based intercrops have been developed for both sorghum and maize. Economic analyses indicate that this strategy is more profitable than both maize mono- and maize-bean inter-crops.

1. Introduction

Striga control by desmodium intercrops was discovered during the development of a 'push-pull' strategy for the control of lepidopteran stemborers in maize in Kenya. The strategy involved creating a 'push-pull' effect using highly attractive trap crops to attract stemborer moths away from the central maize stand, and intercropping between the rows of maize with repellent plants.[1] Highly attractive trap crops such as Napier grass (*Pennisetum purpureum*) are planted as trap plants around maize or sorghum crop (pull) and intercropping between the rows with repellent plants such as *Desmodium* spp. (push) (Fig. 1). In field trials in Suba district of western Kenya where *S. hermonthica* is highly prevalent, these legumes unexpectedly and dramatically reduced infestations by this witchweed.[2]

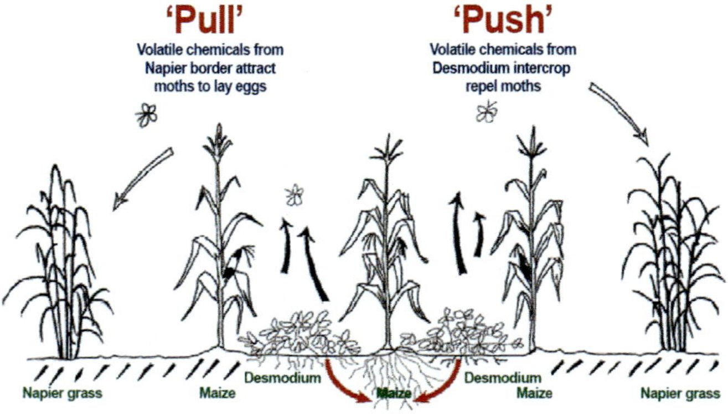

Figure 1. How the 'push–pull' habitat management system works. Chemicals (flavones/isoflavones) secreted by desmodium roots inhibit attachment of germinated Striga *to maize roots and cause rapid depletion of* Striga *seeds in the soil. Adapted from ICIPE Annual Scientific Report*[3].

2. Mechanisms by Which *Desmodium* spp. Control *Striga*

The observed suppression of *S. hermonthica* by *Desmodium* stimulated investigations into its mode of action. A number of mechanisms were proposed, including increase in available nitrogen in the soil, effects of shading, and an allelopathic effect caused by semiochemicals released from *Desmodium* spp. roots.[1,4] The effects of these factors were studied in the field and in a screenhouse in western Kenya using *D. uncinatum*.[4] Field plots of maize intercropped with *D. uncinatum* with or without 120 kg nitrogen/ha, maize monocrop with or without nitrogen, and maize monocrop with artificial ground shading made of maize straw (Hybrid 513) with or without nitrogen were set up. *S. hermonthica* seed levels in each plot, before and after each cropping season, were measured by elutriation.[5] Nitrogen content in each plot, before and after each cropping season was measured by Kjeldahl method.[6] Nitrogen and shading treatments, and the combined nitrogen and shading treatment suppressed *S. hermonthica* compared to the maize monocrop. However, a significantly greater reduction in *S. hermonthica* infestation was achieved by the *D. uncinatum* intercrop and the combined *D. uncinatum* and nitrogen treatments, indicating incremental effects of the intercrop.

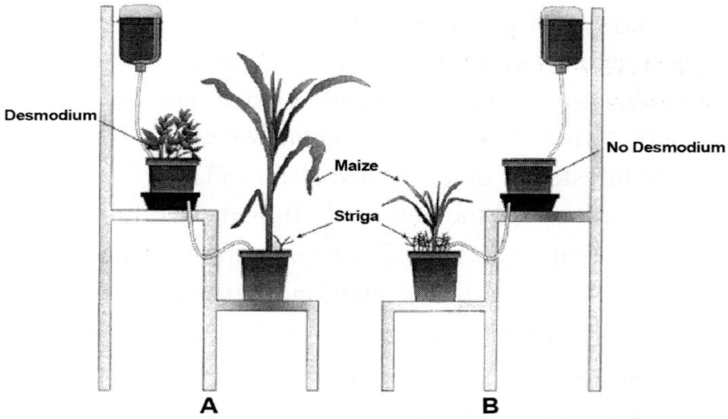

Figure 2. Method of demonstrating an allelochemical mechanism of D. uncinatum *in suppressing* S. hermonthica *infestation of maize. Comparison was made between maize plants irrigated by root eluates of* D. uncinatum *(A) with those irrigated by water passing through pots containing only autoclaved soil (B). From Khan et al.*[4] *with permission of Springer Science and Business Media.*

An allelochemical effect of *D. uncinatum* on *S. hermonthica* was demonstrated in a screenhouse. *D. uncinatum* plants were grown in pots and water dripping from their root systems irrigated maize planted in soil infested with approximately 3000 *S. hermonthica* seeds/pot. *D. uncinatum* was planted with or without the nitrogen-fixing bacterium, *Rhizobium* sp. CB 627, to compare the effect of fixed nitrogen with that of the allelochemicals alone. Autoclaved soil was used in all experiments, and no additional nitrogen was applied. The pots containing *D. uncinatum*, which received distilled water, were placed on shelves, thus allowing the flow of water by gravity through the pots into the maize pots situated below (Fig. 2). Comparisons were made between maize plants irrigated by root eluates from *D. uncinatum* (with or without *Rhizobium* sp.) and those irrigated by water passing through pots containing only autoclaved soil (with or without *Rhizobium* sp.).[4]

The dramatic effect of the aqueous solution of chemicals eluting from pots in which *D. uncinatum* plants were growing on suppression of *S. hermonthica* infestation is illustrated in Fig. 3. In a separate experiment, aqueous samples of chemicals exuded by axenic *D. uncinatum* roots induced germination of *S. hermonthica* as effectively as the maize root exudates, indicating absence of a germination inhibitor.[4] Radicals of germinated seeds exposed to root exudates from *D. uncinatum* 24 and 48 hrs after germination were significantly shorter than of those exposed to maize exudates. Inhibition of the radical growth was observed irrespective of whether the *D. uncinatum* had been grown in *S. hermonthica* infested or in uninfested clean soil.

These observations led the authors to hypothesize that in addition to germination stimulants present in *D. uncinatum* root exudate, there were additional factors affecting the growth and development of germinated *S. hermonthica* and that this prevented normal attachment to host plants.[4] Some of the compounds (e.g. uncinanone B and C) responsible for these observations were later isolated from root exudates of *D. uncinatum*.[7]

Isolated fractions containing one of the compounds (e.g. uncinanone B) induced germination of seeds from *S. hermonthica* and fractions containing another (e.g. uncinanone C) moderately inhibited radical growth. This may result in reduced chances of attachment to the roots of the host plants.[7] Another key post-germination inhibitor was recently characterised (Chapter 5), although full chemical elucidation of all important allelopathic agents is still ongoing. The combined effect of germination stimulants and post-germination inhibitors represents an efficient mechanism of suicidal germination of *Striga* seeds. It leads to effective control of *S. hermonthica* and provides a novel means of continual *in situ* reduction of the *Striga* seed bank in the soil even in the presence of graminaceous host plants in the proximity. Indeed, the density of *S. hermonthica* seeds in the soil of maize–*Desmodium* plots steadily decreased every cropping season, while in maize monocrops and maize-cowpea intercrops the number steadily rose (Fig. 4).

2.1. *Effects of Different* Desmodium *spp. on* Striga

After we had demonstrated control of *Striga* through intercropping maize with *D. uncinatum*, a medium-high altitude species, it was prudent to test whether other *Desmodium* spp. adapted to different agro-ecologies could offer similar levels of control of *S. hermonthica* and enhance comparable grain yields. Four *Desmodium* spp. and a cowpea variety were

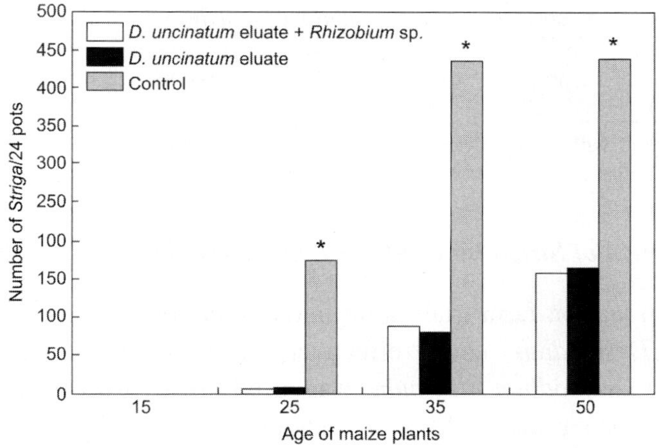

Figure 3. Desmodium *root eluates inhibit emergence of* S. hermonthica *with or without nitrogen-fixing* Rhizobium *sp. bacteria. Within each age group of maize, the treatment marked with an asterisk is significantly different (P<0.05). From Khan et al.*[4] *with permission of Springer Science and Business Media.*

compared. *Desmodium* spp. included: silverleaf, *D. uncinatum* (a medium-high altitude species); greenleaf, *D. intortum* (a low-medium altitude species); Hawaiian tick-trefoil, *D. sandwicense* (a low-medium altitude species), and; pringlei, *D. pringlei*, (a medium-altitude species).[8] *S. hermonthica* counts were significantly reduced in maize-*Desmodium* intercrops (by up to two-fold) compared to the maize monocrop and maize-cowpea intercrop. Similarly, maize plant height and grain yields were significantly higher (by up to two-fold and five-fold, respectively) in maize-*Desmodium* intercrops than in maize monocrop and maize-cowpea intercrop. These results demonstrated that the *Desmodium* spp. assessed had similar effects as *D. uncinatum* on *S. hermonthica* suppression and enhancement of grain yields, indicating comparable phytochemical and physiological attributes in these species (Table 1).[9] Cowpea was ineffective at reducing *Striga* or increasing yield (Table 1)

Table 1. Different species of *Desmodium* equally controlled *S. hermonthica* resulting in significantly taller maize plants and enhanced grain yields. Means represent data averages over four cropping seasons.

	Treatments					
Parameter	Maize monocrop	Maize/*D. pringlei*	Maize/*D. intortum*	Maize/*D. sandwicense*	Maize/*D. uncinatum*	Maize/ cowpea
Striga counts	194.3a	5.5b	3.1b	6.3b	4.9b	144.8a
Plant height	130.7b	192.9a	197.9a	190.8a	193.8a	125.6b
Grain (t/ha)	2.3b	4.8a	5.2a	4.7a	5.2a	2.7b

Means marked with different letters are significantly different (P<0.01). Adapted from Ref. 9.

2.2. *Control of Striga in Sorghum Using Desmodium Intercrops*

On-station and on-farm trials were initiated in western Kenya to assess whether *Desmodium* could effectively suppress *S. hermonthica* in sorghum. *Desmodium intortum* was used as it withstands drought conditions better and wilts less than other species.[10] It also has a relatively higher nitrogen-fixing ability, over 300 kg N/ha/year under optimum conditions[11] and, therefore, would be more appropriate as an intercrop for the degraded environments where sorghum cultivation is widely practiced. In both trials, *S. hermonthica* counts were significantly lower in the intercropped plots of sorghum (commercial hybrid, Gadam Hamam) and *D. intortum* than in the sorghum monocrop plots. These observations were associated with significantly higher grain yields in the intercrop than in the monocrop plots (Table 2).[12]

2.3. *Different Legumes to Control of Striga in Maize and Sorghum*

After we had established and demonstrated the efficiency of *Desmodium* spp. in the control of *S. hermonthica*, we evaluated a number of grain legumes, some of which have been implicated in the control of *Striga*,[13] for similar effects. We assessed their impact on *Striga* alongside *D. intortum*. Maize and sorghum were intercropped with different

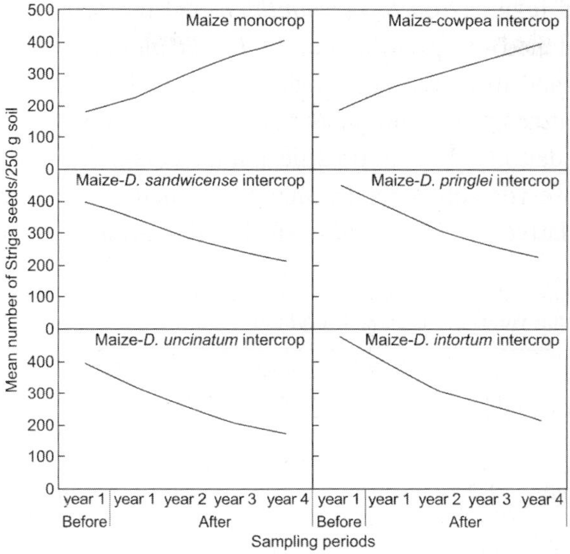

Figure 4. There is a steady depletion of S. hermonthica *seeds in the soil in the maize-*Desmodium *intercrops. Data from long time field trials at ICIPE-Mbita, Western Kenya.*

Table 2. *Desmodium* suppresses *S. hermonthica* and enhances grain yields in sorghum.

Site	Cropping system	Mean no. *Striga* Plants/plot	Mean grain yield (t/ha)
On-station	Sorghum monocrop	467a	1.5b
	Sorghum/desmodium	3b	2.4a
On-farm	Sorghum monocrop	545a	0.9b
	Sorghum/desmodium	60b	1.6a

Within a parameter in a site, means marked with different letters are significantly different ($p<0.05$). (Adapted from Khan *et al.*[12])

legumes (cowpea, crotalaria, beans, groundnuts and greengrams) and *S. hermonthica* counts and grain yields measured. Although crotalaria and cowpea somewhat suppressed *Striga* emergence in sorghum and crotolaria in maize, *D. intortum* had a significantly superior effect in both crops (Table 3), which was reflected in greater reduction in *Striga* and enhanced grain yields relative to the other intercrops.

These results indicated the superiority of *Desmodium* species in the control of *Striga*, with concomitant increases in grain yields.[14] The effect of other food legumes on *S. hermonthica* had previously been assessed but none matched the performance of *D. uncinatum* in the suppression of the weed and in increases in maize grain yields.[4] *Desmodium* fixes 100–180 kg atmospheric nitrogen/ha under Kenyan conditions (Muyekho, unpublished data), increases organic matter content of the soil through leaf fall, conserves soil moisture thereby reducing soil temperature and increasing relative humidity, and is an effective ground cover.[4]

Table 3. *Desmodium* performs better than food legumes in the control of *S. hermonthica* and enhances grain yields in maize and sorghum.

Parameters	Cereal† mono	Cereal /grnt	Cereal /grgm	Cereal /Des	Cereal /crot	Cereal /cowp	Cereal /beans
Sorghum							
Striga counts	579a	358a	104bc	1d	47c	175bc	271ab
Grain yields (t/ha)	1.7b	2.3b	2.4b	3.4a	2.5b	2.1b	2.4b
Maize							
Striga counts	683a	499ab	474ab	2c	185b	385ab	329ab
Grain yields (t/ha)	2.4c	3.1bc	3.1bc	5.4a	3.7b	3.8b	3.1bc

†Represents either sorghum or maize. grnt, groundnut; grgm, greengram; Des, *Desmodium*; crot, crotalaria; cowp, cowpea. Within a parameter (rows) the means marked by different letters are significantly different ($p<0.05$). Means represent averages over four cropping seasons. (Adapted from Khan *et al.*[14])

3. Economics of the *Desmodium* Intercrop in a 'Push-Pull' Strategy

We assessed the economics of the *Desmodium* intercrop in a 'push-pull' strategy by comparing it with two conventional cropping systems, maize mono- and maize-bean inter-crop in five districts in western Kenya via gross margin analysis and returns on labour. There were no significant differences in total variable costs between the 'push-pull' strategy and the two conventional cropping systems (Table 4).

There were six times greater gross benefits with 'push-pull' strategy and more than tripled returns on labour than in the two cropping systems. Similarly, maize-bean intercrop significantly increased gross benefits relative to the monocrop system, although the returns on labour were not different between the two systems. This renders the *Desmodium* in a 'push-pull' intercrop strategy as a more profitable cropping system for smallholder farmers. These results however sharply contrast those of Woomer *et al.*,[15] who reported negative net returns from an intercrop of maize and *Desmodium*, an analysis based on one season of data (compared to ours based on four years) that took into consideration only maize grain yields. These authors also recognised that two main factors influenced their results, poor establishment of *Desmodium* and drought that led to competition for moisture between the two crops.

Table 4. Significantly higher economic returns result from the *Desmodium* intercrop in a 'push-pull' strategy compared to maize monocrop and maize-bean intercropping. Means represent data averages of five districts over four years.

Cropping system	Total variable costs (USD/ha)	Gross benefits (USDA/ha)	Return on labour (USD/person day)
'Push-Pull'	343.3a	598.5a	2.2a
Maize-bean intercrop	347.9a	214.6b	0.7b
Maize monocrop	287.8a	91.5c	0.14b

Within a variable (columns), means marked by different letters are significantly different ($p<0.05$).

3.1. *'Push-Pull' Strategy as a Platform Technology*

The 'Push-pull' strategy is an internally integrated and sustainable habitat management system that addresses all three major constraints on maize and sorghum production (stemborers, *Striga* and soil fertility). It is also a platform technology with the possibility of other forward linkages and associated benefits. Farmers in eastern Africa have embraced the technology with enthusiasm. In addition to improved maize yields, the strategy provides fodder and meets the need for a reliable source of forage, either for their own cattle or for sale. Sales of Napier grass and *Desmodium* herbage to neighbours with stall-fed cattle provides a new source of income. As the forage can be harvested regularly, this brings in money when there are no other crops to sell. Home-grown forage also obviates the need to spend many hours each day gathering fodder for stall-fed cattle or herding animals as they graze. The increase in milk yields add income and also improves the nutritional status of the farming family. *Desmodium* seed is also highly marketable. The net result has been a substantial impact on food security through increased farm productivity.

Environmentally, the practice of the 'push-pull' strategy has long term benefits as well. Improved availability of forage can enhance soil fertility. Instead of feeding crop residues to livestock, farmers can now return them to the soil. If they stall-feed their animals, it is easy to collect the manure and this too can be used to enrich the soil. By introducing a mixture of crop species into the farm environment and reducing the need to use insecticide for stemborer control, this reverses the trend towards monocropping using chemical inputs as a means of increasing productivity. This is much more beneficial to long-term environmental health, enhancing rather than reducing biodiversity. The effect of *Desmodium* on *Striga* is a long-term one.

4. Adoption of the 'Push-Pull' Strategy

Following success of the on-station experiments, dissemination of the technology was initiated among smallholder farmers in Kenya in 1998. Currently, the technology is being practiced by over 7,000 smallholder farmers in Kenya, Uganda and Tanzania. Several dissemination pathways are being evaluated in the promotion of the technology. These include use of farmer-teachers, brochures, field days, tours, farmer groups training, media, radio and television programmes and autonomous diffusion. We are also implementing farmer field schools.

The use of *Desmodium* species to control *S. hermonthica* has been associated with positive crop performance and enhanced grain yields and this has been one of the reasons for its widespread adoption. The high adoption is undoubtedly linked to the farmers' perceptions of the short-term benefits; they can see that they will be better off within one or two seasons, so are willing to invest their time and labour. Indeed, data collected from farmers practicing the technology indicate that their grain yields have increased by up to 100% in some of the areas where stemborers and *S. hermonthica* occur together. The *Striga* seed bank in the soil is almost depleted after about six seasons of continuous practice of the strategy. If the farmer chooses not to continue with the strategy at this point they can then plough out the *Desmodium*.

5. Conclusions and Future Outlook

The 'push-pull' strategy quite uniquely developed from basic science to a practical technology, with farmer take-up and spontaneous technology transfer among farmers. Although the experience to date has been restricted to maize and sorghum-based farming systems, we believe that the general approach is applicable to a much wider range of pest problems in a variety of crops (such as millet) and will be a model for other researchers in their efforts to minimize pest-induced yield losses in an economically and environmentally sustainable manner. We have initiated studies on the potential role of this strategy in the control of other parasitic weeds, particularly the broomrapes, *Orobanche* spp., in eastern Africa. We are also evaluating the technology with imidazolinone-resistant (IR) maize (Chapter 11), especially in the first cropping season before *Desmodium* establishes. Efforts to identify the genes responsible for the phytochemical and physiological attributes of the *Desmodium* spp. relevant in the suppression of *Striga* are being explored with a view to introducing them into edible beans (Chapter 5).

The strategy is now expanding via small-holder farmers into more districts in Kenya, Uganda, Tanzania and Ethiopia. Each region has varying climatic conditions and cultivars, and crops that must be considered. Experience has been gained from pilot studies in various countries. However, wherever these approaches are developed for the specific needs of local farming practices and communities, it is essential that the scientific basis of the modified systems should be completely elucidated. Otherwise there will be a drift from effectiveness and justifiable dissatisfaction on the part of the practising farmers. Every effort will be made to ensure that technology transfer follows the incorporation of these practices into other regions of Africa.

To date, the major constraint to technology diffusion has been availability of *Desmodium* seed. The relative merits of private seed company, community-based seed production, and vegetative propagation by farmers are all being assessed. In addition, the role of different reinforcing interventions such as mass media, information bulletins, field days, farmer teachers, farmer field schools etc. need to be evaluated and the most cost-effective ones identified. The relationship between household socio-economic status and land labour ratio in different areas, and the performance of different diffusion mechanisms is also being studied.

Acknowledgements

We acknowledge the Gatsby Charitable Foundation, Rockefeller Foundation and Kilimo Trust for R&D funds; our institutes for logistical and other support; governments of Kenya, Uganda and Tanzania, NGOs; technicians and farmers in the regions for their enthusiastic contribution in the development and dissemination of the 'push-pull' technology.

References

1. Z. R. Khan, J. A. Pickett, J. Van den Berg, L. J. Wadhams and C. M. Woodcock, *Pest Manage. Sci.* **56,** 957 (2000).
2. Z. R. Khan, J. A. Pickett, L. J. Wadhams and F. Muyekho, *Insect Sci. Appl.* **21,** 375 (2001).
3. ICIPE Annual Scientific Report, (ICIPE Science Press, Nairobi, Kenya, 2005).
4. Z. R. Khan, A. Hassanali, W. Overholt, T. M. Khamis, A. M. Hooper, J. A. Pickett, L. J. Wadhams and C. M. Woodcock, *J. Chem. Ecol.* **28,** 1871 (2002).
5. R. E. Eplee and R. S. Norris, *Monograph series, Weed Science Society of America,* **5,**136 (1990).
6. G. J. Shugar, G. J., R. A. Shugar, L. Bauman and R. S. Bauman, *Chemical Technicians' Ready Reference Handbook,* 2nd ed., (McGraw-Hill, New York, 1981).
7. M. K. Tsanuo, A. Hassanali, A. M. Hooper, Z. R. Khan, F. Kaberia, J. A. Pickett and L. J. Wadhams, *Phytochemistry* **64,** 265 (2003).
8. P. J. Skerman, D. G. Cameron and F. Riveros, Tropical Forage Legumes (FAO, Rome, 1989).
9. Z. R. Khan, J. A. Pickett, L. J. Wadhams, A. Hassanali and C. A. O. Midega, *Crop Prot.* **25,** 989 (2006).
10. H. Ostrowski, *Queen. Agric. J.* **92,** 106 (1966).
11. A. S. Whitney, *Herbage Abstracts,* **38,** 143 (1966).
12. Z. R. Khan, C. A. O. Midega, J. A. Pickett, L. J. Wadhams, A. Hassanali and A. Wanjoya, *Intl. J. Pest Manage.* **52,** 297 (2006).
13. A. Oswald, *Crop Prot.* **24,** 333 (2005).
14. Z. R. Khan, C. A. O. Midega, A. Hassanali, J. A. Pickett and L. J. Wadhams, *Crop Sci.* (2007, in press).
15. P. L. Woomer, J. Tungani, G. Odhiambo, F. M. Mwaura, *Afr. Crop Sci. Conf. Proc.* **7,** 479 (2005).

CHAPTER 19

INTEGRATED *STRIGA* MANAGEMENT TO MEET SORGHUM DEMAND IN TANZANIA

Ambonesigwe M. Mbwaga[a,*], Charles Riches[b] and Gebisa Ejeta[c]

[a]*Uyole Agricultural Research Institute, Mbeya, Tanzania*
[b]*Natural Resource Institute, Chatham UK*
[c]*Agronomy Department, Purdue University, Lafayette IN, USA*
[*]*E-mail: ambwaga@yahoo.co.uk*

Field trials and farmer participatory evaluation resulted in the registration of two sorghum cultivars that perform well on *Striga* infested soils in Tanzania. Cultivars Hakika and Wahi are early maturing and fulfill producer and consumer preferences. Yields can be improved when they are grown in an integrated *Striga* management system with use of animal manure or fertilizer and planted on tied ridges to ensure soil moisture conservation. There is potential for locally produced sorghum to replace the imported grain currently used by commercial processors.

1. Introduction

More than 40 percent of the Tanzanian population lives in chronic food-deficit regions including semi-arid zones where irregular rainfall causes recurring food shortages and consequent malnutrition. Between 1986 and 2005 the area planted with sorghum in Tanzania has ranged from 380,000 to 890,000 ha depending on rainfall, the crop being particularly important for food security in semi-arid districts. Models predict that by 2100 rainfall will decrease by up to 20% in these areas of Tanzania with a fall in national grain production of 10% by 2080, with particularly severe yield reductions in maize.[2,3] Farmers are already responding to climatic variability with the area planted with sorghum in Morogoro region increasing in three of the seasons between 1994 and 2001, when maize production remained static or declined.[3] Increasing sorghum productivity in the semi-arid zone of Tanzania will be a continuing priority for both food security and household income. Sorghum competes strongly with maize on price, particularly when purchased by the National Strategic Grain Reserve.[4]

The semi-arid areas of Tanzania lie in a zone where *Striga asiatica, S. forbesii* and *S. hermonthica* infest cereals.[5] Seventy five percent of farmers interviewed in Shinyanga region considered *Striga* an increasing problem on sorghum, and they were unable to obtain satisfactory advice from extension on effective control strategies.[6] Sorghum, the preferred staple, has been replaced in parts of Missungwi district in Mwanza Region by pearl millet, which is presently not attacked by *Striga* species in Tanzania,[7] but in West Africa it is attacked. Farmers in Dodoma Rural district are well aware that poor sorghum yields are the norm in *Striga* infested fields and that poor crop vigor and *Striga* are associated with declining soil fertility.

Most farmers plant low yielding, drought susceptible traditional landraces, but adoption of improved cultivars, which occupied barely 5% of Tanzania's sorghum area in the early 1990s, had risen to 36% of the area planted by 2002.[8] Cultivar Pato has been widely promoted. It is high yielding on favourable soils but becomes stunted resulting in poor harvests when planted on *Striga* infested soils without addition of manure or fertilizer.[9]

On-farm research was initiated in 2000 to address the *Striga* problem by identifying early maturing drought and *Striga* tolerant/resistant sorghums that fulfill farmer and consumer preferences. Here we summarize the process that resulted in 2002 with new cultivars being registered and promoted to farmers. Sorghum in Tanzania is largely consumed in producing areas with less than 1000 T per year used by commercial processors due to inconsistent quality, high costs of consolidating grain harvests from small growers, transport, and cleaning.[10] The new cultivars provided an opportunity for increased production, so producers and processors were brought together to identify new markets.

2. Methodology

Sorghum lines P9405 and P9406 (80 and 85 days to maturity respectively) bred for resistance to *Striga* (Chapter 7), *Striga* resistant line SRN 39 (maturity 87 days) from Sudan, Weijita (maturity 92 days) a popular brown seeded land race from Mara region in northern Tanzania, the released cultivar Macia (maturity 91 days) developed by ICRISAT), and Pato (maturity 91 days) bred for high yield in Tanzania were all evaluated for resistance on *S. hermonthica, S. asiatica* and *S. forbesii.* These lines were planted on fields naturally infested by one or more *Striga* species in replicated "uniformity" trials at Ukiriguru (Missungwi District in Lake zone – *S. hermonthica, S. asiatica*), Hombolo (Dodoma Rural district in Central zone – *S. asiatica*), Melela (*S. asiatica* and *S. forbesii*) and Ilonga research stations (Morogoro rural and Kilosa districts in Eastern zone respectively). Ilonga provided a *Striga*-free reference site. These trials ran for three seasons and were undertaken and managed according to the specification laid down by the Tanzania Official Seed Certification Institute. The entries were planted on-station in plots of four rows replicated three times. *Striga* counts were from the two centre rows at 9th and 12th week after planting, and again at harvest. Sorghum grain yield was assessed from the two centre rows. The lines were also evaluated by farmer groups. Between 10 and 25 farmers also evaluated the lines in each of Mwagalla, Iteja (Missungwi district, *S. hermonthica* infested), Mvumi, Chipanga (Dodoma Rural district, *S. asiatica* infested) and five growers established plots in Mpalanga (Dodoma Rural district). The plot sizes on-farm were 5 m by 10 m with farm sites used as replicates. Data on *Striga* emergence at 12 weeks after emergence and crop yield were recorded from an area of 5 x 5 m on each plot by village extension officers. In addition to the technical evaluation, mid-season field walks and were organized with each village research group to explore farmer perceptions of the lines.

On-farm trials to validate integrated management options for production of *Striga* tolerant sorghum varieties were also undertaken. Pure stands of sorghum cultivar Pato and P9405 were compared with plots inter-cropped with groundnut (cultivar Nyota) in Mvumi, Dodoma Rural district. The legumes were planted in the same row as sorghum. Composted cattle manure was applied at either 0.5 kg per sorghum planting station on a range of soil types in Dodoma. Manure use was evaluated in 2000 with sorghum lines P9405, P9406, Pato, and Macia. Intercropped and manured plots were planted at five farms in each village, with farms used as replicates. Data were collected on *Striga* emergence and sorghum yield. Group discussions were used to evaluate farmers' perceptions of results of the trials.

The cultivars Hakika (P9405) and Wahi (P9406) were promoted subsequent to official release to farmers in combination with other *Striga* management technologies in Singida rural, Kongwa and Missungwi districts. Farmer groups were provided with seed of the new cultivars for field demonstrations that incorporated tied ridges for water harvesting and application of animal manure. Meetings were facilitated to explore market opportunities and to link sorghum growers with commercial processors.

3. Sorghum Performance

Findings from on-station are presented in Tables 1 to 3. Lines P9405 and P9406 supported lower numbers of emerged *S. asiatica, S. forbesii* and *S. hermonthica* than other lines, particularly the released cultivar Pato. P9405 and P9406 produced higher yield than Pato and Macia at *Striga* infested sites. P9405 produced higher yields than Pato in seven of nine year x location tests ($P < 0.05$). P9406 performed better than Pato in six trials. Pato and Macia have a higher yield potential and perform well under *Striga*-free conditions, as was observed at Ilonga. Although no data were recorded, all lines tested became heavily infested by sorghum midge *(Contarinia sorghicola).* This proved to be a serious pest at

Ukiriguru and village sites in the Lake zone when sorghum was planted after mid-February, during the long rainy season. Farmers' avoid midge damage in this bimodal rainfall area by planting sorghum in the short rainy season (from September to December). P9405, P9405 and Macia were not susceptible to leaf blight (*Exserohilum turcicum*) that was a particular problem with Pato at a number of locations in Dodoma District in 2002. P9406 is somewhat more susceptible than other lines to long smut (*Tolyposporium ehrenbergii*), so it would be better to plant P9405 on *Striga* infested fields in areas where long smut is common. Cultivars P9405 and P9406 had a yield advantage compared to Pato on *S. asiatica*-infested soils in Dodoma Rural District particularly at sites where sorghum productivity is low (Table 4). Yields of Marcia were intermediate between Pato and the P9405 and P9406. SRN39 also produced higher yields than Pato.

4. Farmer Ranking of Sorghum Lines

Farmers ranked the sorghums by their own criteria. Examples of these perceptions are shown in Table 5. P9405 and P9406 ranked highly for a number of important traits including drought and *Striga* tolerance, early maturity and yield. The final ranking exercise was conducted in Chipanga in 2002 by which time some farmers had grown the new lines for four years. In a pair-wise ranking, women ranked Macia first followed by P9405, Pato, and local landraces Lugugu and Mtika. The most significant change from the previous year's evaluation was that modern varieties were ranked more highly overall. Macia appeared to be followed by P9405, Pato, and local landraces Lugugu and Mtika. The most significant change from the previous year's evaluation was that modern varieties were ranked more highly overall. Macia appeared to be particularly favored by women for its early maturity and yield and to some aspects of the ugali (porridge) it produces. P9405 is perceived by women to have a higher yield and better tasting ugali than P9406. Men

Table 1. Multi-location sorghum trials, 2000.

	Ilonga	Melela			Hombolo		Ukiriguru		
Entry	Kg ha^{-1}	*S. asiatica*	*S. forbesii*	Kg ha^{-1}	*S. asiatica*	Kg ha^{-1}	*S. asiatica*	*S. hermonthica*	Kg ha^{-1}
Pato	3200	190.0	17.5	1200	161.8	527	14.3	62.3	233
P9405	2300	21.3	3.5	1700	99.5	1013	26.8	42.8	783
P9406	1800[1]	16.8	8.0	2000	127	567	27.0	9.8	583
SRN 39	2200	114.0	55.5	1600	235	340	9.8	77.5	87
Weijita	3300	117.3	29.5	1600	98.5	420	8.3	122.0	60
Macia	2100	216.8	29.8	1100	149.5	846	6.3	60.0	283
S.E. (d.f. 17)	140	28.3	5.9	90	18.2	108	5.2	11.1	59

Striga number 7.5 m^{-2} at harvest and sorghum grain yield Kg ha^{-1}. (Ilonga is a Striga free site). [1]damaged by American boll-worm at grain-filling.

Table 2. Multi-location sorghum trials, 2001.

	Ilonga	Melela			Hombolo		Ukiriguru		
Entry	Kg ha^{-1}	*S. asiatica*	*S. forbesii*	Kg ha^{-1}	*S. asiatica*	Kg ha^{-1}	*S. asiatica*	*S. hermonthica*	Kg ha^{-1}
Pato	4400	243	75.3	1400	92.3	107	41.8	3.5	0
P9405	2000	2.3	3.3	1900	2.0	453	36.5	2.0	0
P9406	3100	3.5	4.3	2000	7.5	400	32.8	0.0	0
SRN 39	3600	111.3	107.5	900	23.8	180	43.3	2.5	0
Weijita	4300	218.5	61.5	1200	12.0	147	65.3	2.0	0
Macia	3200	263	83.3	1100	19.0	367	35.5	11.3	0
S.E. (d.f. 17)	240	31.6	11.8	120	8.1	38	3.8	1.3	0

Striga numbers 7.5 m^{-2} at harvest and sorghum grain yield Kg ha^{-1}. (Ilonga is a Striga free site). No grain was harvested at Ukiriguru due to sorghum midge infestation following late planting.

Table 3. Multi-location sorghum trials, 2002.

Entry	Ilonga	Melela			Hombolo		Ukiriguru	
	Kg ha^{-1}	*S. asiatica*	*S. forbesii*	Kg ha^{-1}	*S. asiatica*	Kg ha^{-1}	*S. hermonthica*	Kg ha^{-1}
Pato	3700	75.0	4.8	1500	105	900	9.0	823
P9405	3200	0.5	0	1900	25	2000	47.3	933
P9406	3200	0.5	0.3	1200	41	1600	14.0	943
SRN 39	3900	1.0	0.8	2500	169	800	48.0	890
Weijita	2700	67.3	40.8	900	141	700	19.5	953
Macia	4000	24.0	10.3	1300	71	1100	35.0	963
S.E. (d.f.17)	110	14.0	6.3	200	15	150	7.6	55

Striga number per 7.5 m^{-2} at harvest and sorghum grain yield Kg ha^{-1}. (Ilonga is a *Striga* free site).

Table 4. Sorghum performance – means for 5 on-farm trial sites in each in 4 villages in Dodoma District in 2001,2002.

Entry	Village/year									
	Mpalanga 01		Chipanga 01		Mvumi 01		Mvumi 02		Mwagala 02	
	Striga	Yield	*Striga*	Yield	*Striga*	Yield	*Striga*	Yield	*Striga*	Yield
Pato	61.4	0.6	4.0	1.5	7.70	0.34	31.07	0.8	7.70	0.34
P9505	2.9	1.4	0	1.6	2.27	0.60	5.75	2.0	2.27	0.60
P9406	20.3	1.1	0	1.0	1.09	0.78	0.82	1.4	1.09	0.78
SRN39	18.9	1.2	0.32	1.6	-	-	-	-	-	-
Macia	23.1	1.1	0.64	1.4	2.79	0.44	11.07	1.1	2.79	0.44
S.E. (d.f.16)	3.8	0.05	0.61	0.11	0.93	0.12	6.55	0.96	0.93	0.12

S. asiatica number m^{-2} and grain yield t ha^{-1}.

consistently ranked the shorter duration varieties more highly than the tall landraces and ranked P9405 first in both years. This preference according to men is based on earlier maturity, higher yields and greater drought tolerance than tall landraces.

The rankings demonstrate why a range of sorghum types continue to be grown. Macia is ranked highly on yield, early maturity, drought tolerance and ease of marketing. However, the variety is ranked relatively poorly according to post-harvest criteria such as the 'heaviness' and the taste of the ugali. The landrace Lugugu performs well according to almost all the post-harvest criteria, but very low against yield, maturity and drought tolerance. Interestingly though Lugugu appears to have a greater ability to recover if rains come after drought compared to modern varieties. P9405 performs well according to *Striga* tolerance, drought tolerance, yield and it makes good ugali. However, it performs less well in terms of the colour of the ugali and suitability for selling. P9406 ranked highly according to *Striga* and drought tolerance, as well as early maturity, but less well for home consumption.

Many farmers adopted Pato due to high yield and relatively early maturity compared to land races but found that under conditions of drought, *Striga* and foliar disease, its yield potential is not realised. In such situations P9405 and P9406 offer alternative options and these were therefore released for general cultivation as the cultivars Hakika ("certain") and Wahi ("early") in November 2002.

5. Development and Promotion of an Integrated Approach to *Striga* Management

The availability of Hakika and Wahi provided an opportunity to improve sorghum productivity and also a challenge to promote appropriate, affordable sorghum management practices for *Striga* infested fields. Soils in sorghum producing areas of Tanzania are generally low in fertility and the low levels of nitrogen observed on farmers fields favour the growth of *Striga*. With sufficient nitrogen, Pato attained the greatest stem biomass of the cultivars, and typically had the highest yields, but was heavily stunted when infected with *Striga* in both a laboratory and a field study.[9] Wahi and Hakika were not stunted to the same extent, with Hakika retaining the same degree of tolerance despite severe nitrogen limitation.

Table 5. Sorghum variety preference by farmer's criteria in Mvumi Makulu village Central Tanzania. (1 = best, 2 = worst)

Criteria	Tegemeo	Mhuputa	Sandala	Pato	Lugugu	P9406	P9405	Bangala	Lugugu Arusha
High yielding	4	8	5	1	9	2	3	7	6
Withstand drought	4	7	5	3	9	1	1	8	6
Withstand *Striga*	4	9	5	3	8	2	1	7	6
Short plants	3	7	5	4	9	2	1	8	6
Marketing	9	6	3	5	1	6	5	4	2
Birds Resistance	6	-	5	7	2	8	9	1	4
Withstands pests	6	3	5	9	1	7	8	3	4
Not shattering	4	2	5	3	8	2	1	7	6
Storage pest tolerance	9	9	6	5	1	7	8	4	3
Palatability	9	2	7	8	1	6	5	4	2
Total	58	56	51	48	49	43	42	53	45
Ranking	9	8	6	4	5	2	1	3	7

Yields of released sorghum lines can be enhanced when animal manure is available as has been demonstrated in on-farm trials in the Dodoma region (Table 6). Planting legumes provides an alternative approach to soil fertility enhancement. Groundnut intercrops provide farmers in central Tanzania with an opportunity to improve the nutrient content of household diets or cash income without significantly depressing sorghum yield (Table 7). A program for promotion was initiated in three districts after the official release of Hakika and Wahi, focusing on the use of integrated *Striga* management technologies (ISM) to increase sorghum yield in drought and *Striga* prone areas. The ISM

Table 6. Sorghum grain yield and *S. asiatica* emergence at harvest following use of manure, Mvumi Makulu, Central Tanzania. (Source: Pierce *et al.* 2003)

Cultivar	Manure (kg/plant)	*Striga* number (m^{-2})	Yield (Kg ha^{-1})
Pato	0	22	320
	0.5	36	440
P9405	0	38	1360
	0.5	13	1800
Macia	0	476	1000
	0.5	59	1400
Wahi	0	48	1240
	0.5	21	1400

Table 7. Grain yield of sorghum cultivars Pato and Hakika and *S. asiatica* number at harvest, Mvumi Makulu, 2000.

Treatment	*Striga* m^{-2}	Yield kg ha^{-1}
Pato	40	1200
Hakika + groundnuts	5	1560
Hakika	6	1600
Pato + groundnuts	16	1440
S.E.	9	252

Table 8. Performance of sorghum cultivars planted on ridged or non-ridged land with soil fertility treatments. Mean yield (t ha^{-1}) at 12 farm sites, Sepuka Singida, 2004.

	Villages	
Treatments	Msungua	Musimi
Sorghum landrace – no ridges	1.7	1.4
Cv. Hakika (P9405) - no ridges	1.8	1.4
Cv. Hakika + tied ridge + animal manure	2.7	2.0
Cv. Hakika + tied ridge + Urea	2.6	1.9
Mean	2.2	1.7

Table 9. Performance of sorghum cultivars planted on ridged or non-ridged land with soil fertility treatments mean yield (t ha^{-1}) at 12 farm sites, Sepuka-Singida, 2004.

	Villages	
Treatments	Malolo	Mpipiti
Sorghum landrace - no ridges	1.4	0.8
Cv. Wahi (P9406) - no ridges	1.4	1.3
Cv. Wahi + tied ridges	2.6	2.2
Cv. Wahi + tied ridges + Animal manure	2.9	2.9
Mean	1.7	1.4

Note: Farmers used a handful animal manure per hill.

technologies included *Striga* resistant sorghum varieties, tied ridges for moisture conservation and animal manure as compared to the traditional method of growing local sorghum varieties without ridges or manure. The performance of the demonstration plots gave relatively high sorghum yield (Table 8 and 9). The highest yields were produced from short duration sorghum cultivars Hakika/Wahi combined with tied ridges and animal manure (2-2.7 and 2.9t/ha grain respectively).

6. Linking Farmers to the Sorghum Market

A workshop was held to identify new markets for sorghum in the central zone after the introduction of Hakika, and Wahi with ISM technologies to farmers. This workshop was attended by representatives from farmer research groups, extension staff, district cooperative staff, stockiest, traders, processors and exporters. The commercial sector (brewers, food processors and exporters) currently require over 2,500 t sorghum grain per year, but most is imported from South Africa.

Processors and traders are prepared to use locally grown sorghum provided there is a reliable supply of quality grain in sufficient quantities at readily accessible market points. Previous attempts by commercial brewers to substitute local supplies for imported grain have been compromised by grain that is poorly dried and cleaned. Farmers need to be sure that buyers will visit local markets after harvest offering a stable and adequate grain price to ensure sorghum production is profitable before they will expand production.

A number of support activities have been planned to develop the market. These include formation of farmer production and marketing groups, establishment of village-based seed supply mechanisms, training of farmers in post-harvest practices to supply quality grain, dissemination of information on market demand and prices, increased purchase of sorghum by the national Strategic Grain Reserve, rehabilitation of village grain stores and establishment of sorghum marketing points. A task-force has been established to facilitate this process for Kongwa and Singida districts involving district extension staff, farmers and representatives of processors based in Dar es Salaam.

Acknowledgements

Evaluation and registration of Hakika and Wahi was funded by the UK Department for International Development (DFID) and the Government of Tanzania (GoT) with promotion funded by INTSORMIL USA and GoT. The views expressed here are not necessarily those of DFID (Project R7564 Crop Protection Programme) INTSORMIL and GoT.

References

1. WFP *World Hunger – Tanzania.* Web site for UN World Food Programme www.wfp.org/country_brief/ (accessed September, 2006).
2. T. E. Downing in *Managing the Earth: The Linacre Lecture,* Ed. J. C. Brining and T. E. Downing (Oxford University Press, Oxford, UK, 2002), pp 5-34.
3. J. Paavola, *Livelihoods, vulnerability and adaptation to climate change in Morogoro Region, Tanzania.* Working Paper EDM 04-12, (Centre for Social and Economic Research on the Global Environment, University of East Anglia: Norwich, UK, 2004).
4. FEWS *Tanzania Food security update.* Web site for Famine Early Warning System www.fews.net (accessed September 2006).
5. A. M. Mbwaga and A.T. Obilana, *Intl. J. Pest. Manage*, **39,** 449 (1993).
6. S. Reichmann, J. Kroschel, and J. Sauerborn, Distribution and infestation of *Striga* species in Shinyanga region of Tanzania and evaluation of control measures, *Proc. Brighton Crop Protection Conference – Weeds,* 151-156 (1995).
7. A. M. Mbwaga, C. R. Riches and R. Lamboll, The Striga problem in the Dodoma Region and the Lake Zone of Tanzania: Analysis of the problem and research priorities. (Natural Resources Institute, Chatham, UK, 1998).
8. ICRISAT *SA Trends*, Issue15, article 1 (Feb. 2002). http://www.icrisat.org/satrends/feb2002.htm (accessed January 2007).
9. S. Pierce, G. Ley, A. M. Mbwaga, R, I. Lamboll, C. R. Riches, M. C. Press, and J.D. Scholes, *Proc. BCPC International Congress on Crop Science and Technology*, 1017-1022 (2003).
10. D. D. Rohrbach, *Improving the commercial viability of sorghum and millet in Africa.* Web site of South African Grain Information Service www.sagis.org.sa (accessed September 2006).

CHAPTER 20

STRIGA ECONOMICS

Hugo De Groote*

International Maize and Wheat Improvement Centre (CIMMYT),
PO Box 1041-00621 Village Market, Nairobi, Kenya
E-mail: h.degroote@CGIAR.ORG

The economic importance of *Striga* as major pest problem in sub-Sahara Africa, as well as the economic analysis of control methods is poorly documented. Integration of social sciences into the *Striga* research agenda is needed to develop a clear conceptual framework on how *Striga* affects yields through a complex set of interactions. Essential functional relationships between *Striga*, yields, and the natural and socioeconomic environment need to be determined. The key parameters need to be estimated, either from previous research or by integration into on-going research, to compare benefits and costs of control methods over time. The proposed economic analysis consists of seven steps: estimating the extent and intensity of the problem, trials and appropriate economic analysis of new control methods, farmer evaluation of these methods, modeling of the interactions, and impact assessment. Examples are presented for each step from on-going research of *Striga* control measures in maize.

1. Introduction

Social scientists wonder if two decades of efforts in developing technologies for *Striga* control have been efficient. These include intercropping, rotation, weeding, transplanting, soil fertility enhancement, trap crops, and others.[1] Few of these technologies have been adopted by farmers, so social scientists wonder if they really fit the farming systems and are economical.

Few social scientists are involved in *Striga* research, and very little economic analysis is available, especially on-farm. Some information is embedded in agronomic papers, in conference proceedings, project reports and other grey literature, but very little has been published. The available information consists mostly of basic comparisons of benefits and costs, often overstating benefits by using market prices and ignoring transaction costs, and underestimating labor costs and opportunity costs for land. Analysis over time, including an appropriate discounting rate, is generally lacking. Thus, very little debate is being generated on the topic. The major issues concerning *Striga* economics, as well as the appropriate methods of analysis, are not known to most in the *Striga* research community.

This chapter outlines a coherent set of socioeconomic research activities, in seven steps, to accompany technical *Striga* research, all based on experience and literature review. The first step is to determine the extent and intensity of *Striga*. The extent can be measured qualitatively by discussing it with farmers during participatory rural appraisals (PRAs). Quantitatively, the *Striga* area can be determined by direct or indirect geo-referenced observations. The intensity of the problem (step 2) can be determined through direct measurement or by farmer estimation, of the percentage and amount of crop lost to *Striga*.

Once the seriousness of the *Striga* problem is identified and new *Striga* control measures are proposed, they need to be tested in a sequence of trials (step 3). Management by scientists should be gradually decreased, while increasing farmer involvement, ending with trials under full farmer conditions and management accompanied by economic analysis to compare the benefits to the costs (step 4), as well as by farmer evaluation to analyze their interest (step 5).

A dynamic conceptual framework needs to be developed, modeled and estimated to fully understand how a technology works, how it can be improved and what the full benefits and costs are over time (step 6). The model needs to include the pathways through which the technology is likely to work, and two major functions need to be estimated: the effect of control methods on *Striga*, and the effect of *Striga* on yield, including confounding factors. Finally, the impact of new technologies needs to be assessed by comparing the benefits vs. development and disseminating costs (step 7). This can be done ex ante (before the release) or ex post (afterwards, including adoption studies). Assessing the impact of new technologies on the poor is becoming increasingly important.

2. Estimating the Extent of the *Striga* Problem

Before conducting *Striga* research, it is important to determine the extent of the problem, using PRAs, direct observation, farmer surveys, or expert opinion. PRAs allow researchers to interact with farmers who can list the major constraints they face in food production during group discussions, ranked in order of importance, as well as a list of their major pest problems.[2] Other popular PRA techniques used in *Striga* research, include walking transects and village resource mapping.[3] Wealth ranking should be included for poverty analysis. Here, participants discuss the criteria they use to classify farmers as poor, and determine the cut-off points for different wealth classes. These criteria can than be used to select technologies useful to the poor.

PRAs were organized in all maize agroecological zones in Kenya for the Insect Resistant Maize for Africa (IRMA) project.[2] Where possible, group discussions were held separately with men and women. Participants were asked to list and rank the constraints they face in maize production, and the pest problems. Country-wide, stemborers and storage pests were considered the first pest problem in maize, except for the moist mid-altitudes zone, the area around Lake Victoria.[2] In this area, five villages were selected for the PRAs in three districts.[4] The major constraints were low soil fertility, cash, farm implements and *Striga*. *Striga* was consistently ranked first among pest problems in Butere-Mumias and Homa Bay districts (Table 1). In the third district, Busia, *Striga* is considered the second or third pest problem, after stemborers and weevils.

Once *Striga* is identified as a serious problem, the area at risk can be geo-referenced. Direct observations are often considered by to be more precise. However, given the high variability in occurence, and the limited time period that direct observations are possible, it is quite

Table 1. *Striga* ranks high as a major maize pest problem by farmers in participatory rural appraisals in five western Kenya villages.[4]

	Butere-Mumias	Homa Bay District		Busia District	
Pest Problem	Ebubala	Kayambo	Koyolo	Bulemia	Sirisia
Striga	1	1	1	3	2
Weevils	2	NM	NM	2	4
Stemborer	3	2	3	4	1
Termites	5	3	NM	1	5
Maize streak	NM	7	2	NM	NM
Head smut	6	9	4	5	3

NM: Not mentioned among the first 10 pest problems. Homa Bay data mixed men and women, otherwise data are from women farmers

expensive to repeatedly cover a wider area. It is therefore more efficient use farmers' observations of *Striga* incidence in their fields, obtained during farmer surveys.

In a western Kenya survey, 367 farmers were interviewed and their farm geo-referenced in 1993-1994, clearly revealing the *Striga*-prone area (Figure 1). Farmers with *Striga* in their fields are marked by a circle, those without *Striga* with a triangle. All farms between Lake Victoria (1,150 m) and the 1,500 m contour (thick line) faced *Striga* problems. Between the 1,500 m and 1,600 m, some farmers also had *Striga*, but there were none above the 1,600 m contour. The *Striga* zone largely overlaps with moist mid-altitude maize production zone (grey area) containing a population estimated at 5.8 million people (1.3 million households) on 16,000 km^2, with an average density of 359 people /km^2. Maize production data provided by ILRI provide a production estimate for 1994-1999 of almost 0.5 million tons of maize on 212,000 ha, or 14% of Kenya's average maize area during that period (1.5 million ha).

Expert opinion is a fast and cheap alternative to farmer surveys. Maize breeders, extension officers, and other knowledgeable people often have a good idea of *Striga* occurrence in a broader geographical area. They can easily identify and map administrative units with *Striga*. Maps of *Striga*-prone areas, using different techniques, are available for Nigeria,[5] Tanzania,[6,7] and Ethiopia (Chapter 15).

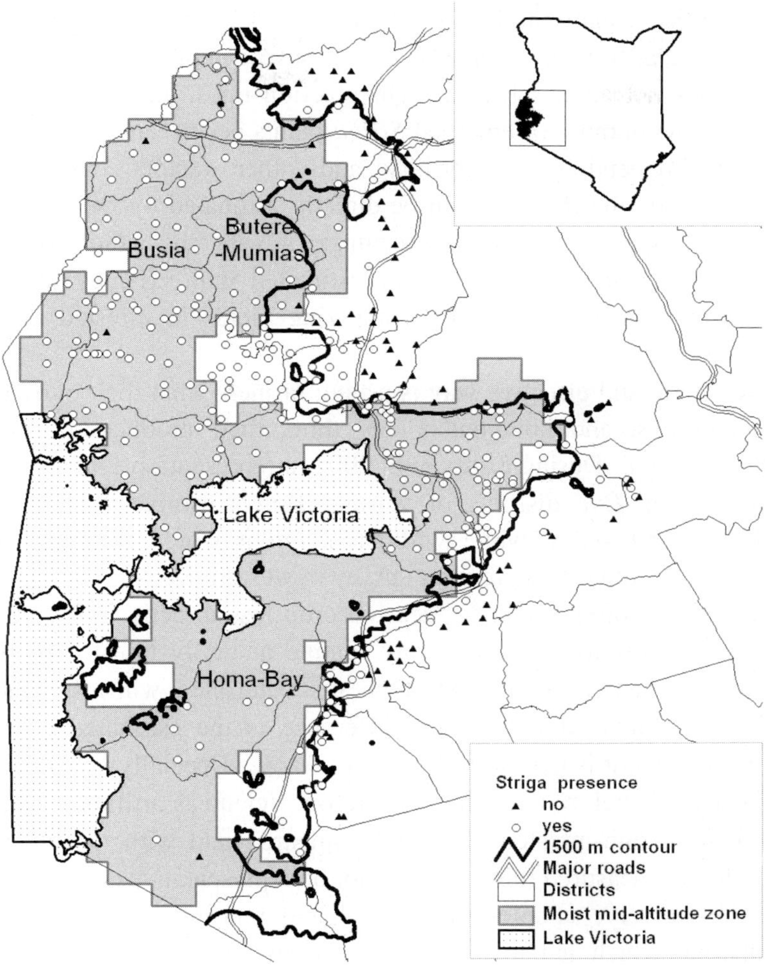

Figure 1. Location of Striga *area and the moist mid-altitude zone in western Kenya.*

3. Estimating the Intensity of the Problem

Once the extent of a pest problem is determined, the intensity of a pest problem needs to be assessed, usually by measuring infestation levels and their impact on yield, as convenient indicator of *Striga* infestation, especially related to yield, are rare. Two common indicators are *Striga* seed density (usually in the top 10-20 cm) and *Striga* emergence counted at a particular time after planting. Both have complications: a high seed bank does not always lead a high infection rates, and the relationship between the number of emerged *Striga* plants per m^2 and yield is highly variable, depending on fertility and other factors (Chapter 11). Alternatively, the intensity can be directly estimated by measuring crop loss, usually expressed as a percentage of what the yield would have been without *Striga*. This can be estimated indirectly, by farmers or expert opinion, or directly by comparing yields in infested and non-infested fields.

A simple and effective way is asking farmers what their current crop production is, and how much they think their production would be without *Striga*. During a survey in 2004, a representative sample of 111 farmers from five districts of western Kenya estimated their current yields at 550 kg/ha, while without *Striga* it would be 1,200 kg/ha, remarkably similar to a previous survey in western Kenya.[8]

Biologists often prefer to measure crop loss directly in the field. A standard procedure is to randomly sample naturally infested plots, and divide them in two. In one half the pest is controlled, while the other one is left untreated, and the yield difference is the estimated crop loss. Unfortunately, it is hard to fully control *Striga*, especially as it does most of its damage before emerging. Therefore, scientists artificially infest of one half of non-infested plots and compare yield with the other half. Crop losses measured this way may not be representative.

Alternatively, samples of infested and non-infested plants can be compared, which led to an estimated crop loss of 68% in West and Central Africa.[9] This assumes that there are no confounding factors influencing both yield and probability of being infested. A similar method is to take a sample of naturally infested plots and estimate the relationship between yield and infestation levels using regression analysis. Using this function, expected yields can be calculated at zero and at average infestation, the difference being an estimation of yield loss.[10] This method also ignores confounding factors that influence both yield and the level of infestation, such as soil fertility, although such factors could be incorporated in a larger model.

Biologists often use expert opinion, usually that of their colleagues, when unable to directly measure crop loss. Such an exercise was conducted with CIMMYT and BASF staff to estimate the proportion of *Striga* infested maize area in the major producing countries of SSA. Currently, SSA produces about 37.3 million tons of maize on 25.2 million ha (FAOSTAT, 2004 data), with an estimated value, at US$115/ton, of $4.94 billion. Based on the expert opinion of proportion infested, the maize area infested with *Striga* in SSA is estimated at 3.64 million ha infested (14% of total area), producing 5.4 million ton, valued at $610 million. Crop loss was estimated at between 30 and 50%, or between 2.3 and 5.4 million tons, with an estimated value of between $305 million and $622 million. The accuracy of such estimates depends, of course, on the reliability of expert opinion.

4. Testing *Striga* Control Methods in the Field

Pest control methods need to be systematically tested in a sequence of trials that typically start under full scientific control and management, and end with testing under full farmer conditions and management. Early trials typically take place on-station to establish if and how a technology works. Imadazolinone resistant (IR) maize was first tested in on-station trials in western Kenya, where it was demonstrated that the herbicide resistant gene could be incorporated in locally adapted germplasm, and that coating the seed with the herbicide showed good control of *Striga* (Chapter 11).[11]

These trials were followed by on-farm but researcher-managed trials., These trials were held for IR maize in 2002 on three farms, with three repetitions, both without fertilizer and with fertilizer. Without fertilizer, the yield increased from 1.0 to 3.7 tons/ha, more than tripling yield. Within the IR plots, however, there was no difference between the fertilized and unfertilized treatments (Figure 2). This was unexpected, as *Striga* is supposed to be linked to low soil fertility. Probably the soil fertility in these plots was high, but its effect was not realized due to prior *Striga* infestation.

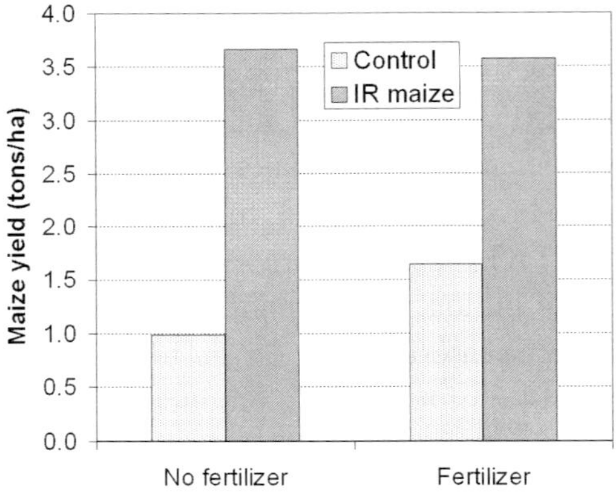

Figure 2. Yields of IR maize and the control, with and without fertilizer, during the 2002 trials in western Kenya.[12]

The sample size was very small and the high soil fertility levels not representative, so the trials were repeated in 2004, but now on a larger scale and under farmer management. Farmers were given a short explanation on the technology, 250 g of IR maize seed intended for about 25 m^2, and asked to plant it next to their own preferred variety for yield comparisons. When moving trials into farmers' conditions, scientists inevitably lose some control, and confounding factors and unexpected events complicate the interpretations of the results.

In this trial, for example, the farmers were visited by officials from an NGO, who told farmers to take particularly good care of the plots with the new varieties, out of respect for the scientists. As a result, many farmers applied more fertilizer in the IR plots than in the controls. Fortunately, the difference was captured in the input/output data sheets farmers were using. A second complication was that IR did not control *Striga* well in plots in one of the three districts, where the heavy rainfalls at the beginning of the season might have washed away the herbicide. Still, yields of IR maize were still significantly higher overall, doubling yields from 0.6 tons/ha to 1.3 tons/ha. Interpretation is, however, complicated by the fact that *Striga* counts were only reduced in 2 of the 3 districts, and fertilizer application was higher in many IR plots.

5. Economic Analysis of Control Methods in Trials

On-farm trials provide data for technical efficacy and, more importantly, for economic analysis. Partial budget and marginal analysis are used to calculate if the extra benefits of the technology, as compared to current farming practices, outweigh the extra costs.[13] A spreadsheet is created to compare the variable factors that are different in the new technology.

For example, the IR maize yielded 2.7 tons/ha more than the control (c omparing the plots without fertilizer), so at the local maize price of $202 tons/ha at that time, the extra revenue is valued at $741/ha. The extra cost of the technology is estimated at $4/ha, the cost of the herbicide, while all other factors are constant and therefore not included. The marginal rate of return (MRR) is defined as as the ratio of the extra benefit, in this case 737/ha (revenue minus costs), over the extra cost, $4/ha, resulting in 135:1. This means that for each extra dollar invested in the IR technology, the farmer receives an extra benefit of $135, suggesting that IR maize offers a very neat return to the investment under these conditions. The use of fertilizer in this trial was not very interesting. In the control, the fertilizer increases yield by 659 kg/ha, valued at $133, but at an extra cost $125/ha. The extra benefit is therefore on $4, and the MRR is only 4%, much less the recommended 100-150%.

6. Farmer Evaluation

Several methods are currently used in farmer evaluation of new technologies. Farmers are typically invited to visit trials to evaluate the new technologies, using ranking or scoring methods. The evaluation should start with a presentation and some group discussions in which farmers rank constraints and pest problems. This sets the stage and confirms the importance of *Striga* in the participants' farms. After an introduction explaining the objectives and the lay-out of the trials, farmers are invited to inspect and evaluate the different treatments in the trial. One popular method is to rank the treatments, but analysis of ranking across sites and trials is problematic. Therefore, the use of scoring, also called rating, has become more popular.

For scoring, farmers are first asked, in the group discussion or in a preceding PRA, to list the criteria they use in selecting new maize technologies.[2] Each treatment is then evaluated by the most important criteria, on a scale of 1 (very poor) to 5 (very good), and also given an overall evaluation. The appropriate method of analysis is ordinal regression as such scores are not continuous variables.[14] Using the score as dependent variable, and binary variables for the different treatments as independent variables, ordinal regression coefficients can be calculated with standard software packages such as SPSS. These estimated coefficients represent the logarithm of the odds ratio, the ratio of the probability that a farmer prefers this treatment to the control, over the probability that farmers prefer the control over the treatment.

In the *Striga*, stemborer and soil fertility management project in western Kenya, push-pull,[15] IR maize, and soybean rotation were compared in on-farm trials. In 2004, 263 farmers evaluated the trials at four sites, using scores, on *Striga* resistance (among other criteria).[16] Analysis using ordinal regression showed significant coefficients for all treatments, indicating that farmers prefer them, for this criterion, over the control, monocrop of local maize. The combination push-pull with IR maize scored best, with push-pull with local varieties coming in second. IR is clearly preferred to local maize in the push-pull and in the monocropping, but not in the soybean rotation.

The best way to predictively evaluate new technologies is clearly to let farmers test it out, in their own fields under their own particular conditions. To reach a sufficient number of representative farmers over a larger area is expensive and hard to manage. The 2004 IR trials took place on 60 farms. During the mid-season evaluation, IR maize scored significantly better on all criteria using the same techniques.[12] Fewer farmers were available at harvest and IR did not control *Striga* well in one district. As a result, IR only scored significantly better for maturity period and disease resistance, but not for yield.[12]

At least 50 farmers are needed to provide significant results in a scoring exercise. The number can be increased by inviting neighboring farmers to evaluate new technologies at on-farm trials. This requires substantial involvement of the scientist to assure that evaluations are based on timely group discussions, and that farmers understand the trial and the evaluation methodology. Collaboration with farmer groups, extension officers, extension officers and NGOs are proposed.

A simpler and cheaper alternative, the contingent valuation method, can be considered if a technology can be explained in a straightforward manner; farmers can be asked if they would be willing to purchase the technology. In western Kenya 123 farmers were first given a short presentation on IR maize technology, and then asked if they would be interested in buying IR maize seed at current seed prices ($1.67/kg). Almost all were interested, and would, on average, like to purchase 4 kg. Subsequently, they were asked how much of the seed they would be willing to purchase at different prices levels and the responses indicated a strong reduction of demand with increasing prices (Figure 3). At a seed rate of 25 kg/ha, IR seed for one hectare would cost about $40 at current prices, for an increase in crop loss of 0.5 tons, valued at $100. If the IR seed price would be double the current price, there would not be much incentive to purchase, especially given the risk of yield and price variation farmers face.

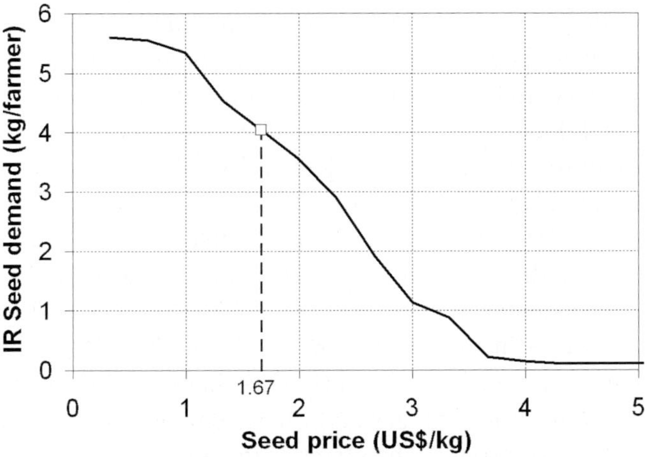

Figure 3. Willingness to pay for IR maize seed (farmer survey in western Kenya, 2004).

7. Modeling and Econometrics

Conventional economic analysis of agricultural technologies is based on the estimation of the production function, which measures the effect of inputs on outputs. This relationship is usually positive but decreasing, so the optimal input level is where the marginal benefit equals the marginal cost. This assumption does not always hold for pest control technologies, as yields do not necessarily increase with the level of control. A yield benefit is only realized when there is an infestation, and the input only works through a reduction of crop loss. Confounding factors can influence both yield and infestation to complicate matters.

An appropriate conceptual framework should be developed to help focus the empirical work. Based on theory and available information, the model should be based on two equations. First, the effect of the pest problem on production has to be assessed, with all the confounding factors. Secondly, the effect of the control measure on the pest should be re-evaluated with the factors that influence infestation. As the outcome of this season influences the outcome of next season, both equations need to be dynamic. In the final analysis, the properly discounted net present value of benefits and costs must be compared.

The conceptual framework then needs to be translated in an empirical model where the functional forms are determined and the parameters estimated. Such a model is lacking for *Striga*, although many separate relationships have been estimated. The effects of seed bank, fertilizer, soil fertility, weeding, and other factors on both yield and infestation have been well documented. It is now time to gradually include them in economic analyses, while developing the appropriate models.

Fertilizer and different control levels were clearly confounding factors in the 2004 IR trials. A regression analysis was therefore performed, using the yield difference between IR maize and the control (kg/ha) as dependent variables, and the difference in *Striga* infestation and fertilizer use as independent variables (Table 2). The intercept, 370 kg/ha, can be interpreted as the difference between IR maize and the control, controlling for fertilizer and *Striga* emergence, or the effect of the superior IR germplasm. The effect of the herbicide can be calculated indirectly: each extra *Striga* plant per m^2 reduces yield by 49 kg/ha, and as the IR maize reduced *Striga* by an average of 4 plants/m^2, the herbicide effect is an increase in maize yield of about 200 kg/ha.

Table 2. Yield enhancement with the IR technology (results of regression analysis, with yield difference between IR and control as dependent variable, and differences in N application and *striga* plants as independent variables).[12]

Variables (differences)	Coefficients	Std. Error	P
Intercept	370	253	0.152
Nitrogen application (kg N/ha)	7	4	0.113
Striga emergence (plants/ m^2)	-49	15	0.002
R^2	0.27		
N	40		

8. Impact Assessment

The economic impact of an agricultural technology can be estimated before it is disseminated (ex ante) or afterwards (ex post). Ex ante impact is often estimated by multiplying the average yield increase by the crop area on which it is adopted and the average output price. This ignores, however, the price depressing effect of production increases. Most technical developments in agriculture have a higher impact through a price reduction, benefiting the consumers, than through revenue increases for farmers. Therefore, impact is better assessed through the economic surplus (ES) method, which combines both producer and consumer benefits.[17] The latter is particularly important since many small-scale farmers are net-consumers of maize. ES can be conveniently calculated using the Dynamic Research Evaluation for Management (DREAM) software, developed by the International Food Policy Research Institute (IFPRI). The total benefits thus calculated are compared to the development and dissemination costs. Applications to pest problems include impact assessment of Bt maize against stemborers.[18]

The economic surplus method does not provide any quantitative analysis of the benefits to the poor, be it producers or consumers. Poverty analysis in agricultural research is still at the early stages, although several methods have become available. Livelihood analysis provides a framework to improve on the PRAs and include factors such as poverty and vulnerability. Poverty mapping helps to assess if the problem under study is relevant to the poor. By overlaying the *Striga* map with the poverty map of Kenya, for example, 61% of people in the area were found to live below the poverty line. Wealth ranking and classification criteria from the PRAs can also be used to analyze farmers' preferences by different wealth categories.[16]

Impact assessment in the larger sense should also include evaluating externalities, or the impact of an activity on other people. Negative externalities are commonly assessed in pesticide use. Positive externalities of *Striga* control, for example, would include the effect of controlling *Striga* in one farmer's field on reducing the likelihood of its spreading to a neighbor's field.

An ex post impact assessment usually starts with adoption studies. They assess the level of adoption, estimating the number of adopting farmers and the crop area under the new technology.[19] These parameters, with a farmers' estimate of increased yield, can be incorporated in the ES model. This method is frequently applied to different pest control strategies, water hyacinth for example,[20] but not yet on *Striga* control measures. Adoption studies allow an analysis of the factors influencing adoption levels, such as extension and credit, and then provide recommendations for policies to improve adoption.[21] Analyzing adoption by wealth category also provides insights on the impact on the poor.

9. Constraints and Integration

The major constraints to the economic analysis of *Striga* control methods is the lack of integration of social sciences in *Striga* research and lack of communication between the two groups of scientists. Social scientists have not been very active in *Striga* research. Most research institutes involved in *Striga* research have few resources, so it is hard for them to take the initiative. Social scientists could do a better job explaining their methods to the biologists, who mostly are not very knowledgeable on social science methods and economic analysis. Biologists mostly drive the *Striga* research agenda, and should include social scientists to understand what should and could be done, and budget the necessary resources.

These teams should then develop a clear conceptual framework to determine the key functional relationships between *Striga*, yields, and the natural and socioeconomic environment in a dynamic framework, and estimate the key parameters. The seven steps laid out here provide a guideline: estimating extent and intensity, trials and appropriate economic analysis, farmer evaluation, modeling and impact assessment. Institutes should review these steps and see how they can be integrated in their *Striga* research with adaptations based on specific experience and research results. *Striga* scientists should do a better job in publishing their economic analyses, build a body of scientific knowledge, provide empirical evidence of the problem and possible solutions, and improve the methodology for economic analysis.

Acknowledgments

The author thanks the Rockefeller Foundation, the Syngenta Foundation for Sustainable Development, and the UK Department for International Development (DFID) for funding the economic research on *Striga* and stemborers on which this paper is based. Its content was strongly influenced by elaborate discussions with colleagues from the bio-physical sciences, in particular in particular Fred Kanampiu, Dennis Friesen, Alpha Diallo, Bernard Van Lauwe, Jonathan Gressel, Zia Khan, John Pickett, Lester Wadhams and Tom van Mourik, which is greatly appreciated. Special thanks go to research assistants Lucy Wangare and Esther Rutto for their help with field work and literature review, and to Hugo Frost and Dana Berner for sharing their data and information.

References

1. A. Oswald. *Crop Prot.* **24**, 333 (2005).
2. H. De Groote, J. O. Okuro, C. Bett, L. Mose, M. Odendo and E. Wekesa, in *Participatory Plant Breeding and Participatory Plant Genetic Resource Enhancement: An Africa-wide Exchange of Experiences. Proceedings of a workshop held in Bouake, Ivory Coast. May 7-10, 2001*, Es. L. Sperling, J. Lancon and M. Loosvelt (CGIAR Systemwide Program on Participatory Research and Gender Analysis, Cali, Colombia, 2004), p. 148.
3. J. B. Werner, *Participatory development of agricultural innovations: Procedures and methods of on-farm research*, (GTZ, Eschborn, Germany, 1993).

4. M. Odendo, H. De Groote and O. M. Odongo, in *Proceedings of the 5th African Crop Science Conference (Lagos, Nigeria October 21-26, 2001)*, (African Crop Science Association, Kampala, 2001), p. 769.
5. I. Y. Dugje, A. Y. Kamara and L. O. Omoigui. *Agriculture, Ecosystems and Environment* **116**, 251 (2006).
6. A. M. Mbwaga, in *SADC/ICRISAT regional sorghum and pearl millet workshop,* Ed. K. Leuschner and C. S. Manthe (ICRISAT, Pantancheru, 1996), p. 195.
7. A. M. Mbwaga and A. T. Obilana. *Intl. J. Pest. Manag.* **39**, 449 (1993).
8. R. Hassan, J. K. Ransom and J. Ojiem, in *Maize Research for Stress Environments, Proceedings of the Fourth Eastern and Southern Africa Regional Maize Conference, held at Harare, Zimbabwe, 28 March-1 April, 1994*, Eds. D. C. Jewell, S. R. Waddington, J. K. Ransom and K. V. Pixley (CIMMYT, Mexico D.F., 1994), p. 250.
9. S. K. Kim, V. O. Adetimirin, C. The and R. Dossou. *Int. J. Pest. Manag.* **48**, 211 (2002).
10. P. T. Walker, in *Crop loss assessment and pest management*, Ed. P. S. Teng (The American Phytopathological Society Press, St. Paul, Minnesota, 1991), p. 133.
11. F. K. Kanampiu, V. Kabambe, C. Massawe, L. Jasi, D. Friesen, J. K. Ransom and J. Gressel. *Crop Prot.* **22**, 697 (2003).
12. H. De Groote, L. Wangare and F. Kanampiu. *Crop Prot.* (forthcoming).
13. CIMMYT, *From Agronomic Data to Farmer Recommendations: An Economics Training Manual.* (CIMMYT, Mexico, D. F., 1988).
14. R. Coe, in *Quantitative analysis of data from participatory methods in plant breeding,* Ed. M. R. Bellon and J. Reeves (CIMMYT, Mexico, DF, 2002), p. 46.
15. Z. R. Khan, J. A. Pickett, L. J. Wadhams, A. Hassanali and C. A. O. Midega. *Crop Prot*. 25, 989 (2006).
16. E. Rutto, H. De Groote, B. Vanlauwe, F. Kanampiu, G. Odhiambo and Z. Khan, *Participatory Evaluation of Integrated Pest and Soil Fertility Management Options Using Ordered Categorical Data Analysis.* Paper Presented at the IAAE Conference, Gold Coast, Australia (2006).
17. J. M. Alston, G. W. Norton and P. G. Pardey, *Science Under Scarcity: Principles and Practices for Agricultural Research Evaluation and Priority Setting* (CABI, Wallingford, U.K., 1998).
18. H. De Groote, W. Overholt, J. O. Ouma and S. Mugo. Paper presented at the IAAE Conference, Durban, South Africa (2003).
19. CIMMYT Economics Program, *The Adoption of Agricultural Technology: A Guide for Survey Design* (CIMMYT, Mexico, D.F., 1993).
20. H. De Groote, O. Ajuonu, S. Attignon, R. Djessou and P. Neuenschwander. *Ecol. Econ*. 45, 105-117 (2003).
21. C. Doss, W. Mwangi, H. Verkuijl and H. De Groote, *Adoption of Maize and Wheat technologies in East Africa: Synthesis of the Findings of 22 Case Studies,* (CIMMYT, Mexico, D.F., 2003).

Section VI

Biocontrol: Untapped Potential?

CHAPTER 21

BIOCONTROL USING *FUSARIUM OXYSPORUM*; A CRITICAL COMPONENT OF INTEGRATED *STRIGA* MANAGEMENT

Fen D. Beed[a,*], Steven G. Hallett[b], Julien Venne[c] and Alan K. Watson[c]

[a]*International Institute of Tropical Agriculture, Box 7878, Kampala, Uganda*
[b]*Department of Botany & Plant Pathology, Purdue University, 915 West State Street, West Lafayette, IN 47907, USA*
[c]*Department of Plant Science, McGill University, 21111 Lakeshore Road, Ste-Anne-de-Bellevue, QC H9X 3V9 Canada*
[*]*E-mail: f.beed@iita-uganda.org*

Striga can be controlled in a safe, environmentally friendly and cost effective manner, using soilborne microorganisms that can be readily grown, stored, formulated, and deployed. Several groups have come to strikingly similar conclusions with *Fusarium oxysporum* f. sp. *striga* selected as the optimal candidate. The potential and constraints of *F. oxysporum* based bioherbicides for the control of *Striga* are discussed in detail. Clearly, a bioherbicide will only be adopted if field efficacy is proven to farmers / policy makers, and will only provide significant value if integrated with other techniques for control of *Striga*. Here we discuss preliminary results from multi institutional bioherbicide field trials in West Africa using *Striga* susceptible and resistant varieties of sorghum and maize. *Striga* parasitism and the efficacy and persistence of bioherbicides are affected by the dynamic biotic and abiotic environment of the rhizosphere. Rhizosphere studies can be used to identify agronomic practices that are synergistic with biocontrol agents across the range of biophysical environments used for cereal production. A set of experiments is reported demonstrating the dynamic nature of the rhizosphere environment, showing that nutrient input and crop variety selection influence the microbiology of the cereal-*Striga* rhizosphere.

1. Introduction

Controlling *Striga* early in the season, at the initiation of parasitism, is essential for effective Integrated *Striga* Management (ISM). Control must be effected in the rhizosphere; the zone of soil under the influence of the plant root, which is generally nutrient enriched, and harbours a microbial flora different from the rest of the soil. The rhizosphere contains signalling molecules, such as those that induce *Striga* germination, and may support naturally occurring pathogens, biocontrol agents and their antagonists.[1] The biology of the cereal rhizosphere during parasitism by *Striga* is poorly understood although a diverse array of microbes can suppress *Striga.*[2-10] Furthermore, pasteurization suppresses the ability of certain soils to suppress *Striga* .[11] The link between soil degradation and *Striga* intensification may be explained, in part, by a loss of soil microbial diversity or reduced inoculum levels or activity of key microbes in degraded soils. The incorporation of organic matter into impoverished soils can reduce the impact of *Striga*, presumably due to enhanced microbial diversity and biomass as mediated by increases in nitrogen, aeration and water retention.[12,13] The SIPWEMA (Sustainable Integrated Parasitic Weed Management in Cereal Legume Production Systems in Africa; 2003) working group has therefore identified biocontrol of *Striga* with soilborne microbes as its first cross cutting issue.

Biological control of *Striga hermonthica*, has gained considerable attention in recent years as a viable supplement to other control methods within an integrated approach. Intensive surveys on the occurrence of micro-organisms pathogenic to *S. hermonthica* were conducted in Sudan,[2] Burkina Faso, Mali and Niger,[3,6] Ghana,[9,14] and Nigeria.[10] *Fusarium* species have received the most attention for the biological control of *Striga hermonthica* ever since the first isolation and pathogenicity testing of *F. equiseti*, from diseased *Striga* in 1977.[15] Biocontrol researchers have increasingly focused their attention on isolates of *Fusarium oxysporum* that have been independently isolated from diseased *Striga* by different groups working in several countries of the African Sahel and Savanna.[2-10] The isolates attacked all growth stages of *Striga*, including seeds, germlings, seedlings, and flowering shoots, thus affecting the target prior to the onset of yield loss in addition to reducing the soil seedbank. *Fusarium oxysporum* has the potential to be combined with other *Striga*-suppressive microbes such as *Pseudomonas* spp.[4,16] Conversely, some soilborne microbes, also including *Pseudomonas* spp., could be antagonistic to *F. oxysporum* based bioherbicides.[17-22] This chapter reviews the history and potential of a *Fusarium oxysporum* f. sp. *striga* based bioherbicide. Furthermore, it highlights studies required to facilitate the development of a bioherbicide suitable for implementation as a critical component of integrated *Striga* management.

2. Disease Surveys

Twenty-eight fungal and two bacterial pathogens of *S. hermonthica* were isolated in a 1989 survey in Sudan.[2] A *Fusarium nygamai* isolate reduced *Striga* emergence by up to 100% in greenhouse trials when incorporated into the soil before sowing and killed emerged *Striga* following spore applications. The potential use of *Fusarium nygamai* as a bioherbicide is, however, seriously compromised because of fumonisin B_1 production. This powerful phytotoxin against *Striga*, is also a carcinogenic mycotoxin.[23-24]

Over two hundred fungi were collected from over a hundred diseased *S. hermonthica* shoots in Burkina Faso, Mali, and Niger in 1991,[6] One isolate of *F. oxysporum* (M12-4A) from Mali, grown on sorghum straw and incorporated into potted soil, prevented all emergence of *S. hermonthica* and resulted in a four fold increase in sorghum yield.

Another 13 fungal species were isolated from diseased *S. hermonthica* in northern Ghana.[9,14] The pathogenicity of 12 isolates, including *Fusarium equiseti*, *F. equiseti* var. *bullatum*, *F. solani* and *F. oxysporum*, were evaluated against *S. hermonthica* under controlled environmental conditions. All isolates were pathogenic, but one *F. oxysporum* (Foxy 2) was highly virulent, reducing the emergence of *S. hermonthica* by 98% and increasing sorghum yield by 26%.[9]

Other surveys in Burkina Faso, Mali, and Niger,[3,25-26] recovered *Fusarium* from more than 90% of diseased *Striga* samples. When *Fusarium* isolates were applied pre-sowing at a rate of 5 g kg^{-1} of soil infested with *S. hermonthica* in pots, emergence of *Striga* was reduced and shoot biomass and grain yields of millet and sorghum were increased. One *F. oxysporum* isolate from Burkina Faso, 4-3-B, was particularly virulent.[25-26]

Virulence tests for *S. hermonthica* of several *Fusarium* spp. from a fourth survey, in the Nigerian Savanna, led to the selection of another *F. oxysporum* isolate (PSM-197).[10] When this isolate was grown on sorghum grain and incorporated into soil, emergence of *S. hermonthica* was completely inhibited.

These surveys encompassed a substantial portion of Africa infested by *S. hermonthica* and showed the abundance of *F. oxysporum* recovered from *Striga* plants with vascular wilt symptoms. Research therefore focused on the selection of isolates of *F. oxysporum* (namely M12-4A, Foxy-2, 4-3-B, and PSM-197) having the greatest biocontrol potential to suppress *S. hermonthica.*

3. Host Specificity and Safety

Some *Fusarium* spp., such as *Fusarium nygamai* mentioned above, are notorious for their broad host ranges and production of mycotoxins, which negate their potential use as biocontrol agents. However, under all conditions tested, isolate M12-4A of *F. oxysporum* did not produce mycotoxins, and hence it does not constitute a known health hazard to humans or livestock.[27] The host specificity of most *Fusarium* spp. are at the genus or species level, leading to the *formae specialis* taxonomic classification.[28-29] The *F. oxysporum* f. sp. *striga* isolates tested (M12-4A, Foxy-2, 4-3-B, and PSM-197) were host specific to *Striga* spp.[6,9,10,30-31] and non pathogenic on sorghum, pearl millet, maize, rice, fonio, cotton, groundnut, cowpea, or okra. The genetic variability of *Striga* spp. is high, and this may impact the efficacy of the *F. oxysporum*-bioherbicide. Isolate Foxy 2 controlled both *S. hermonthica* and *S. asiatica* in greenhouse trials,[8] but the efficacy of *F. oxysporum* f. sp. *striga* isolates requires evaluation against multiple *Striga* populations from different hosts across varied environments. Similarly, *S. hermonthica*, an out-crossing species, demonstrates intra and inter population variation with differing degrees of virulence on host plants,[35] which is likely to impact on *F. oxysporum*-bioherbicide efficacy.

Generally, *Fusarium* spp. isolates differ in their vegetative compatibility grouping (VCG) pattern in accordance with their host range.[29] There is a high degree of genetic similarity among the various isolates of *F. oxysporum* from *S. hermonthica*.[32] All *F. oxysporum* f. sp. *striga* collected from Kenya, Niger, and Mali are in one VCG. Random amplified polymorphic DNA (RAPD) assays have identified markers restricted to a set of *F. oxysporum* strains isolated from *Striga*. Two SCAR primers (FUN001 and FUN002) amplified a single band of 157 bp in all isolates tested from *Striga*, which included M12-4A from Mali and Foxy 2 from Ghana. Thus, PCR-based assays confirm the VCG results, indicating *F. oxysporum* isolates from *Striga* are genetically similar suggesting co-existence of *F. oxysporum* f. sp. *striga* with its host across the Sahel and the Savanna. Conversely, an isolate of *F. oxysporum* from Burkina Faso was not amplified by the SCAR primers and isolates from Benin and Burkina Faso were in a different VCG group than M12-4A (J. Venne, unpublished), although the presence of a few VCG groups in a single *forma specialis* is coherent with data gathered from other *F. oxysporum* subspecies.[29,32]

Some countries may not approve the deployment of a bioherbicidal fungus that was isolated from outside their national boundaries. We note, however, that VCG analyses and preliminary molecular analyses indicate commonality among isolates of *F. oxysporum* from *Striga*.[32-36] That *F. oxysporum* is host specific and has no known sexual stage, supports its potential as an extremely safe *Striga* control option. Similarly, studies with a saprophytic isolate of *F. oxysporum* under development for the control of pathogenic isolates of the same species also identified no risks.[29,37] Further research is required, but the findings to date are promising, suggesting that a *F. oxysporum* f. sp. *striga* bioherbicide may have utility across the entire region.

4. *Fusarium*-Based Bioherbicide for *Striga* Control in the Field

Isolate M12-4A applied in chopped or ground sorghum straw resulted in a 60% reduction of emerged *Striga* at 82 days after sowing and doubled sorghum biomass compared to the control.[38] *S. hermonthica* emergence was completely suppressed by applications of a chlamydospore powder in the planting hole, or as a seed coating in the field, (Table 1).[39] Only 80 g of the chlamydospore powder were needed to treat 1 ha.

Table 1. Effect of *Fusarium oxysporum* M12-4A on *Striga hermonthica* emergence in the field.

Inoculum treatments per seed planting hole	Striga plants / plot[a]		
Control (no straw incorporated)	32.1	(17.3)	a
Sterilized straw control (10g)	16.8	(6.2)	ab
Sterilized ground straw control (2.6g)	21.3	(12.5)	ab
Ground colonized straw inoculum (2.6g)	7.9	(4.5)	b
Chlamydospore powder (0.5g)	6.9	(4.9)	b
Chlamydospore powder (0.5g) + sterilized straw (10 g)	3.6	(1.9)	b
Chlamydospore powder (1.0g)	2.7	(1.7)	b
Chlamydospore powder (1.0g) + sterilized straw (10 g)	2.5	(1.4)	b

[a]Mean number of *S. hermonthica* in plots; Values in parentheses are standard errors; Values having the same letter are not significantly different at = 0.05 according to the Student-Neuman-Keuls multiple range test. Source: Ciotola *et al.*[39]

Isolate PSM 197 was applied in combination with a *Striga*-resistant sorghum cv.Samsorg 40 and a *Striga* tolerant landrace Yar'rurukain in on-farm trials in savanna of Nigeria.[36,40] Five g of *Fusarium*-colonized grains were added in each planting hole, equivalent to 167 kg/ha of biocontrol product. *Striga* counts were significantly reduced by around 95%, crop stands were significantly higher and sorghum yields were 50% higher in plots where the bioherbicide was applied and the resistant variety was grown. The bioherbicide alone increased the yields of the *Striga* tolerant landrace cultivar by 20-40%. These results illustrate the potential of an integrated management strategy that incorporates host plant resistance and biological control using *F. oxysporum* as an effective means of *Striga* control.

Field trials were initiated in 2006 to compare *F. oxysporum* isolates and application methods using experimental sites in Benin, Burkina Faso, Nigeria and Togo in a collaboration between scientists at IITA, INRAB, INERA, ITRA, Ahmadu Bello University, McGill University, and the University of Hohenheim. Uniform inoculum of chlamydospores of each isolate was produced on aerated water containing 5% sterile sorghum glumes. Preliminary analyses of the results indicated significant differences in the efficacy control of *Striga* emergence, between isolates tested and untreated controls. However, there was no difference in control efficacy among all the isolates of *F. oxysporum* f.sp. *striga* tested: PSM197; Ahmadu Bello, Foxy 2; Hohenheim, M12-4A; McGill; and the most virulent isolates recovered from Bénin; IITA and Burkina Faso; INERA). Therefore, the data for all isolates were combined to show mean treatment effects for these isolates of *F. oxysporum* (Figs. 1 and 2).

Two types of formulation were prepared: 1) "Pesta", a granular formulation containing semolina, kaolin and sucrose inoculated with aqueous chlamydospores suspension of *F. oxysporum*, kneaded and extruded into a sheet through a hand operated pasta maker, air dried and blended into 1- 2 mm granules,[30,36,40-1] and 2) crop seeds coated with chlamydospores in gum arabic.[39] Three thousand *Striga* seeds mixed in soil were applied to each planting hole. Fields were managed following normal farmer practice, except that crops were thinned to a single plant per hole. A series of both locally favoured and recently bred varieties of maize and sorghum that are either susceptible or resistant to *Striga* were cultivated.

Despite our attempts to find *Striga*-free sites, control treatments showed a low background level of *Striga* infestation, but infestation was much lower than in the experimentally-infested plots. Emergence of *Striga* ten weeks after sowing was greater for maize than sorghum for the varieties tested (Fig. 1). There was greater *Striga* emergence on susceptible vs. resistant maize varieties, but emergence was similar on the resistant and susceptible sorghum varieties (Fig. 1). Both bioherbicide formulation controls, with granules and seed-coating (prepared without *F. oxysporum* inoculum), had high levels of emergence and flowering of *Striga*, possibly through the supply of exogenous nutrients to the parasite (Figs. 1 and 2). Bioherbicide treatments of both formulations provided statistically significant levels of *Striga* control, with improved control being provided by granules compared to seed coating. The degree of *Striga* suppression in maize provided by susceptible cultivars following Pesta applications of *F. oxysporum* was similar to that provided by resistant germplasm (Fig. 1). The granular formulation provided significantly better control for sorghum in combination with resistant germplasm than did resistant germplasm alone (Fig. 1). The number of flowering *Striga* ten weeks after sowing was similarly reduced by the cultivation of resistant varieties and also by applications of both forms of bioherbicides for maize (Fig. 2).

5. Bioherbicide Production and Delivery to Target

Various substrates have been used to produce *Fusarium* inoculum for greenhouse and field trials, including cereal grains, sorghum straw and liquid fermentation, and the fungus has been deployed as a seed coating and "Pesta" granules.[42] An inoculation production system was developed for *F. oxysporum* f. sp. *striga* M12-4A utilizing a liquid suspension of finely ground sorghum straw as the substrate, fashioned on a cottage industry model.[39] The material was dried, ground and stored at room temperature. Gum arabic was used to stick the bioherbicide powder, predominately chlamydospores (1×10^7 g $^{-1}$) onto sorghum seeds prior to planting. The concept was tested through training village-level producers of *Striga* bioherbicide in four villages in Mali.[43] The production strategy was constrained by contamination of preparation utensils, when this was tested on farm.

The commercial sales of a *F. oxysporum* f. sp. *striga* bioherbicide on treated seed is needed to fully ascertain the value of the technology.[40] Production of the bioherbicide could be carried out regionally by local entrepreneurs or farmer cooperatives with scientific capacity and facilities, but quality control would best be attained through production of dry powder inoculum for seed coating, or a "Pesta" formulation at one or more central facilities with shipment to other locations. For example, many commercial seed companies exist in Nigeria producing seeds of cowpea, soybean, groundnut, maize, sorghum, rice, millet and vegetables and they also have facilities for seed treatment. The national annual market requirement in Nigeria for improved sorghum seeds is about 110 000 t, 100 000 t for millet, 96 000 t for maize and 280 000 t for rice.[40,44]

Further development of improved and certified seed production in *Striga* infested regions would not only improve crop production but would certainly aid the delivery of *F. oxysporum* f. sp. *striga* bioherbicides. Presently in sub Saharan Africa, 75% or more of the farmers are subsistence farmers, without access to improved crop seed.

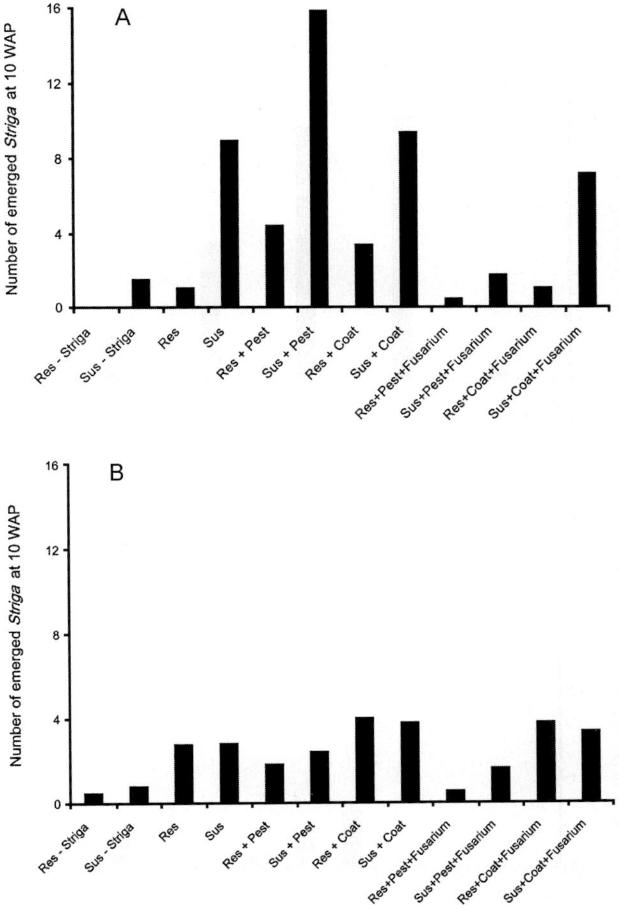

Figure 1. Fusarium oxysporum *reduced* Striga *emergence to a greater extent than resistant varieties of A. maize and B. sorghum, 10 weeks after planting. Res =* Striga *resistant varieties and Sus = susceptible. –* Striga *= without* Striga *seed inoculation, all other treatments received 3000 seeds mixed in sand per planting hole. Pest = Pesta application and Coat = seed coating using gum arabic. Fusarium = combined results for PSM197, Foxy 2, M12-4A, isolates from Benin and Burkina Faso.*

Innovative means have to be developed to bring the *Fusarium* bioherbicide strategy to the majority of farmers while waiting for certified seed systems to be established, unless they can be developed simultaneously.

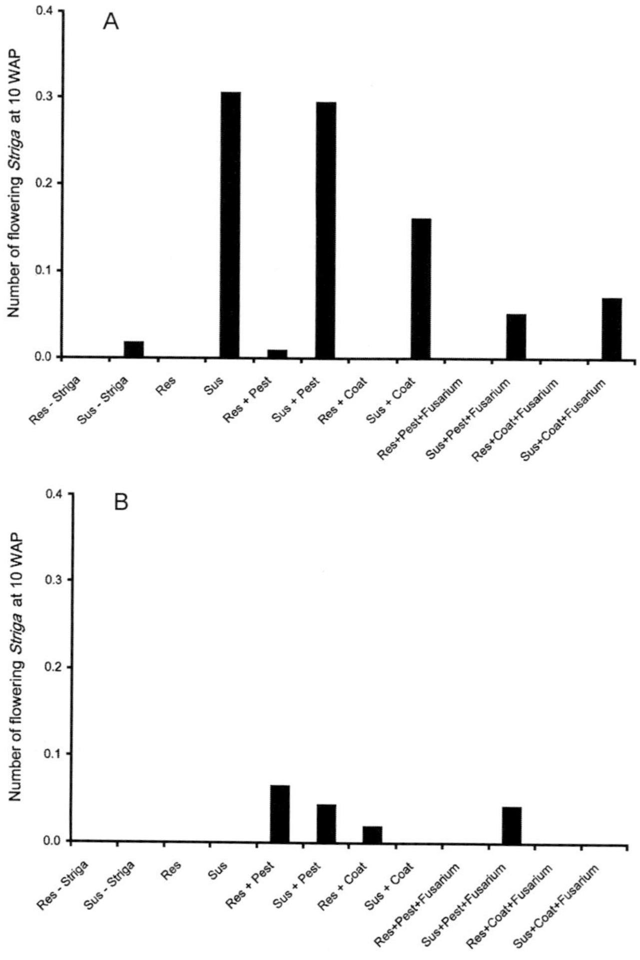

Figure 2. Number of flowering Striga per plant for A. maize and B sorghum, 10 weeks after planting. Res = Striga resistant varieties and Sus = susceptible. – Striga = without Striga seed inoculation, all other treatments received 3000 seeds mixed in sand per planting hole. Pest = Pesta application and Coat = seed coating using gum arabic. Fusarium = combined results for PSM197, Foxy 2, M12-4A, isolates from Benin and Burkina Faso.

6. Understanding the Microbiology of the Cereal-*Striga* Interface

Soil microbial communities can have different impacts upon the parasitism of cereals by *Striga* and the modification of microbial communities may ameliorate or exacerbate parasitism. One way of modifying soil microbial communities may be to plant different crop species or different varieties of a given crop. In order to test this hypothesis, the effect of different varieties of sorghum on the composition of microbial communities of the rhizosphere was examined. Four varieties of sorghum were grown in the same soil for 45 days while some soil was maintained without sorghum as a control. Two of the varieties have some resistance to *Striga* (Framida and P9401) and two are fully susceptible (Shanqui Red and P954603). After the 45 d, plants were removed, and the fungal and bacterial communities in the remaining soil analyzed by PCR-DGGE.[45-48] Briefly, DNA was extracted from soils, Bacterial 16S rDNA and fungal ITS rDNA fragments were amplified by PCR,[49-50] and PCR products separated on DGGE gels. Gels were stained with SYBR Green and photographed over an UV transilluminator. Banding patterns were analyzed by Shannon-Weaver diversity index and Principal Component Analysis to ascertain similarity and diversity indices as demonstrated by their degree of clustering.

Sorghum influenced the composition of rhizosphere microbial communities in variety-specific ways (Figs. 3 and 4). The diversity of fungi was generally decreased following the growth of sorghum, whereas the diversity of bacterial communities was generally increased (Table 2). We hypothesize that sorghum exudates are more toxic to fungi than to bacteria.

Fertilizer applications may also affect the composition of microbial communities of the rhizosphere. To test this hypothesis, two varieties of maize (one susceptible: 5057, one tolerant: 9450 STRS) were grown in containers either infested or not infested with *Striga*, and fertilized at three different levels. Fungal and bacterial communities in the root zone of the maize were sampled from each container four weeks after planting and analyzed by PCR-DGGE as described above.

Distinct microbial communities were found in soils fertilized at different levels and planted with different varieties of maize, demonstrating that the microbiology of soils can be influenced by multiple manipulations in cereal cropping systems (Fig. 5).

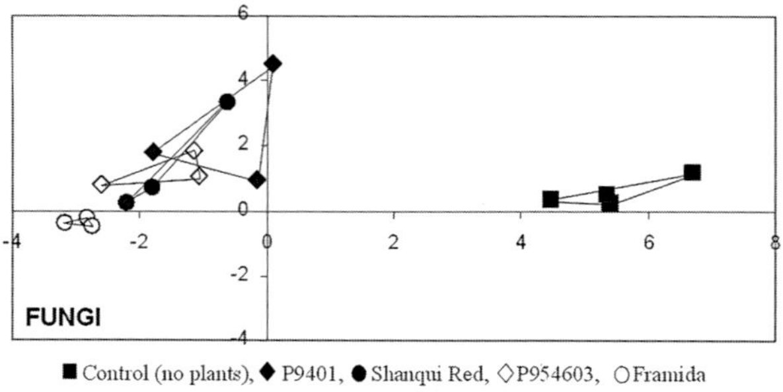

Figure 3. Growth of different sorghum varieties in the same soil for 45 d alters the composition of soil fungal communities. Principal component analysis of fungal ribosomal ITS DNA collected from the rhizospheres of sorghum. Values on the axes indicate the percentage of the total variation explained by the axes. The distance between symbols is a measure of the dissimilarity in the composition of the microbial communities represented.

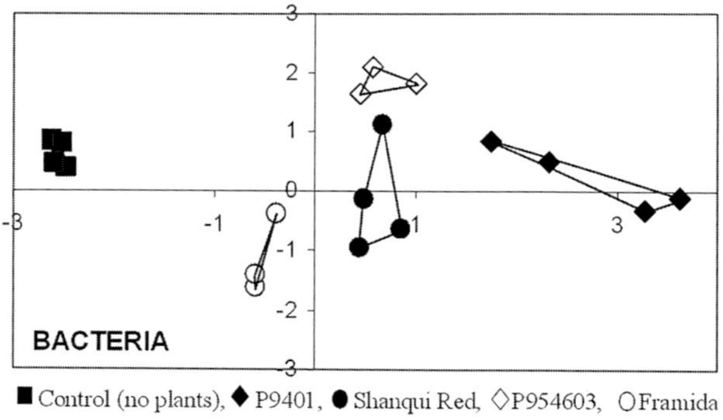

Figure 4. Growth of different sorghum varieties in the same soil for 45 d alters the composition of soil bacterial communities. Principal component analysis of the bacterial ribosomal 16S rDNA collected from the rhizospheres of sorghum. Values on the axes indicate the percentage of the total variation.

These findings illustrate the dynamic nature of the microbial communities in the cereal rhizosphere. Simple techniques available to farmers in *Striga* afflicted regions, including variety selection and fertilization, can affect the composition of microbial communities. Further research is required to determine if microbial communities can be modified by these techniques in a way that will decrease the incidence or impacts of *Striga* parasitism.

Table 2. Growth of different sorghum varieties in the same soil for 45 d alters the diversity of soil fungal and bacterial communities.

Variety	Fungal Diversity	Bacterial Diversity
Untrained (Control)	3.2 (0.05)	2.4 (0.03)
Framida (Resistant)	2.2 (0.03)	2.5 (0.20)
P9401 (Resistant)	2.9 (0.15)	2.6 (0.09)
Shanqui Red (Susceptible)	2.5 (0.12)	2.9 (0.07)
P954603 (Susceptible)	2.6 (0.11)	2.8 (0.01)

Shannon-Weaver indices of fungal and bacterial diversity calculated from banding patterns on DGGE gels in soils trained by different varieties of sorghum, and an untrained control (standard error of the means from four replicates).

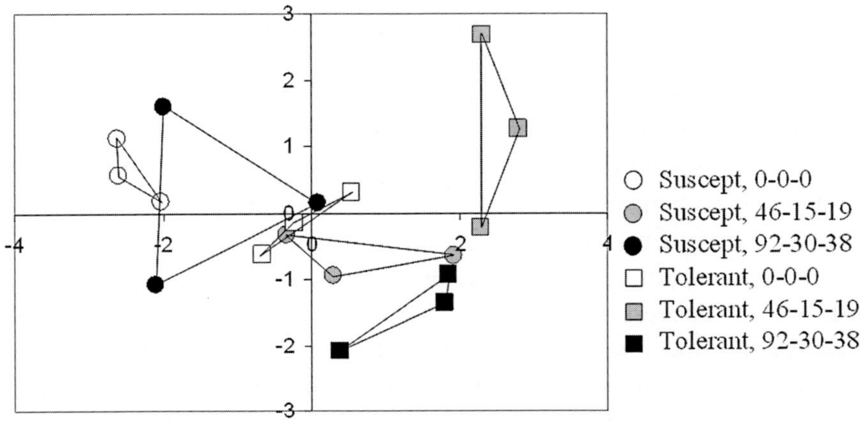

Figure 5. Growth of two maize varieties in the field with different levels of fertilization alters the composition of soil fungal communities. Principal component analysis of the fungal ribosomal ITS region collected from the rhizospheres of maize. Values on the axes indicate the percentage of the total variation explained by the axes. The distance between symbols is a measure of the dissimilarity in the composition of the microbial communities represented.

7. Conclusions and Outlook

The preliminary results of the field trials with *F. oxysporum* f. sp. *striga* are encouraging, with significant activity observed under a range of different conditions. Generally, the granular formulation (Pesta) was more efficient than seed-coating applications. We speculate that seed coating with *F. oxysporum* was less effective on sorghum than on maize due to the reduced inoculum dose delivered by the smaller surface area of the sorghum seed. Pesta granules may have the added advantage of increased shelf life, though they may be more problematic to distribute to farmers, and to apply.[30,32] The seed coating methodology provided significantly lower efficacy but could be used in combination with resistant germplasm or with acetolactate synthase (ALS) resistant maize seed (Chapter 11).[51] An imazapyr (acetolactate synthase) resistant *F. oxysporum* f. sp. *striga* mutant, jointly applied as a seed dressing with imazapyr to ALS resistant maize seeds, could effectively reduce the amount of herbicide use and extend the duration of protection from *Striga*. Seed coating is a technology that could be readily delivered to farmers on cereal seed,[39,52] although the general lack of efficient seed companies in sub- Saharan Africa is a barrier, currently forcing most farmers to use farm-saved or locally available seed.

Rhizosphere colonisation and persistence are essential characteristics of a soilborne biological control agent.[53-54] Variation in control efficacy across environments could be caused by a range of abiotic and biotic factors. Competitive saprophytes and antagonists may reduce control efficacy or, conversely, *F. oxysporum* f. sp. *striga* may protect cereals from pathogenic fungi. Extensive field testing under a wide range of environmental conditions is now required to determine the range of conditions under which efficacy can be expected, and to identify potential limiting factors. The finding that the microbiology of the cereal rhizosphere can be manipulated by nutrient amendments and variety selection, has implications for the management of *Striga,* as these interventions are among the few that are available to African subsistence farmers. The soil microbial communities are crop variety-specific, which might indicate that some of the resistance of varieties may be linked to indirect effects mediated by microbes in the rhizosphere. Possibilities may thus exist to manage *Striga* by planting varieties that promote the growth of *Striga*-suppressive soil microbial species.

More research is needed to understand the impacts of these interactions and the influence of other components of ISM e.g. trap crops,[55-57] and variance in the *Striga* seedbank.[58-62] Such research might include the identification of microbes in the cereal-*Striga* rhizosphere by sequencing,[63-65] and the tracking of bioherbicide penetration in the rhizosphere and the soil by fluorescent tagging (if the isolate was transformed to deploy a green fluorescent protein). The manipulation of soil microbial communities may be of considerable value when integrated with other control interventions, in particular biological control with soilborne fungi such as *Fusarium* species.

It is unlikely that a single control option will be effective across the wide diversity of biophysical and socioeconomic environments in which *Striga* is a problem,[12,66] but manipulating the rhizosphere so that it becomes favourable to the proliferation and persistence of a biocontrol agent such as *F. oxysporum* is a realistic aim.[40,44] This requires an understanding of the agronomic and ISM practices that may favour conducive rhizosphere conditions, and this in turn can only be achieved through an improved understanding of the rhizosphere. There needs to be a concerted effort to distinguish between the impact of annual applications of *F. oxysporum* f.sp. *striga* compared to producing a stable inoculum level in the soil and for the two concepts to be harmonised to provide large scale control.

Despite some remaining knowledge gaps, the outlook for the integration of effective biocontrol agents into ISM systems is promising. Effective control of *Striga* has been demonstrated in the field using a safe, environmentally benign organism that can be readily grown, stored, formulated and deployed. Biocontrol of *Striga* is therefore a methodology that is perfectly suited to the socioeconomic needs and realities of African cereal farmers. The opportune time is now for more research on *Striga* biocontrol, coupled with vigorous implementation of the various other components of ISM as they become available, through farmers' participatory techniques. The search for a biocontrol of *Striga* has a leading candidate in *F. oxysporum*. Challenges and opportunities remain. Mass production and delivery strategies need to be optimized and controlled for consistent quality and effect. Could the pathogenic *Fusarium oxysporum* Fo47[67] be jointly applied with the bioherbicide to replace the fungicide in seed treatment? Diversity amongst the different *F. oxysporum*-bioherbicide isolates needs to be examined relative to *S. hermonthica* diversity in the regions. These challenges can be met and farmers will have another management component in their struggle against *Striga*.

Acknowledgments

This work was supported by the International Development Research Centre, Ottawa, Canada and by the Consultative Group for International Agricultural Research (CGIAR)-Canada Linkage Fund in collaboration with the International Institute for Tropical Agriculture (IITA) in Cotonou, Benin. Thanks to Adolphe Avocanh, Yao Kolombia, Djibril Yonli, Afiniki Zarafi, Chabi Yallou, Abuelgasim Elzein, Flavien Zinsou, and Andre Aivodji for production of inoculum, management of field trials, data collection and amino acid experimentation. Thanks to Jean-Sébastien Boulanger for critical assistance with Figures and Peter Neuenschwander for editorial comment. Thanks also to Kathy Anderson and Karen Mitchell at Purdue University for technical assistance.

References

1. R. P. Larkin and D. R. Fravel, *Plant Dis.* **82**, 1022- (1998).
2. A. A. Abbasher and J. Sauerborn, *Biol. Control* **2**, 291 (1992).
3. A. A. Abbasher, D. E. Hess, and J. Sauerborn, *Afri. Crop Sci. J.* **6**, 179 (1998).
4. M. O. Ahonsi, D. K. Berner, A. M. Emechebe and S. T. O. Lagoke, *Biol. Control* **24**, 143 (2002).
5. O. Ahonsi, D. K. Berner, A. M. Emechebe, A. M. Sanginga and S. T. O. Lagoke, *Afr. Crop Sci. J.* **10**, 145 (2002).
6. M. Ciotola, S. G. Hallett and A. K. Watson, *Weed Res.* **35**, 303 (1995).
7. M. Ciotola, A. DiTommaso and A. K. Watson, *Biocontrol Sci. Techn.* **10**, 129 (2000).
8. E. M Elzein and J. Kroschel, *Weed Res.* 44, 433 (2003).
9. J. Kroschel, A. Hundt, A. Abbasher and J. Sauerborn, *Weed Res.* **36**, 515 (1996).
10. P. S. Marley, S. M. Ahmed, J.A.Y. Shebayan and S. T. O. Lagoke, *Biocontrol Sci. Techn.* **9**, 159 (1999).
11. D. K. Berner, R. J. Carsky, K. E Dashiell, J. G. Kling and V. M. Manyong, *Outlook Agri.* **25**, 157 (1996).

12. L. Franke, J. Ellis-Jones, G. Tarawali, S. Schultz, M. A. Hussaini, I. Kureh, R. White, D. Chikoye, B. Douthwaite, B. D. Oyewole and A. S. Olanrewaju, *Crop Prot.* **25**, 868 (2006).
13. E. Gacheru and R. M. Rao, *Intl. J. Pest Manag.* **47**, 233 (2001).
14. A. A. Abbasher, J. Kroschel and J. Sauerborn. *Biocontrol Sci. Techn.* **5**, 157 (1995).
15. N. Zummo, *Plant Dis. Reptr.* **61**, 428 (1977).
16. D. K. Berner, N. W. Schaad and B. Völksch, *Biol Control* **15**, 274 (1999).
17. L. M. Miché, L. Boillant, R. Rohr, G. Sallé and R. Bally, *Eur. J. Plant Pathol.* **106**, 347 (2000).
18. O. Becker and R. J. Cook, *Phytopathology* **78**, 778 (1988).
19. J. Deacon, *Modern Mycology* (Blackwell Publishing Ltd, Oxford, UK. 303 pp. 1997).
20. J. Leong, Ann. Rev. Phytopathol. **24**, 187 (1986).
21. H. S. Lim and S. D. Kim, *J. Micro. Biotech.* **7**, 13 (1997).
22. J. B. Neilands and S. A. Leong, *Ann. Rev. Plant Physiol. Plant Mol. Biol.* **37**, 187 (1986).
23. P. G. Thiel, W. F. Marasas, E. W. Sydenham, G. S. Shephard, W. C. Gelderblom, and J. J. Nieuwenhuis, *Appl Environ Microbiol.* **57**, 1089 (1991).
24. J. Kroschel and A. Elzein, *Biocontrol Sci. Tech.* **14**, 117 (2004).
25. D. Yonli, H. Traore, D. E. Hess, A. A. Abbasher and I. J. Boussim, *Biocontrol Sci. Tech.* **14**, 417 (2004).
26. D. Yonli, H. Traoré, D. E, Hess, P. Sankara and P. Sérémé, *Weed Res.* **46**, 73 (2006).
27. M. E. Savard, J. D. Miller, M. Ciotola and A. K. Watson, *Biocontrol Sci. Tech.* **7**, 61 (1997).
28. P. E. Nelson, T. A. Toussoum, and R. J. Cook (Eds.), *Fusarium: disease, biology and taxonomy.* (Pennsylvanian State University Press. University Park, PA, 1981).
29. J. F Leslie,. and B. A. Summerell, *Fusarium Laboratory Manual. First Edition,* (Blackwell Publishing, Ames, Iowa, United States. 388 pages, 2006).
30. A. Elzein, *In: Tropical Agriculture – Advances in Crop Research 12(2)* Ed. J. Kroschel, (Margraf Verlag, Weikersheim, Germany, 2003).
31. A. Elzein and J. Kroschel, *Weed Res.* **44**, 433 (2004).
32. A. Watson, J. Gressel, D. Sands, S. Hallett, M. Vurro, and F. Beed, *In: Novel biotechnologies for biocontrol agent enhancement and management.* Springer-Verlag, Dordrecht (p. 213, 2007).
33. M. Ciotola, S. Jabaji-Hare, C. Leclerc-Potvin, G. Taylor and A.K. Watson, *7th Int. Parasitic Weed Symp.* 5-8 June 2001, Nantes, France (2001).
34. A. Elzein and J. Kroschel, *J Plant Dis. Prot.* **Special Issue XX**, 875 (2006).
35. A. Elzein and J. Kroschel, *J Plant Dis. Prot* **Special Issue XX** 889 (2006).
36. P. S. Marley and J. A. Y. Shebayan, *BioControl* **50**: 389 (2005).
37. M. L. Gullino, Q. Migheli and M. Mezzalana, *Plant Dis.* **79**, 1139 (1995).
38. C. Diarra, M. Ciotola, S. G. Hallett., D. E. Hess and A. K. Watson. *Nuis. Pests Prag.* **4**, 257 (1996).
39. M. Ciotola, A. DiTommaso and A. K. Watson, *Biocontrol Sci. Techn* **10**, 129 (2000).

40. P. S. Marley, D. A. Aba, J. A. Y. Shebayan, R. Musa and A. Sanni, *Weed Res.* **44**, 157 (2004).
41. B. I. G. Haussmann, *Proc. 8th International Parasitic Weeds Symposium*, (Durban, South Africa, June 24-25, p. 16, 2004).
42. A. Elzein, J. Kroschel and D. Muller-Stover, *Biol. Control* **30**, 203 (2004).
43. C. Bastian, M.Sc. Thesis, McGill University (2002).
44. B. Schaub, P. Marley, A. Elzein and J. Kroschel, *J. Plant Dis. d Prot.* **Special Issue XX**, 691 (2006).
45. I. M. Head, J. R. Saunders and R. W. Pickup, *Microbial Ecol.* **35**, (1998).
46. M. H. Nicolaisen, and N. B. Ramsing,. *J. Microbiol. Meth.* **50**, 189 (2002).
47. G. O'Donnell, and H. E. Gorres, *Curr. Opinion in Biotechnol.* **10**, 225 (1999).
48. J. F. Salles, F. A. de Souza and J. D. van Elsas, *Appl. Environ. Microbiol.* **68**, 1595 (2002).
49. S. Muyzer, A. Hottentrager, A. Teske and C. Wawer, *Mol. Microb. Ecol.* **1**, 23. (1996).
50. E. J. Vaino and J. Hantula, *Mycol. Res.* **104**, 927 (2000).
51. F. F. Kanampiu, J. K. Ransom and J. Gressel, *Crop Prot.* **20**, 885 (2001).
52. D. K. Berner, J. G. Kling and B. B. Singh, *Plant Dis.* **79**, 652 (1995).
53. P. S. Ojiambo and H. Scherm, *Phytopathology* **96**, 1168 (2006).
54. J. Whipps, *J. Exp. Bot.* **52**, 487 (2001).
55. R. Carsky, L. Singh and R. Ndikawa, *Exp. Agr.* **30**, 349 (1994).
56. R. Carsky, D. K. Berner, K. Dashiell, B. Oyewole and S. Schulz, *Int. J. Pest Manag.* **46**, 115 (2000).
57. C. Parker, *Crop Prot.* **10**, 6 (1991).
58. A. Van Ast and L. Bastiaans, *Weed Res.* **46**, 264 (2006).
59. J. G. Geth, M. E. Smith, S. E. Mitchell and S Kresovich, *Weed Res.* **45**, 64 (2005).
60. G. Gbèhounou, A. H. Pieterse and J. A. C. Verkleij, *Weed Sci.* **51**, 940 (2003).
61. R. Matusoua, T. A. Van Mourik and H. J. Bowmeester, *Seed Sci. Res.* **14,** 335 (2004).
62. T. A. Van Mourik, T. J. Stomph and A. J. Murdoch, *J. Appl. Ecol.* **42**, 299 (2005).
63. T. I. Arie, I. Kaneko, T. Yoshida, M. Noguchi, Y. Nomura and I. Yamaguchi, *Mol. Plant-Micro. Interact.* **13**, 1330 (2000).
64. S. Teunissen, M. Rep, P. M. Houterman, B. J. C. Cornelissen and M. A. Haring, *Mol. Gen. and Genomics*, **269**, 215 (2003).
65. S.-H. Yun, T. Arie, I. Kaneko, O. C. Yoder and B. G. Turgeon, *Fungal Gen. and Biol.* **31**, 7 (2000).
66. D. Chikoye, J. Ellis-Jones, G. Tarawali and A. F. Lum, *Proceedings of the second Striga management stakeholders' conference,* (Zaria, Kaduna State, Nigeria. ISBN 978-131-250-5, 114 p, 2004).
67. C. Olivain and C. Alabouvette, *New Phytol.* **137,** 481 (1997).

CHAPTER 22

GENETICALLY ENHANCING THE VIRULENCE AND EFFICACY OF PLANT PATHOGENS FOR BIOLOGICAL CONTROL OF PARASITIC PLANTS

David C. Sands[*], Alice L. Pilgeram, Brian M. Thompson and Matthew M. Kirkpatrick

Plant Sciences and Plant Pathology, Montana State University, Bozeman, MT, 59717 USA

[*]*E-mail: dsands@montana.edu*

A considerable number of plant pathogens have been studied for their possible use in weed control. Some have proven virulent enough to control weed species and to compete commercially with chemical herbicides. However, most pathogens of weeds are not useful in their wild form because they are not sufficiently host-specific and/or are not sufficiently virulent. We believe that these barriers are not inescapable and may be overcome. Pathogens can be selected for greater virulence. Our research has focused on the inhibitory effects of certain amino acids on the growth and development of specific plants. Pathogens that overproduce these selected amino acids can have increased virulence on the target weed and enhanced field performance. We report enhancement of virulence in three separate pathogen-host systems, two with *Fusarium* and one with *Pseudomonas*. In this report we will suggest that the same technology might be used to enhance the virulence and efficacy of the pathogens that attack parasitic plants.

1. Plant Disease Epidemics

Biocontrol researchers have exerted tremendous efforts to find naturally-occurring pathogens capable of controlling noxious weeds.[1] There are pathogens that will attack weeds; however, there are very few pathogens that suppress weed expansion, much less actually eradicate an established weed population. In pathogen-host interactions, virulence is expensive and eradication of the host is suicidal. Therefore, parasitism becomes the more beneficial interaction for the pathogen, ensuring longer-term survival. As a result, severe disease epidemics are rarely observed in native plant or dispersed weed populations. Epidemics are more frequently observed in monocultures that lack genetic diversity and distance between susceptible plants. Even small changes in the fitness or susceptibility of a plant or changes in the virulence of a pathogen can drastically alter the severity of a plant epidemic. Multigenic changes in crop plant resistance can occur rather rapidly due to breeding and more recently genetic engineering. In contrast, changes in pathogen populations must result from random mutations or from natural genetic recombination. Plant disease, regardless of severity, may be contained simply by distance between susceptible hosts. Early in a weed infestation, plants are dispersed. However, when weed infestations progress to a density approaching monoculture, they tend to provide a uniformly susceptible or resistant host population dependent on its competitiveness in a new environment and the inability of pathogens and insect pests to match its rapid expansion.

In the special case of diseases of parasitic plants, several additional layers of complexity need to be explored. Firstly, parasitic plants are physically attached to a crop plant and suppression must be specific to the parasite. Secondly, both *Orobanche ramosa* and *Striga hermonthica* build up substantial long-term seed banks in the soil and to be effective the pathogen would have to be very effective in reducing the seed population.

Parasitic weed seeds usually become vulnerable to pathogen attack only after they break dormancy in response to the specific host root exudates described in Chapter 4. Thirdly, in order to protect the crop plant from serious damage, a biocontrol agent must kill the parasitic plant in the soil before it penetrates the root of the host crop or in an early stage of parasitism. Thus the pathogen itself must be soil-borne and/or colonize the rhizosphere of the crop plant. We will focus on controlling these devastating weeds by selecting pathogens with increased virulence combined with appropriate dissemination and crop management.

2. Amino Acid Inhibition of Plants, an Exploitable Weakness

In our research, we have found that every weed, in fact every plant so far examined, is inhibited by at least one amino acid. This observation leads to the conclusion that weeds have a weakness that can be readily exploited. The fitness of a weed, and therefore its resistance to plant pathogens and other plant pests can be reduced by direct application of inhibitory amino acids. Furthermore, our studies have found that the virulence and efficacy of bioherbicides can be greatly enhanced by selecting weed pathogens that overproduce and excrete specific amino acids that are inhibitory to a target plant.[2] At least in this case of a valine excreting *Fusarium oxysporum* against *Cannabis sativa,* the host range of the enhanced pathogen remained. Plants within a population that are tolerant of the amino acid imbalance have not been observed. Thus, amino acid supplementation is a viable strategy for the development of biocontrol agents for suppression of parasitic weeds.

3. Enhancement of Bioherbicides

3.1. *Criteria for Selection of Biocontrol Agents*

Classical biocontrol has proven successful in a few situations including biocontrol of rush skeletonweed with *Puccinia chondrilla*[3] in Australia and *Acacia saligna* by the rust fungus *Uromycladium tepperianum*[4] in South Africa. These successes utilized obligate pathogens that are highly host-specific, highly virulent, and capable of naturally spreading from a focal inoculation point. The overall success of these biocontrol agents has been tempered by the genetic diversity of the target weed. For example, *P. chondrilla* attacks the 'broad-leaved' form of rush skeletonweed.[5] The 'narrow-' and 'intermediate-leaved' forms of rush skeletonweed were not susceptible to the pathogen.

Highly virulent, host-specific pathogens are few and far between. However, as virulence of a pathogen can be increased, we can greatly expand our pool of potential biocontrol agents. Our selection foci include host specificity, nature of dissemination, and ease of production. There are a number of genera of plant pathogens where there are forma speciales or pathovars that display narrow host specificity including fungi (*Fusarium oxysporum*, several species of *Phomopsis* and *Colletotrichum*, and the rust fungi) and bacteria (*Ralstonia*, *Pseudomonas syringae* and *Xanthomonas*). *Fusarium oxysporum* is at the top of our list for control of both *Striga* and *Orobanche*.

Fusarium oxysporum is host-specific, perhaps even to a fault. It is a well characterized fungus from its morphology to its DNA. It is soil-borne and disseminated by water. It has good knockdown, is easy to culture on minimal media, is prototrophic for all amino acids, makes three kinds of spores including long-lived chlamydospores. Most importantly, it saprophytically colonizes the rhizosphere of numerous non-host species and establishes high populations in the soil. It has a spontaneous mutation frequency that permits selection of mutants without need for mutagenic agents. The literature on the fungus is extensive because it encompasses over 200 forma speciales. Each forma speciales is specific for a particular host species of plant or a group of closely related host species. Forma speciales specific for *Striga* spp.[6-12] and *Orobanche* spp.[13-17] have been characterized. Also, there are non-pathogenic forms that have saprophytic rhizosphere competence, and these may be useful in biological control.

3.2. *Selection of Biocontrol Agents that Excrete Target Amino Acids*

The virulence and efficacy of bioherbicides are enhanced by selection of variants of the pathogens that overproduce and excrete amino acids that are inhibitory to the target plant.[2,18] Our approach is modeled after "frenching disease", a naturally occurring disease of tobacco.[19] Steinberg *et al.*[20] discovered that saprophytic bacteria growing on the roots of symptomatic plants overproduced a single amino acid, isoleucine. Isoleucine is synthesized in plants via the branched chain amino acid pathway. The end products of the pathway (valine, leucine, and isoleucine) allosterically regulate the activity of acetolactate synthase (ALS). ALS is differentially inhibited by these amino acids in different plant species. In "frenching disease", overproduction of isoleucine by the saprophytic bacteria inhibited the activity of ALS in the tobacco, shutting down synthesis of valine and leucine, which in turn disrupted

Table 1. Valine excretion and virulence of wild type and valine overproducing variants of *F. oxysporum* f. sp. *cannabis*.[2]

Strain	Description	Valine Excretion[b] (mg/l)	Mortality Rate[c]	%Kill
C95	Wild-type	0-0.18	6-8 weeks	25
4nv	Norvaline resistant[a]	2.84	2-3 weeks	70
6pa	Penicillamine resistant[a]	2.48	2-3 weeks	90
8pa	Penicillamine resistant[a]	9.93	2 weeks	90

[a] Spontaneous mutant strains were selected for their resistance to successively higher levels of valine analogs. Strain 4nv is resistant to norvaline and strains 6pa and 8pa are resistant to penicillamine.

[b] Valine excretion was bioassayed by spectrophotometric analysis of growth of *Pediococcus cerevisiae* ATCC 8042 in culture supernatant.

[c] Mortality rate is the duration between inoculation and the first appearance of severe disease symptoms or death (greenhouse studies).

essential protein synthesis. Interestingly, several modern chemical herbicides mimic this strategy by inhibiting single biosynthetic enzymes in plants, rendering treated plants incapable of producing a metabolite essential for plant growth.[21]

Cannabis sativa, an illicit crop and a noxious weed, is inhibited by the amino acid valine. We isolated variants of *F. oxysporum* f. sp. *cannabis* that were resistant to valine analogs.[22] When analyzed, these variants excreted 10-55 times more valine than their wild type parent (Table 1). The valine-excreting strains of *F. oxysporum* f. sp. *cannabis* were more virulent to *C. sativa* than the wild type parent. The wild type strain caused 25% mortality in the target plant. Mortality in plants treated with the valine mutants ranged from 70-90%. In addition, development of wilt disease was more rapid in the plants infested with the valine overproducers. Limited studies on fourteen other plant species did not reveal a change in host range.

Thus, overproduction of an essential amino acid provided a highly effective means of enhancing the virulence of a biocontrol agent and has been used to enhance the virulence of *F. oxysporum* f. sp. *cannabis*,[22] *F. oxysporum* f. sp. *papaveris*,[2] *Pseudomonas syringae* pv. *tagetis* (N. Zidack, personal communication), *F. oxysporum* for control of *Orobanche*,[18] and *Xanthomonas campestris* pv. *poae* (A. Pilgeram, personal communication).

3.3. *Inhibition of Weeds by Amino Acids*

Amino acids, when applied to plants or seeds, have a measurable effect on plant health. In all cases where noxious weeds have been analyzed for amino acid sensitivity, an amino acid has been found that negatively affects the health of the plant. Inhibitory effects vary and include necrosis, wilting and stunting of growth. Certain other amino acids selectively enhance the growth and vigor of plants. Amino acids can be applied to the soil at the base of the plant or drenched over the entire plant. For example, when lysine was applied to *Cirsium arvense*, necrosis was observed on the leaves within days. Application of methionine plus lysine to *Cirsium arvense* resulted in yellowing on new leaf buds as well as necrosis. Other amino acids had little or no effect on the plants.

4. General Methodology

4.1. *Determination of Amino Acids or Combinations of Amino Acids that are Most Inhibitory to Target Weeds*

Surface sterilized seeds are placed on plates of water agar supplemented with a single amino acid (2-5mM l-form). Inhibitory effects were observed utilizing amino acids in the branched chain pathway (valine, leucine, and isoleucine), the aspartate pathway (lysine, threonine, and methionine) and the aromatic pathway (tyrosine, tryptophan, and phenylalanine). Inhibitory effects include reduced seed germination, inhibition of shoot growth and/or necrosis. Effects were seen with single or combinations of amino acids depending on the plant species involved.

The lowest inhibitory concentrations of amino acids that are inhibitory to a target plant are determined by placing surface-sterilized seeds on water agar that has been supplemented with decreasing concentrations of the selected amino acid(s).

4.2. *Selection of Variants of the Bioherbicide Resistant to Analogs of the Selected Amino Acid*

Amino acid overproducing and excreting strains of each fungus or bacterium can be selected by exposure to specific amino acid analogs.[23] For example, if the target weed is inhibited by lysine, then potential pathogens for control of that weed are exposed to lysine analogs (e.g. S-aminoethyl-cysteine or lysine hydroxamate) to select for lysine-producing mutants. Resistant colonies are selected using a well zone–diffusion assay on CUTS minimal medium. The zone diffusion plates are prepared by cutting a plug from the center of the CUTS plate with a sterile cork borer. The plates are then spread with 10^6 - 10^7 fungal spores, a suspension of 10^3 - 10^5 mycelial fragments, or a suspension of 10^7 - 10^8 bacteria. A sterile solution of the amino acid analog (0.1 ml of a 100 mM solution) is then added to the well. The plates are incubated in a laminar flow hood for 4 hours. An additional 0.1 ml of the analog solution can be added to the well. The plates are then incubated at 28C and monitored daily for the appearance of zones of inhibition and resistant colonies within the zone. Resistant colonies are isolated and analyzed for amino acid excretion. This selection may need to be repeated several times using increasing concentrations of analog and/or different analogs.

4.3. *Assay for Amino Acid Excretion*

In our laboratory amino acid excretion is assayed by growth of a bacterial auxotroph.[23] The auxotroph is seeded into media lacking the amino acid required for growth. Subsequent growth of the auxotroph in the medium is dependent upon and proportional to the quantity of added amino acid. For example, in order to assay valine, a valine auxotroph of *E. coli* (strain CAG18431) is seeded into CUTS media. The auxotroph will not grow unless exogenous valine is added to the media. Colonies of the plant pathogenic fungi or bacteria that are resistant to a valine analog are sub-cultured onto the seeded media. The plates are incubated at 28 C for 2-3 days. If the resistant variants excrete valine, there will be a zone of auxotroph growth surrounding the sub-cultured colony. The size of the zone is an indication of the magnitude of valine excretion. A standard dose-response can be determined by placing discs containing various levels of amino acid onto the auxotroph seeded agar. This straight forward bioassay is a tool that enables the researcher to screen numerous mutants in search of variants that excrete a desired amino acid.

4.4. *Testing Virulence and Host Range of the Amino Acid Overproducing Variants in Growth Chamber Studies*

The virulence (rate of kill and % mortality) of amino acid producing variants of each pathogen should first be evaluated in environmental growth chambers in order to eliminate external factors that may influence experimental results. In the initial studies, target weed plants are inoculated with each amino acid excreting variant and its respective wild type parent. Amino acid excreting variants that are more virulent than the parent are further evaluated in host range and scale-up experiments.

5. Improving Dissemination

A soil-applied pathogen will not be an efficacious mycoherbicide, even if it has specificity, sufficient lethality, and long-term soil survival, unless it can be delivered in a cost-effective manner. Fungi grown in liquid or solid-phase fermentation are inherently expensive when applied to large areas at 10^4 spores per gram of soil. Conventional formulation methods with spore suspensions and food-based formulations did not provide enough spores in the root zone of the target weed.[6,24-26] However, plant pathogenic fungi such as *F. oxysporum* saprophytically colonize the roots of many non-host plants[26] and thus, *F. oxysporum* mycoherbicides could be delivered to farmer's fields on non-host seed such as crops or grass, positioning the mycoherbicide directly in the rhizosphere of target weed.[2,10] The multiplication of fungal biomass in the rhizosphere of the carrier seedling allows for application of low levels of the mycoherbicide, greatly reducing the cost of inoculum production.

6. Possible Applications for the Control of *Striga* and *Orobanche*

Parasitic plants are not easily controlled, partially because of their survival as seeds in large numbers in soil for years, and their stealth-like ability to cause damage to their hosts before emergence above ground. Approaches to biological control of these plants may have to contend with the following:

- determining which amino acids inhibit the parasite and not its crop host;
- finding a pathogen and/or a saprophyte that is rhizosphere competent and capable of rapid multiplication in soil;
- obtaining appropriate amino acid excreting mutants of the selected pathogen;
- choosing a "carrier seed" system for inexpensive soil inoculation;
- developing an appropriate production system for the biocontrol fungus or the bacterium.

Additional control strategies include:

- developing soil microbes that produce germination stimulants to artificially induce *Striga* or *Orobanche* germination, or;
- developing strains that produce enzymes that destroy secreted germination stimulants prior to diffusion from the crop plant rhizosphere;
- isolation of strains that inhibit *Striga* or *Orobanche* germination;[27]
- selecting strains that overproduce fusaric acid;[15]
- genetically modifying strains with virulence genes (Chapter 23 and Refs. 28-29);
- co-applying a biocontrol agent with a chemical herbicide;[30]
- integrating biocontrol agents and parasitic plant resistance[31] or transgenic herbicide tolerance[32].

Numerous other approaches might be successful, but one thing seems certain. The approach that succeeds will have to combine the expertise of several research disciplines (plant pathology, weed biology, agronomy, chemistry, molecular biology). This is clearly a situation where collaboration by sharing of findings, timely communication and comparative field plots will all play crucial roles. Critical decisions will have to be made to select pathogens; to use or not use genetic recombinant techniques; and to develop reliable and economic production and dissemination capacity.

7. Conclusions

Over the last thirty years, numerous pathogens have been investigated as potential bioherbicides. Despite this intensive research effort, few pathogens have been successfully developed as biocontrol agents. The inherent constraints associated with biological species are largely responsible for this failure, yet our preconceived ideas about these agents are also at fault. We believe that a paradigm shift must occur if bioherbicides are to enjoy wider success as a weed control method. In the past, researchers focused on lethality and host specificity as initial screening requirements for successful agents. However, many pathogens that do not meet these criteria could be enhanced by synergistic additions or genetic modification. Embracing new methodologies may allow many "unsuitable" pathogens to be developed into successful biocontrol agents. Likewise, embracing collaborations amongst scientists with diverse approaches to biocontrol may provide the necessary synergy to implement a successful biocontrol project. Success in controlling a serious agricultural parasitic weed such as *Striga* would do wonders for farmers, farm-based economies, and biocontrol researchers.

References

1. R. E. C. McFadyen, *Annu. Rev. Entomol.* **43**, 369 (1998).
2. K. Tiourebaev, Montana State University Ph.D. Thesis (Bozeman, MT) (1999).
3. S. Hasan and A. J. Wapshere, *Ann. Appl. Bio.* **74**, 325 (1973).
4. M. J. Morris, A. R. Wood, and A. den Breeÿen, *African Entomol. Mem.* **1**, 125 (1999).
5. M. E. Hanley, and R. H. Groves, *Weed Res.* **42**, 370 (2002).
6. M. Ciotola, A. DiTommaso, and A. Watson, *Biocontrol Sci. Technol.* **10**, 129 (2000).
7. P. S. Marley, S. M. Ahmed, J. A. Y. Shebayan, and S. T. O. Lagoke, *Biocontrol Sci. Technol.* **9**, 159 (1999).

8. P. S. Marley and J. A. Y. Shebayan, *BioControl* **50**, 389 (2005).
9. A. Elzein and J. Kroschel, *Weed Res.* **44**, 433 (2004).
10. D. Yonli, H. Traoré, D. E. Hess, P. Sankara, and P. Sérémé, *Weed Res.* **46**, 73 (2006).
11. A. Elzein, J. Kroschel, and V. Leth, *Biocontrol Sci. Technol.* **16**, 3 (2006).
12. D. Yonli, D. E. Hess, A. A. Abbasher, P. Sérémé, and P. Sankara, *Afr. Crop Sci. J.* **13**, 41 (2005).
13. Z. Amsellem, Y. Kleifeld, Y. Kerenyi, L. Hornok, Y. Goldwasser, and J. Gressel, *Biol. Cont.* 21, 274 (2001).
14. Z. Amsellem, N. K. Zidack, P. C. Quimby, Jr., and J. Gressel, *Crop Protect.* **18**, 643 (1999).
15. B. Bouizgarne, H. el Maarouf-Bouteau, K. Madiona, B. Biliqui, M. Monestiez, A. M. Pennarun, Z. Amiar, J. P. Rona, Y. Ouhdouch, I. el Hadrami, and F. Bouteau, *Mol. Plant–Microbe Interact.* **19**, 550 (2006).
16. D. Müller-Stöver, H. Thomas, J. Sauerborn, and J. Kroschel, *BioControl* **49**, 595 (2004).
17. A. Boari and M. Vurro, *Biol. Cont.* **30**, 212 (2004).
18. M. Vurro, A. Boari, A. L. Pilgeram, and D. C. Sands, *Biol. Cont.* **36**, 258 (2006).
19. R. A. Steinberg, *Science* **103**, 329 (1946).
20. R. A. Steinberg, J. D. Bowling, and J. E. McMurtrey, Jr., *Plant Physiol.* **25**, 279 (1950).
21. N. Amrhein, *Rec. Adv. Phytochem.* **20**, 83 (1986).
22. K. S. Tiourebaev, G. V. Semenchenko, M. Dolgovskaya, M. K. McCarthy, T. W. Anderson, L. D. Carsten, A. L. Pilgeram, and D. C. Sands, *Biocontrol Sci. Technol.* **11**, 535 (2001).
23. D. C. Sands and L. Hankin, *Appl. Microbiol.* **28**, 523 (1974).
24. D. R. Fravel, S. K. Stosz, and R. P. Larkin, *Phytopathology* **86**, 236 (1996).
25. B. A. Bailey, K. P. Hebbar, M. Strem, L. C. Darlington, and R. D. Lumsden, *Biocontrol Sci. Technol.* **7**, 423 (1997).
26. W. J. Connick, Jr., D. J. Daigle, A. B. Pepperman, K. P. Hebbar, R. D. Lumsden, T. W. Anderson, and D. C. Sands, *Biol. Cont.* **13**, 79 (1998).
26. L. W. Burgess, in *Fusarium: Diseases, Biology, and Taxonomy*, Ed. P. E. Nelson, T. A. Toussoun, and R. J. Cook (Pennsylvania State University Press, University Park, 1981) p. 225.
27. A. Andolfi, A. Boari, A. Evidente, and M. Vurro, *J. Agric. Food Chem.* **53**, 1598 (2005).
28. B. A. Cohen, Z. Amsellem, R. Maor, A. Sharon, and J. Gressel, *Phytopathology* **92**, 590 (2002).
29. Z. Amsellem, B. A. Cohen, and J. Gressel, *Nature Biotech.* 20, 1035 (2002).
30. D. Müller-Stöver, H. Buschmann, and J. Sauerborn, *Eur. J. Plant Path.* **111**, 193 (2005).
31. P. S. Marley, D. A. Aba, J. A. Y. Shebayan, R. Musa, and A. Sanni, *Weed Res.* **44**, 157 (2004).
32. T. Surov, D. Aviv, R. Aly, D. Joel, T. Goldman-Guez, and J. Gressel, *Theor. Appl. Genet.* **96**, 132 (1998).

CHAPTER 23

TRANSGENIC BIOCONTROL AGENTS TO OVERCOME EVOLUTIONARY BARRIERS

Jonathan Gressel[a,*], Sagit Meir[a], Yoav Herschkovitz[a], Hani Al-Ahmad[a,b], Olubukola Babalola[a,c] and Ziva Amsellem[a]

[a]Plant Sciences, Weizmann Institute of Science, Rehovot, 76100, Israel
[b]Present address: Department of Biology and Biotechnology, An-Najah National University, Nablus, Palestinian Authority
[c]Present address: Microbiology Department, Olabisi Onabanjo University, Ago-iwoye, Ogun state, Nigeria
[*]E-mail: Jonathan.Gressel@weizmann.ac.il

Inundative mycoherbicides have not been successful in weed control in row crops, probably due to evolutionary barriers. Adding virulence factors was considered essential. Exogenous addition of the products of various genes was used to ascertain synergy as a prelude to adding them transgenically. Transgenically over-expressing single 'soft' genes (host lytic enzymes such as pectinase, cellulase, and expansins, or natural hormones such as IAA), or 'hard' genes encoding toxins such as NEP1 and CP1, has enhanced virulence, but not enough. We have studied these in three weed/pathogen systems: *Abutilon / Colletotrichum*; *Senna* (*Cassia*) / *Alternaria*; and the root parasitic relative of *Striga, Orobanche / Fusarium* spp. We deal below with the *Orobanche*, as a model for *Striga*, a species one does not work with where it might flourish. Gene stacking to obtain synergies among the various genes is considered a top priority, both to achieve sufficient virulence and to delay the evolution of weed resistance.

1. The Need for Enhancement — Exogenous Synergists vs. Endogenous Transgenes

Inundative mycoherbicides have rarely been commercialized in row crop agriculture, where they must compete with conventional herbicides.

That is not to say there is no need for them; there are many row crop situations where no conventional herbicide can selectively distinguish between crop and related weed. The barrier is often evolutionary: if the specific pathogen had the extreme virulence needed in row crops, it would kill all host plants, and both might become extinct. Thus the need to enhance the potential of mycoherbicides with virulence factors from other sources.

1.1. *Synergists as Gene Models*

Synergists that help overcome host defenses can assist in enhancing virulence of a biocontrol agent. This has a cost of the synergists, and they cannot always be used except in a laboratory situation; e.g. when a biocontrol agent is to be soil applied, the likelihood of its persistence long enough to be effective is minimal. Thus, it is suggested that the synergist be made by the biocontrol agent, by genetically engineering the appropriate genes. Exogenously added synergists have a biosafety advantage over engineered synergists insofar as the biocontrol agent is only hypervirulent when the synergist is present.

When a synergist does provide hypervirulence, it provides an inkling about what genes might be transformed to provide hypervirulence. One way to choose putative synergists for testing is to scan the literature on characterized mutants that lost virulence and ask whether adding the missing gene product to the wild type enhances virulence. The literature is replete with reports that fungal mutants losing the ability to produce auxins, various cell wall and middle lamellae hydrolases, oxalate biosynthesis, callose biosynthesis, as well as phytoalexin biosynthesis have less virulence than the wild type. There were no previous reports that engineering overproduction of these compounds enhance virulence. This concept led to the demonstration that an antimetabolite preventing phytoalexin biosynthesis,[1] and agents that complex calcium, a key co-factor in callose biosynthesis[2] could serve as synergists. This led to engineering enhanced oxalate (a calcium complexing agent) biosynthesis into *Colletotrichum* (data not shown). Adding pectinase or cellulase to fungal inocula could synergize virulence (Fig. 1), leading to using genes for overproducing cell wall/middle lamellae degrading enzymes.

There are times where there can be an apparent failure from engineering genes for overproduction, based on synergies. Adding genes encoding IAA overexpression to the *Colletotrichum coccodes* specific to *Abutilon theophrasti* did not increase virulence (Amsellem and Gressel, unpub.), even though the same gene enhanced *Fusarium* spp. on *Orobanche*.[3] As the requisite enzymes were expressed, we hypothesized that they had insufficient endogenous substrate, and added tryptophan, which greatly enhanced the activity of the transgenic fungi, but not the wild type (Amsellem and Gressel, unpub.). Thus, you can even synergize a transgenic biocontrol agent. One could also genetically (Chapter 22), or transgenically enhance tryptophan production.

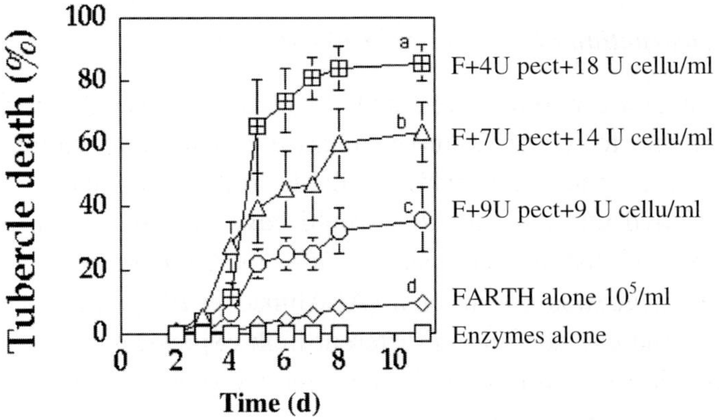

Figure 1. Exogenous applications of cellulase and of pectinase increase the virulence of Fusarium sp. CNCM I-1621 *(FARTH) on* Orobanche. Orobanche aegyptiaca *tubercles attached to the roots and parasitizing tomato were sprayed with chopped mycelia amended with 10 units/ml of Cellulysin added to the mycelial suspension (2.2x10^6 propagules/ml) or 1.4 units/ml pectinase added to the mycelial suspension (4x10^6 propagules/ml). The results are averages of 4 replicates ± SE.*

1.2. *Concept of 'Soft' Genes vs. 'Hard' Genes*

We divide the genes that are being engineered into mycoherbicidal agents as 'soft' and 'hard', based on their modes of action, prevalence, and virulence. Those genes whose products are already present in the human food supply and would have "generally regarded as safe (GRAS) toxicological status, would be considered 'soft', e.g. carbohydrases, auxin, and oxalate. Their effects are also not expected to be as dramatic as 'hard' genes, such as those encoding phytotoxins. Organisms with hard genes may be harder to get through regulatory channels, but their greater efficacy requires that they be considered.

Nature rarely uses a single solution for a problem, unlike too many of the single 'stand alone' solutions used for pest control. It is advisable to learn from nature, and combine genes for hypervirulence. This should give synergistic interactions (or at least additive ones) allowing getting closer to cost effective weed control. A multitude of genes will also make it harder for weeds to evolve resistance to the transgene products in the hypervirulent biocontrol agent.

1.3. *Construction of a Universal Cassette*

All the genes we wished to test had already been isolated and cloned. It was necessary to prepare a universal cassette with many cloning sites so that the genes graciously made available to us by colleagues could be easily inserted into a vector that would have the same high expression trpC promoter that we have successfully used on previous occasions.[4] Such a cassette was constructed (Al-Ahmad *et al.*, unpublished) along with a second cassette with a different high expression toxA promoter.[5] The protoplast transformation system that we routinely use allows us to co-transform many genes simultaneously. We have both hygromycin and bleomycin selectable markers so that we can transform strains that have previously been transformed with other genes, and the other selectable marker.

2. 'Soft' Genes

The various genes that we have obtained from a variety of sources are being transformed into the three model systems described above.

2.1. *Auxins*

The two genes responsible for bacterial biosynthesis of auxin from tryptophan, *IAA*H, and *IAA*M were transformed into the *Fusarium* spp.

When the fungus was pre-cultured on tryptophan prior to preparing inocula, the level of virulence was doubled.[3] While this was statistically significant, it was far from the orders of magnitude increased virulence that was necessary.

2.2. *Pectinase*

Pectinases (polygalacturonidases) are typically used by fungi to separate the host cells following penetration, and adding pectinases enhances virulence. Pectinase genes originating from higher plants have no sequence homology to those of fungi. Thus, we inserted an apple pectinase gene[6] into our universal cassettes, with a feeling of surety that there would be no co-suppression of the gene due to homology with the fungal gene. Fungal virulence increased (Figure 2).

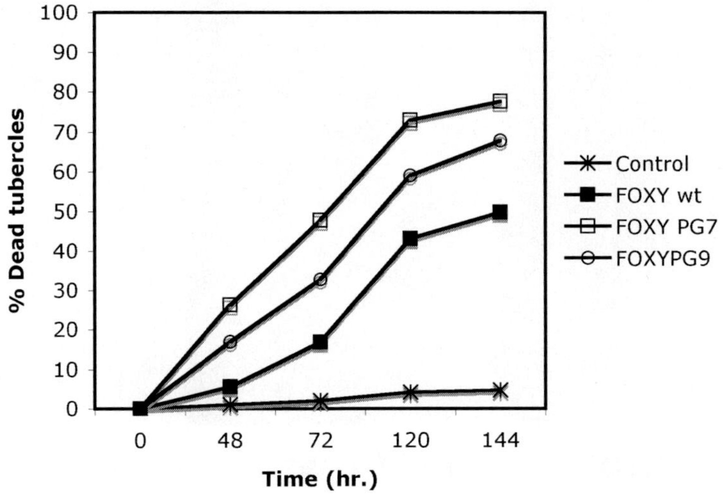

Figure 2. Transformation of pectinase (PG) gene into Fusarium oxysporum *(FOXY) enhances the death of* Orobanche aegyptiaca *tubercles attached to the roots and parasitizing tomato.. The pectinase gene*[6] *was transformed into the fungus under the control of the ToxA promoter (transformants UC2PG7 and UC2PG9). The tubercles were sprayed with chopped mycelia. Each treatment is an average of 5 plants with about 180 tubercles (total). The experiment was repeated three times.*

2.3. *Expansins*

Expansins are similar to pectinases insofar as they separate cell walls. They are secreted by nematodes upon penetration into plant tissue, allowing them to slither between the cells. We inserted the nematode expansin *Gr-Exp1* gene[7] into our universal cassettes and transformed them to our model fungal systems. The virulence of the *Fusarium oxysporum* transformants (Fig. 3) and of the *Colletotrichum coccodes* transformants (not shown) increased significantly towards their hosts.

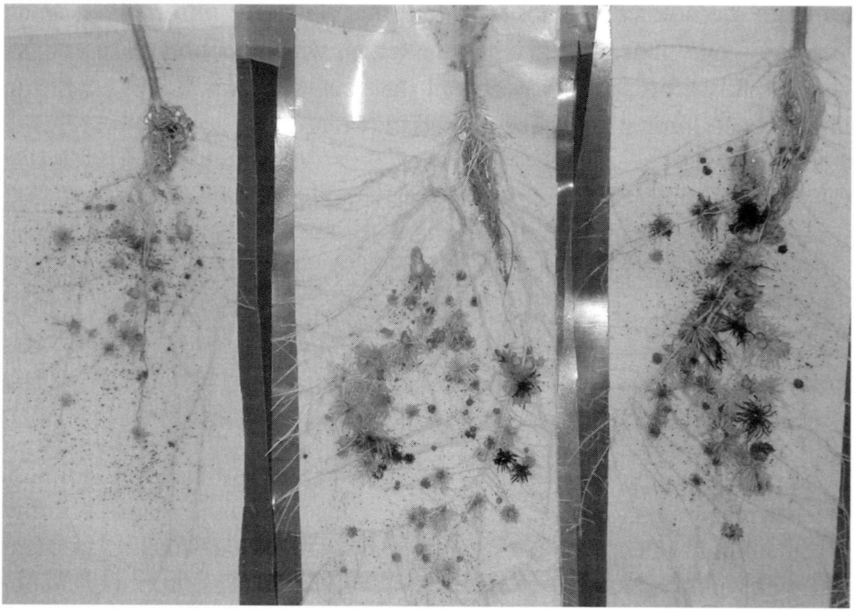

Figure 3. Transformation of expansin GR-Exp1 *(Exp) into* Fusarium oxysporum *(FOXY) causes rapid death of* Orobanche aegyptiaca *tubercles attached to the roots and parasitizing tomato. The nematode* GR-Exp1 *gene*[7] *was transformed into* Fusarium oxysporum *(FOXY). The* Orobanche *tubercles attached to and parasitizing tomato roots, were sprayed with chopped mycelia. Photograph taken 5 days after infection. Each treatment is an average of 5 plants with about 180 tubercles (total). The experiment was repeated 3 times. Note that tomato itself was unaffected by the transformed fungus.*

2.4. *Ethylene*

It has long been known that ethylene stimulates the germination of *Striga*, but not of *Orobanche*.[8,9] Bacteria,[10,11] free living soil[12] and mycorrhizal[13] fungi that produce ethylene all stimulate the germination of *Striga*. The genes of ethylene synthesis from methionine are well known, and have been sequenced and cloned from a variety of sources and could easily be transformed into mycoherbicidal organisms that attack *Striga*, giving them a dual function of inducing *Striga* and then devouring the germlings.

2.5. *Cellulases*

Cellulases are routinely secreted by fungi to assist in dissolving cell walls, releasing free sugars and allowing fungal penetration into cells. Bacterial cellulase genes have little sequence homology to the fungal genes, and thus the bacterial cellulases *cel*Y and *cel*Z[14] have been cloned into the universal cassettes, with the hope that there would be no co-suppression of the fungal gene upon transformation. Indeed, cellulae biosynthesis by *Colletotrichum* was increased (data not shown).

3. Hard Genes

3.1. *Nep1*

NEP1 is a *Fusarium oxysporum* gene encoding a 'necrosis enhancing protein', which was once considered to be a potential natural herbicide.[15] It was rapidly realized that it could hardly be made to penetrate plants when used as a stand-alone. We utilized this gene with the high-expression cassette provided by Bailey and found it to be exceedingly potent in enhancing virulence of *Colletotrichum* on *Abutilon,*[4] of *Alternaria* on *Senna* (Safran *et al.* unpublished) and *Fusarium* sp. CNCM I-1621 on *Orobanche* (Meir *et al.* unpublished). It did not enhance the virulence of our forma specialis of *Fusarium oxysporum* that attacks *Orobanche*. We rapidly discovered that all forma speciales of *F. oxysporum* that we checked bear the gene, but express it at very low levels. For this reason we are excising the native gene, and are reinserting the high expression gene, hoping to obtain hypervirulence with this weed/fungus pair.

The over-expressed *NEP*1 transgene in *Colletotrichum* expanded the host range beyond its high specificity to *Abutilon* and it was pathogenic to species such as tomato and tobacco.[4] This is probably because the fungus caused minor injury that allowed the phytotoxin to enter leaves, causing a necrotic lesion that allowed the fungus to attack as a heterotrophic organism, i.e. not a true pathogen. When *Fusarium* sp. CNCM I-1621 overexpressing *NEP*1 colonized tomato roots "waiting" for *Orobanche* to attack the tomato, it had no deleterious effects on the tomato plants. Thus, when the fungus does not scar the plant, the NEP1 protein does not affect it.

The *Fusarium* sp. CNCM I-1621 with *NEP*1 is still insufficiently virulent for commercial use and will need to be stacked with other transgenes.

3.2. *Cerato-platanin*

The phytotoxic protein cerato-platanin is produced by the plant pathogenic fungus *Ceratocystis fimbriata f. platani.*[16] This fungus attacks *Plantanus* species (London plane, oriental plane and American sycamore) and causes a canker stain disease. The disease is characterized by foliar wilting and spreading lesions that involve phloem, cambium and extensive regions of sapwood. Cerato-platanin shares some structural and functional characteristics with fungal hydrophobins.

We inserted the cerato-platanin gene into our universal cassettes and transformed them into our model fungal systems. The cerato-platanin transformants showed virulence enhancement in *Colletotricum coccodes* (not shown) and *Fusarium oxysporum* (Fig. 4). Overexpression of cerato-platanin alone did not enhance the virulence in *Fusarium* sp. CNCM I-1621, thus we co-transformed the cerato-platanin gene with *NEP*1 gene to obtain hypervirulence strain (Fig. 4).

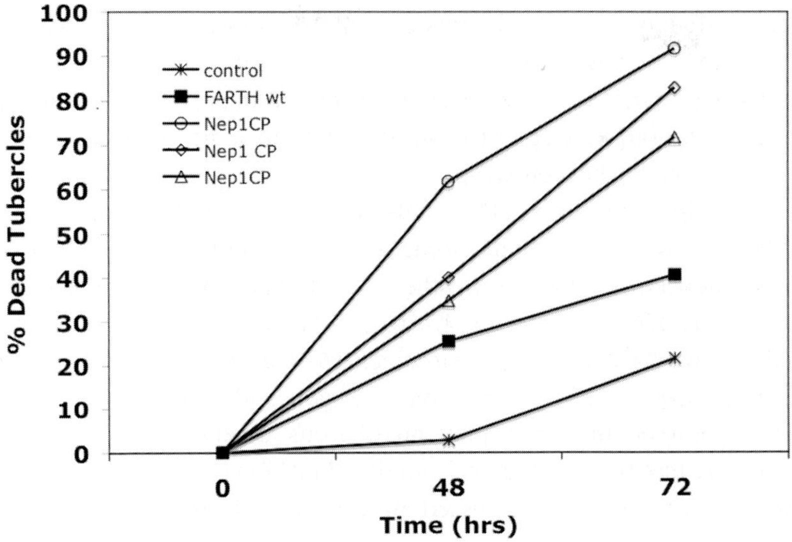

Figure 4. Co-transformation of NEP*1 and Cerato-platanin (CP) genes enhanced the virulence of* Fusarium sp. CNCM I-1621 *(FARTH) on* Orobanche aegyptiaca *tubercles. The CP gene*[16] *under the control of the trpC promoter was transformed into* NEP1 *transformant of* Fusarium sp. CNCM I-1621. *The* Orobanche *tubercles were sprayed with chopped mycelia (10^5 propagules/ml). Each treatment is an average of 5 plants with about 180 tubercles (total). The experiment was repeated 3 times.*

4. Transgenically Overcoming Host Defenses

We had hypothesized that *Orobanche* would utilize phytoalexins to ward off attach by pathogenic fungi, yet in an exhaustive attempt to find such pathogen-induced small inhibitory molecules, we were quite unsuccessful.[17] This does not mean that the parasites may not have evolved constitutive mechanisms to make it harder for pathogens to attack. Peculiarly, root parasitic weeds accumulate high levels of two compounds, which we thought might be part of a defense system. The tubercles of *Orobanche* have a specific mannose phosphate reductase that immediately converts the hexoses stolen from the crop to mannose and then to its alcohol - mannitol;[18] and they accumulate asparagine[19] in their tissue. They could be part of an evolutionary trick to ward off pathogens by sequestering storage materials in a form inaccessible to attackers, limiting their growth.

Asparagine constitutes 80% of the free amino acids in underground portions of *Striga.*[20] There are some past findings that may point to such effects: induced asparagine synthesis in tomato was correlated with reduced *Fusarium oxysporum* f. sp. *lycopersici* infection. We have tested the two *Fusarium* spp. that attack *Orobanche* and found that they can utilize asparagine as both sole nitrogen and sole carbon sources. Thus, asparagine does not play an obvious defense role. It is still conceivable that the asparagine from the plant tissue causes some type of imbalance (of the type discussed in Chapter 22) that reduces fungal growth.

Fusarium oxysporum grew on mannose at the same rate as on glucose, with ammonium as the sole nitrogen source. Thus, the mannose also has no clear cut effect on *Fusarium.*

Once the reasons are elucidated if/why mannose and asparagine affect *Fusarium* as a biocontrol agent in situ, genes can be found that encode enzymes that rapidly convert these utilizable forms by the fungus, while overcoming any feedback inhibitions. This would deplete the parasites of both metabolic resources as well as defenses. Another defense could be direct, a small RNAi produced by the tomato host that could be transposed into the parasite that would inhibit the mannosylphosphate reductase, depleting mannose at the source.

Acknowledgments

The technical assistance of Adi Maoz and Mayan Shaviv at various stages of this project is acknowledged. Bryan Bailey and Mary Strem, USDA, kindly supplied the *NEP*1 gene construct and the polyclonal antibody against the gene product. Linda Ciufetti, David Straney, Amir Sharon, Luigia Pazzagli, Hans Helder, Ross Atkinson, Peter Schaap, and Lonnie Ingram kindly provided the genes used in our research. This research was supported as part of the EU 6th Framework Priority 5 - Food Quality and Safety Project: Enhancement and Exploitation of Soil Biocontrol Agents for Bio-Constraint Management in Crops (Contract no. FOOD-CT-2003-001687). The information/opinions provided in the paper do not necessarily represent the official position/opinion of the European Commission.

References

1. A. Sharon, Z. Amsellem and J. Gressel, *Plant Physiol.* **98**, 654 (1992).
2. J. Gressel, D. Michaeli, V. Kampel, Z. Amsellem and A. Warshawsky, *J. Agric. Food Chem.* **50**, 6353 (2002).
3. B. Cohen, Z. Amsellem, R. Maor, A. Sharon and J. Gressel, *Phytopathologye* **92**, 590 (2002).
4. Z. Amsellem, B. A. Cohen and J. Gressel, *Nature Biotech.* **20**, 1035 (2002).
5. L. M. Ciuffetti, R. P. Tuori and J. M. Gaventa, *Plant Celle* **9**, 135 (1997).
6. R. G. Atkinson, *Plant Physiologye* 105, 1437 (1994).
7. L. Qin, U. Kudla, E. H. A. Roze, A. Goverse, H. Popeijus, J. Nieuwland, H. Overmars, J. T. Jones, A. Schots, G. Smant, J. Bakker and J. Helder, *Naturee* **427**, 30 (2004).
8. A. G. T. Babiker, L. G. Butler, G. Ejeta and W. R. Woodson, *Physiol. Plant.* **89**, 21 (1993).
9. F. F. Bebawi and R. E. Eplee, *Weed Sci.* **34**, 694 (1986).
10. M. O. Ahonsi, D. K. Berner, A. M. Emechebe, S. T. Lagoke and N. Sanginga, *Biol. Cont.* **28,** 1 (2003).
11. D. K. Berner, N. W. Schaad and B. Volksch, *Biol. Cont.* **15**, 274 (1999).
12. N. E. Ahmed, Y. Sugimoto, A. G. T. Babiker, Y. Q. Ma, S. Inanaga and H. Nakajima, *Weed Sci.* **49**, 354 (2001).
13. V. W. Lendzemo, T. W. Kuyper, M. J. Kropff and A. van Ast, *Field Crops Res.* **91**, 51 (2005).
14. S. G. Zhou and L. O. Ingram, *J. Bact.* **182**, 5676 (2000).
15. B. Bailey, R. Collins and J. Anderson, *Weed Sci.* **48**, 776 (2000).
16. L. Pazzagli, G. Cappugi, G. Manao, G. Camici, A. Santini and A. Scala, *J. Biol. Chem.* **274**, 24959 (1999).
17. B. A. Cohen, Z. Amsellem, S. Lev-Yadun and J. Gressel, *Ann. Bot.* **90**, 567 (2002).
18. P. Delavault, P. Simier, S. Thoiron, C. Veronesi, A. Fer and P. Thalouarn, *Physiol. Plant.* **115**, 48 (2002).
19. P. Simier, P. Delavault, E. Demarsy, J. B. Pouvreau, K. Pageau, B. Le Bizec, A. Fer and P. Thalouarn, *Physiol. Plant.* **123**, 9 (2005).
20. K. Pageau, P. Simier, B. Le Bizec, R. J. Robins and A. Fer, *J. Exp. Bot.* **54,** 789 (2003).

Section VII

Epilogue

CHAPTER 24

EPILOGUE — WILL THERE BE INTEGRATED *STRIGA* CONTROL?

Jonathan Gressel[a,*] and Brhane Gebrekidan[b]

[a]*Plant Sciences, Weizmann Institute of Science, Rehovot, Israel*
[b]*AMAREW Project, Bahir Dar, Ethiopia*
[*]*E-mail: jonathan.gressel@weizmann.ac.il*

1. The Concept

In conceiving this meeting of minds and people, Prof. Gebisa Ejeta took a great risk. He understood that no single stand-alone technology could sustainably conquer the scourge of *Striga* in Africa, and that there must be an integration of approaches. But, could he convince all the practitioners of such a need? Two basic problems stood in his way: (1) Scientists have egos, and each thinks s/he has found the "only true solution"; (2) biologists are taught to vary single parameters in experiments, and integrating technologies leads to very messy experiments that not all journals will consider bear all the necessary controls. Indeed, initially as the discussions started with each presenter tooting his/her own horn, it became apparent that each solution that worked in the field had drawbacks, and the realization soon hit home that integration of technologies was essential. The transformation in the group was indeed palpable and collaborations were begun. This chapter will deal more with thoughts about what we heard, than a regurgitation of what has been written and said; for that the preceding chapters exist. Indeed, the emphasis is on the drawbacks of each technology, and why integration is imperative, lest we forget.

1.1. *Striga the Witch*

Gebisa Ejeta set the scene by trying to demonstrate with numbers how bad the situation actually is with *Striga*. This can be accentuated in that *Striga* doubly fits the abbreviation "HIV". *Striga* is truly "Highly Invasive Vegetation", especially *Striga hermonthica*, thought to have evolved from *Striga aspera* after the introduction of maize to Africa from Mexico centuries ago.[1] It evolved to be a pest of crops with few if any wild hosts. Whereas sorghum, an African native, had been in some sort of balance with *S. aspera, S. hermonthica* was devastating to sorghum. Was that due to the huge seed bank left after maize? No one knows for sure what happened, and the issues of diversity have yet to be fully worked out (Chapter 6). Overpopulation pushed agriculture from a rotation with fallow to continuous monocropping. As a result, productivity declined, and *Striga* thrived best when soil productivity is low. This in part caused men to seek livelihoods in cities, leaving families in the villages to scratch an existence in sub-subsistence agriculture. Such migrations of men throughout human history, on all continents, have always been responsible for epidemics of sexually transmitted diseases, where the other HIV comes in.

The scourge of *Striga* becomes even more important in light of a major shift in world agriculture, which brings "good news/ bad news". This is the shift to using good food/feed crops as substrates for producing biofuels.[2] The bad news is that all excess future world grain is going to biofuel, and future commodity prices are increasing in line with petroleum prices. This is bad news for Africa as it means no food for famine relief nor cheap grain purchases by governments to keep the masses happy.[2] This is indeed bad news. The good news is that subsidized grain will not be dumped (using the economists' term) on African markets i.e., below production costs of African farmers. The inability to compete with dumped grain is one reason for subsistence agriculture in Africa vs. production agriculture, where production agriculture is defined as having considerable excess above what is needed for the family to subsist, an excess that can be commercialized. Production agriculture is only possible with improved crop management along with inputs of fertilizer, pesticides, and especially quality seed.

These can be affordable if the infrastructures will be put in place by African governments with public and private investors. Governments can no longer afford to depend on donors to fulfill the governments' obligations to assure food security of their citizens. The additional good news is that the African farmer will be motivated and forced to be more productive and self-reliant. In the case of *Striga* affected cereals, sorghum, millets, and maize throughout Africa, as well as legumes in West Africa, the needed shift to more productive and market based agriculture will not be possible without conquering *Striga* with integrated technologies, a window of opportunity for those with the technologies to integrate.

1.2. *Basic Science and* Striga *Control*

The excellent results with some technologies, and promising results with others presented in this book indicate how wrong the initial approach to *Striga* was. When a small patch of *Striga* established itself on US shores, multi-millions of dollars were spent on eradication, over decades, but hardly a penny for basic research. More recently, various bodies have been willing to invest in providing extension for farmers on how to deal with *Striga*, but what information did extension workers have to "extend" to farmers? Not much. The single component solutions promoted by researchers for decades (e.g. long term rotations into non-food crops, or heavy manuring with orders of magnitudes more manure than is available) have neither been practical nor economical for the average African farmer. Many researchers have yet to learn that solutions must be practical, if they wish to have the satisfaction of seeing them adopted.

The ancient exhortation of "know thine enemy" was not adhered to. Such information comes from basic research. Researchers had to scrounge for funds, against odds, to perform the necessary basic reconnaissance to find the enemy's weak spots. It is clear that much that is presented in this book emanates from basic research in the last couple of decades (Chapters 3, 4, 5, 7, 9, 13), a credit to the persistence of researchers who found parasitic weeds and their interaction with plants a fascinating research topic. We owe much to these colleagues, and their job must continue. As the information in these chapters shows, we are just beginning to scratch the surface, but have come up with gold. We still know far too little on the physiological interactions between host and parasite. What we already know proved to be important for breeding. What will be found out will be necessary for the micro-RNAi approach (discussed in Chapters 3 and 14). We must know much more about *Striga* evolution, its rates and variability so that we can know best how to deal with the propensity for it to evolve resistance to the technologies that are developed and will be developed. It is important not to repeat the mistakes with *Striga's* cousin *Orobanche*. Single gene *Orobanche*-resistant sunflower lines were bred only to quickly fall by the wayside as *Orobanche* sequentially evolved resistance strains to each gene thrown in its path.[3] This example provides another reason for integration of technologies. Considering the almost incalculable *Striga* seed bank size, there are a lot of individuals to choose among for evolution, and if the variability among them is great, the sustainability of any stand-alone technology is jeopardized.

Just to further ensure that strigologists continue to view the necessity of integration, and to assure that their successes do not go to their heads, we focus below mainly on the drawbacks of each technology to accentuate the needs to further integrate practices.

2. Solutions to Integrate

There is much we still do not know about *Striga* that would be useful for obtaining more and perhaps better solutions (see Chapter 1, section 5 for an elaboration). Still, many solutions are available to integrate, and integrate we must, as none are without blemishes.

2.1. *Breeding for* Striga *Resistance*

The crops being bred for *Striga* resistance can be divided into two groups; native African crops that co-evolved with *Striga:* (sorghum, millets, various legumes), and the introduced crops (maize, rice) that fall easy prey to *Striga,* possibly exacerbating the evolution of *Striga* to being more pernicious, as discussed above.

The sorghum breeders abandoned the old paradigm of just recombining genes and looking at *Striga* stands, and have been focusing on pyramiding multiple genes addressing different mechanisms of resistance. This facilitated breeding cultivars with resistance to different stages of *Striga* development (Chapter 7). This in turn has proven to be highly successful as we heard, and should delay evolution of resistance. Still, it is and will be cumbersome to cross these multiple genes, on different chromosomes into cultivars that are locally adapted. This is somewhat easier with marker assisted breeding, but there are not yet enough cloned markers for all the sorghum genes (Chapter 12), and local breeders are not yet attuned to marker assisted breeding.

We saw the success in the field, where the new *Striga* resistant sorghum cultivars looked great, except in the very worst of *Striga* hot-spots. Still, even with the apparently good yields there was still *Striga* in the fields, so the seed bank will not be substantially reduced, which could be dangerous for other crops, and for the evolutionary future. For these reasons, and the incomplete control in "hot-spots", integration with more resistance genes and other technologies is needed.

The maize breeders continue with the old selection, field assay breeding, and have had some local successes (Chapter 8). These successes may not always carry over to distant locales and other soil types. Indeed some researchers point that resistance in the maize types with this modicum of resistance is due to a root growth pattern that leads to avoidance; that the roots quickly grow beneath the surface layers where *Striga* is most effective. Even this resistance, polygenic in nature, and without markers, will be hard to move to locally-adapted varieties. Should we expect there to be *Striga* resistance in a Mexican crop that only recently met (in evolutionary terms) with *Striga*, and has no similar pests in Mexico? Some genetic resistance is there, but probably less than is in sorghum, which co-evolved with *Striga*

2.2. *Transgenic Crops*

We must look to genetic engineering in cases like maize where there is a good likelihood that endogenous genes for resistance are scarce. Genes can be transformed into maize from sources wherever else they may be found. Indeed, the multiple genes from sorghum (Chapter 7) may well be appropriate for maize and other crops. Genes conferring resistance to herbicides that are less prone to evolution of resistance than the mutants described in Chapter 11 are also good candidates, as might be genes encoding allelochemical biosynthesis (Chapter 5). As more becomes known about the metabolic differences between host and parasite, micro-RNAs that utilize these differences may help, as they have with protecting crops from nematodes.[4]

There were breeders in the group who propounded that breeding is preferable to genetic engineering in all crops, forgetting that if the genes do not exist in a species, recombining genes *ad infinitum, ad nausea* quickly reaches diminishing returns. The genes can only be brought from elsewhere by genetic engineering after a crop has reached its "genetic glass ceiling",[5] and has no further genes to recombine.

How long will single genes from genetic engineering last until *Striga* overcomes them with evolutionary tricks? This depends on how transgenic technologies are integrated with other management practices.

The necessity to integrate biotechnological solutions and breeding was heavily discussed informally in between the various talks, leading to incipient collaborations, but also more formally in a panel discussion on 'integrating crop breeding and biotechnology solutions for *Striga* control'. Any novel gene combination, whether genetic or transgenic, will have to be in a form of a cultivar the farmer will value and can use. A gene in the wrong genetic background is useless, which surprisingly is not always realized. Marker assisted selection should streamline the conversion of landraces or elite varieties to *Striga*-resistant types.

2.3. *Intercropping*

The push-pull technology using *Desmodium* as a perennial intercrop gave very good results (Chapter 18). This legume intercrop cannot be grown everywhere, is hard to establish, and there must be animals to feed the legume to. In many of the subsistence zones there are no cattle to benefit from this technology. The poor are trying to grow enough grain to feed themselves, and cattle are a luxury. It is not known whether *Desmodium* can be grown in western Africa, where legume-attacking

Striga gesnerioides is rampant, as susceptibility to this *Striga* has yet to be tested (Z. Khan, pers. comm.). The ones profiting the most from *Desmodium* at present are the first to start, and have legume seeds to sell to the other farmers.

The basic researchers are stepping in, and will possibly allow eliminating the legume "middle man". They are isolating the allelochemicals that suppress *Striga* (Chapter 5); once the genes are found, and if they are not too many for the gene jockeys to deal with, and if their products are not too "expensive" for the crop to produce, we may have another tool to integrate in a new manner.

2.4. *Herbicide Resistant Crops*

The herbicide seed-treated resistant maize gave excellent field results with short season maize, except in a few hot spots for yet un-explained reasons (Chapter 11). Single gene (albeit recessive at the herbicide levels used on the seed) is easy to transfer from one variety to other locally-adapted varieties. With its total control, the *Striga* seed bank could become exhausted, especially with the integration with push-pull already underway (Chapter 18). Resistance should not evolve quickly in short season maize, but will probably evolve in long season maize as semi-dominant resistant *Striga* could easily evolve late in season when part of the herbicide has dissipated.[6] Recombination will quickly generate recessive, highly resistant material. Thus, the technology is an excellent "stand-alone" stop gap, but long term sustainability will only be maintained with integration with other technologies.

Transgenic glyphosate resistant maize, released so far only in South-Africa, should be useful for *Striga* control, as glyphosate can translocate from leaves to roots, killing attached parasitic weeds.[7]

2.5. *Biocontrol*

Biocontrol requires no need for crop breeding, and the indigenous agents can be applied to any seed of any crop, a huge advantage (Chapter 21). So far, biocontrol has not proven to be good enough, and genetic and transgenic improvements of the biocontrol agents are sought (Chapters

22, 23). Clearly biocontrol is an excellent integrative tool to augment other tricks.

2.6. *Integrations with Agronomy*

The most promising approach to tackling the *Striga* menace in Africa to date has been the deployment of integrated *Striga* management (ISM), which includes among other things, good seed of a resistant variety, fertilizer application, moisture conservation measures, and good agronomic practices (Chapter 15). The critical item in the ISM approach is good crop management, which should pay attention not only to *Striga* but also to a host of other biotic and abiotic factors, which limit crop productivity. It is only when all the good practices are integrated under good and integrated agronomy that the farmer can realize good harvest. Our visits to sorghum farmers' fields in eastern Ethiopia, Ghelemso and Fedis, convincingly demonstrated that it is possible to harvest high sorghum yields even under *Striga* hot-spot situations where *Striga* still flourishes. The real challenge in the future is to scale up and out these success stories elsewhere in Ethiopia and indeed Africa. As the seed bank is not sufficiently lowered by the resistant sorghums, especially in the hot spots, one wonders how long it will be until *Striga* evolves means to overcome the modicum of resistance in these varieties.

3. Other Constraints

Striga is not the only problem farmers have, and the others must be addressed for production agriculture to sustain the growing population of Africa. African grain yields are a third of world averages, which they actually bring down. Why is this? *Striga* is important, but in addition to the *Striga*-related technical factors enumerated above, there are a number of institutional and organizational issues, and biotic factors that constrain integration.

3.1. Striga *is Not the Only Biotic Constraint*

In addition to *Striga* a wide range of pests, diseases, and weeds attack African grains. Biotic stresses such as stem-borers and grain storage moths and weevils may not be amenable to breeding and require transgenic or chemical interventions.[8] Sorghum also has a bird predation problem that was one of the major reasons for the shift to maize and other crops.[9] Here too breeding has not proved that effective.

It is unfortunate that there are those promoting farmer-saved seed as a necessity for Africa. Almost every study performed in the world has shown that the quality of farmer-saved seed gradually worsens, and contamination with pathogens and weed seeds increases, in the hands of the vast majority of farmers, and they regularly need seed from a reliable source. Indeed, one study showed that much of the non-certified maize seed sold in western Africa markets was contaminated with *Striga* seed.[10] The transgenic solutions to some of these constraints cannot be implemented until African countries have a regulatory system in place to scientifically deal with the biosafety regulations.

3.2. *Institutional Constraints to Integration*

The continuing expansion of *Striga* in Africa, instead of its decline, is a sign of institutional neglect, and a lack of realization of the gravity of the situation. Besides its direct impact, *Striga* is a major indicator that soils are being degraded and poverty is increasing, and the national institutions are not sufficiently dealing with their national food security issues. Among such principal constraints that are inadequately being dealt with are underdeveloped and poorly supported national research and extension, weak linkage between research and extension organizations and farmers, shortage of trained manpower in agriculture, poorly developed or the absence of an effective national seed industry, and poorly developed and ineffective commodity market systems. These institutional and organizational issues collectively are major obstacles to integrating the technical factors to make a difference in the productivity and production of African farmers. Research and extension organizations are often understaffed and poorly supported. The poor linkage and working relationships between researchers, extension specialists, and farmers are long standing problems. A shortage of trained manpower in all sectors of agriculture is dominant throughout the continent.

Many of the simplest solutions to *Striga* and other biotic and abiotic constraints that plague African food production are through transgenic crops.[5] Africa is woefully behind the developing regions of Asia and South America in instituting national biosafety regulatory frameworks that can scientifically deal with the risks and benefits of transgenics, keeping these solutions from their farmers and consumers. The Republic of South Africa is the only African nation where transgenic crops are cultivated, including glyphosate resistant maize, which could be useful to control *Striga* (see Section 2.4).

Then there are the infrastructural problems that the developing countries in the southeast Asia have learned to deal with, but Africa lays way behind: the availability of fertilizer at world prices; a viable seed industry that provides good seed, and modern, inexpensive grain storage facilities that buffer prices and prevent infestation by storage pests. The presence of strong regional or national seed industries, public or private, particularly hybrid seed based ones, could serve as good integrators of the technical factors thus serving as the main catalyst in the development of commercial agriculture, which is not yet the case in much of Africa. Notable examples of successful seed industries in sub-Saharan Africa are those of Kenya, Zimbabwe, and South Africa, which could be emulated.

On top of all these constraints, the poorly developed commodity market systems exacerbate the problem. The governments of Africa will have to rapidly change their policies to deal with these infrastructural issues. A major shift may be required from a dependency on donors for infrastructure, often with strings attached, to self-financing the infrastructural needs to meet food security requirements.

These issues were discussed in illuminating panel discussions on 'integrating agricultural R&D with technology deployment in Africa the ultimate integration', and 'aligning agricultural technology with markets and development policy'. Many of the messages from these panel discussions were alluded to in the final versions of the chapters, updated in light of the meeting, especially in Chapter 20.

Until all the infrastructural issues are dealt with, the farmers will prefer to remain at subsistence, hoping to succeed to produce just enough for their families. They lack the incentives and wherewithal to produce much more to feed the growing urban populations.

Thus, clearly not only is the will of strigologists to integrate needed, the governments must be willing to integrate the needed infrastructures into their systems so that their farmers will be willing to switch to production agriculture. This is an issue of self-preservation for Africa, and the time has come for the Africans to deal with the issues, with less dependence on donors, and more on local and regional resources and resourcefulness.

References

1. E. I. Aigbokhan, D. K. Berner and L. J. Musselman, *Phytopathology* **88**, 563 (1998).
2. J. Gressel, *African Crops News Service Newsletter,* www.africancrops.net/news/march06/gressel.htm, (2006).
3. S. Echevarria-Zomeno, A. Perez-de-Luque, J. Jorrin and A. M. Maldonado, *J. Exp. Bot.* **57**, 4189 (2006).
4. G. Z. Huang, R. Allen, E. L. Davis, T. J. Baum and R. S. Hussey, *Proc. Natl. Acad. Sci. USA* **103**, 14302 (2006).
5. J. Gressel, *Genetic Glass Ceilings: Transgenics for crop diversity.* (Johns Hopkins University Press, Baltimore, 2007)
6. J. Gressel, *Crop Prot.* **24**, 1007 (2005).
7. D. M. Joel, Y. Kleifeld, D. Losner-Goshen, G. Herzlinger, and J. Gressel, *Nature* **374**, 220 (1995).
8. J. Gressel, A. Hanafi, G. Head, W. Marasas, A. Obilana, B., J. Ochanda, T. Souissi and G. Tzotzos, *Crop Prot.* **23**, 661 (2004).
9. H. Doggett. *Sorghum* (Longman, Harlow UK, 1988).
10. D. K. Berner, K. F. Cardwell, B. O. l. Faturoti, F. O. Ikee and O. Williams, *Plant Dis.* **78**, 402 (1994).

INDEX

ABA ... 28
Abshir ... 200
Abutilon ... 319
Acacia saligna ... 303
acetolactate synthase ... 155
acetyl-CoA carboxylase ... 187
adapted germplasm ... 101
additive genetic effects ... 110
adenylo-succinate synthase ... 187, 188
adoption ... 250, 267
advanced backcross ... 130, 131
AFLP ... 74, 77, 108, 122, 170
African landraces ... 129
afrormosin ... 64
after-ripening ... 20, 21
Agalinis ... 36
agar gel assay ... 94
agents ... 284
Alectra vogelii ... 115, 116
alfalfa ... 64
alfalone ... 64
allele-specific primers ... 169
allelic frequencies ... 76, 108
allelochemical ... 62, 243, 332
allozyme markers ... 72, 74
Alternaria ... 319
AM fungi ... 22, 53, 56
amino acids ... 303, 304
aminoethyl-cysteine ... 307
apigenin ... 65–67
apocarotenoids ... 56
Arabidopsis ... 38, 39
arbuscular mycorrhizal ... 22, 50, 205
asparagine ... 322
attachment ... 25, 149
auxin signaling ... 178
auxotroph ... 307, 308
backcross inbred lines (BILs) ... 178
bacterial communities ... 295
basic science ... 329
benefits and costs ... 266
bewitching ... 5
bioassays ... 89
biochanin ... 64
biocontrol ... 12, 301–304, 333
biofuels ... 328
bioherbicide ... 287, 288
bioinformatics ... 37
biosafety ... 314, 335
bird predation ... 335
BLASTX ... 39
breeder seed ... 207
breeding strategies ... 101
brewers ... 263
broad sense heritability ... 135, 136
broad-based resistance ... 100
broadcast planting ... 237
Cajanus cajan ... 229, 233
callose biosynthesis ... 314
calycosin ... 64
Cannabis sativa ... 303, 305
carbohydrases ... 315
carotenoid ... 53, 54, 57
catch crops ... 54
caudatums ... 130
cDNA Arrays ... 41, 42
cellular communication ... 181
cellulose ... 314, 315, 319
center of origin ... 7
Ceratocystis fimbriata f. platani ... 320
cerato-platanin ... 320, 321
certified seed ... 290, 292
chemical cues ... 173
chemical fertilizer ... 210

chemical signaling 48
chemotropism 22
chlamydospore 288, 290, 304
chlorsulfuron 223, 224
Cirsium arvense 306
classical biocontrol 303
Colletotrichum coccodes 304, 318, 319
combining ability 109
composites 102
conditioning 19, 21, 217
consistency of resistance 104
containment 11
Contarinia sorghicola 256
contingent valuation 275
conventional breeding 88
cortex 26
cotton 49, 286
cowpea 4, 93, 115, 231–233, 235–236, 245, 247, 286
credit 278
crotalaria 247
crop damage 10, 270
crop phytotoxicity 153
crop rotation 205, 234
crop-management 214
CTase 187
2,4-D 224, 229, 237, 238
daidzein 64
day neutral 29
DEAE cellulose 148
delay in germination 147
demonstration 203
deoxystrigol 50, 52
Desmodium 61–68, 153, 156, 222, 226, 241–251, 332, 333
dicamba 219, 223, 224
diffusion 208, 210
dihydrosorgoleone 23, 92
2, 6-dimethoxy-*p*-benzoquinone (DMBQ) 24, 38, 92
di-*O*-methyldiadzein 64
disaster 15
disease epidemics 301
dispersal 8
dissemination 308
diversity arrays technology (DArT) 175
diversity assessment 108
DMBQ 39, 41, 42
DNA microarrays 180
dominant inheritance 119
Dowex-1 148
drought tolerance 166, 204
durra sorghums 130
ecological niche models 78, 81
econometrics 275
economic analysis 273, 276, 278
economic surplus 277
emergence 223, 270, 291
empirical selection 90
endodermis 176
enoyl-ACP reductase 188
environment 137, 217
epidermis 26
EPSP synthase (5-enoylpyruvylshikimate 3-phosphate synthase) 187
eradication 11, 13, 82, 214, 329
EST 39, 66, 67
ethylene 50, 218, 221, 319
Euphorbia 80
evolution of resistance 154
evolutionary barriers 313
expansin 318
extended agar gel assay 93
extension 335
fallow 205, 230
famine relief 328
farmer evaluation 202, 268, 273
farmer participation 208
farmer-saved seed 335
fecundity 76
fertility management 206

fertilizer..........220, 225, 226, 235, 238, 271, 272, 336
field resistance 107, 136, 161
field tests/trials................ 131, 148, 162
fine mapping 179
Florida.. 75
flowering 29
fluorescent tagging.......................... 297
fluridone.. 55
fonio 66, 286
food legumes 248
food security 329
foreground screening....................... 167
formulation..................................... 308
frenching disease............................. 304
full-length cDNA library................... 37
full-sib family selection 101
fumonisin 285
fusaric acid 309
Fusarium equiseti............................ 285
Fusarium nygamai 285
Fusarium oxysporum f. sp. *striga*.... 283
Fusarium oxysporum 219, 303, 317
gancaonins 64
gene candidates 42
gene expression 30, 180
gene for gene 175
gene pyramiding............................. 168
genetic distance................................. 75
genetic diversity 71, 77, 100
genetic engineering 331
genetic glass ceiling 332
genetic mapping 122, 160, 163
genetic polymorphism..................... 163
genetic resistance 119, 219
genetic structure 74
genetic uniformity 75
genetic variability....................... 78, 88
genistein 64
genomic technologies...................... 178
genotype dependency 163
genotype x environment.......... 104, 138
genotypic profiling................. 117, 167
geographic distributions.................... 72
germination 22, 35
germination bioassay......................... 50
germination stimulants................ 49, 50
germplasm screening....................... 160
glycosyl transferases 66, 67
glycosylated flavones 65
glyphosate resistant maize............... 333
glyphosate 155
Gobiye.. 200
GR24................................. 50–52, 132
grain quality 200
grain yield 110, 112, 150, 204, 247–249, 334
gray leaf spot.................................. 148
greengrams 247
groundnut 247, 261, 286
growth habit 121
GTZ.. 14
Guinea sorghums 130
gum arabic.............................. 289, 291
hand weeding 215, 216, 229, 237
hard genes 315, 319
haricot bean.................................... 232
haustoria....... 20, 24, 25, 35–38, 40, 41, 89, 95, 176, 189
haustorial initiation factor (LHF) 92
head smut 268
herbaria 81
herbicide resistant crops.......... 147, 333
herbicides 332
heritability 101, 135
high capacity ion exchangers 147
host defenses 321
host forward (HF) library 38
host plant resistance 178, 201
host preference 117
host signal 35
host specificity 52, 57, 80, 286
host-parasite interactions.................. 19
hyaline bodies 27

hydrophobins 320
hydroquinones.................................. 23
hypersensitive response...... 93–95, 118, 165, 166, 175, 176
hypervirulent......................... 314, 316
*IAA*H and *IAA*M.............................. 316
imazapyr 147–150, 152, 154, 218–220, 222, 296
imidazolinone-resistant (IR) maize................ 147–157, 271
impact assessment.......................... 277
inbred lines.................................... 161
incipient infestations...................... 217
incompatible response (IR)........ 94, 95, 137, 166
infestation 6
informal seed exchange.................. 208
injera 209
institutional constraints.................. 335
integrated pest management.... 214, 256
integrated *Striga* control............ 96, 236
integrated *Striga* management........ 199, 334
integrative approach......................... 15
intercropping.......... 113, 229–232, 265, 332
inundative mycoherbicides............. 313
invasive potential....................... 78, 80
investments 13
ion exchangers 152
IR maize............................... 272–275
isoangustone A................................ 64
isoflavone synthase.......................... 64
isoleucine...................................... 305
isomollupentin 66
Kendall's coefficient....................... 104
laboratory assays.................... 132, 139
landraces 254, 332
large gene families........................... 39
leaf rust 148
legume intercropping...................... 146
legume shrubs 234
leucine.................................. 304–306
leucine rich repeat (LRR)............... 181
libraries 67
licorice *Glycyrrhiza pallidiflora*........ 64
life cycle............................. 19, 48, 89
likelihood ratio 180
linkage analyses 169
livelihood analysis 277
long smut....................................... 257
long term sustainability.................. 154
low germination stimulant (*lgs*) 92, 132, 137, 161
low haustoria initiation factor ... 95, 165
lupiwighteone.................................. 64
lysine... 306
lysine hydroxamate 307
maize............ 3, 99, 100, 225, 286, 328
maize streak 148, 268
management options 11
mannitol 28
mannose 322
mannosylphosphate reductase......... 322
manure 250, 253–256, 261, 262, 329
marginal rate of return.................... 273
marker assisted selection........ 122, 124, 159–167, 174, 331, 332
master-switches.............................. 182
mechanism of resistance 91, 118
methionine 24, 306, 319
microbial communities................... 293
modeling 71, 78–82, 132, 154, 275
moisture conservation 206
molecular markers.......... 123, 160, 167, 169, 179
moths... 335
multi-location......................... 218, 258
multivariate analysis 108
mutants.. 304
mycoherbicide................. 205, 308, 314
mycotoxin 285
napier 242, 249

necrosis enhancing 319
nematode 318, 322
*NEP*1 .... 319, 321
nitrogen application 277
norvaline 305
ODA 14
okra 286
on-farm demonstrations 112, 153
open pollinated varieties 104, 149
ordinal regression 274
Orobanche 302, 304, 309, 317, 321
orobanchol 52
osmoprotectants 28
oxalate 314
paired crosses 168
pallidiflorin 64
pandemic 3
parasite transcripts 41
parental lines 107
participatory field-testing 151
participatory rural appraisals
(PRAs) 199, 266, 268
pathogenesis related (PR)
genes 181
PCR 37, 72, 183, 187, 287, 293
pearl millet 3, 8–10, 66, 254, 286
pectinase 314, 315, 317
penicillamine 305
Pennisetum purpureum 242
peonidin 38, 41
perception 51
performance trial 105
persistent seed bank 72
pesta 289–292, 296
phenotypic variance 164, 180
phenotyping 161
phenoxy herbicides 146
phenylalanine 306
phloem connections 28
Phomopsis 304
phosphate 53, 56, 201
phosphomannose isomerase 190
physiological specialization 9
phytoalexins 321
pollen mutagenesis 156
polyA RNAs 37
polygalacturonidases 317
polygenic resistance 100–104
popularization 205
population 6
post-attachment mechanism 137
post-attachment resistance 132, 137,
174, 176, 189
poverty analysis 267, 277
preconditioning 49, 51
primary haustoria 27
principal component analysis 108,
293, 294
private seed industry 9
production agriculture 155, 328
proof of function 181
prunetin 64
Pseudomonas 285, 304, 306
Puccinia chondrilla 303
push-pull 11, 241, 242, 248, 249,
274, 332
pyramiding 155, 175, 331
pyrithiobac 147
QTL 162, 163, 167, 169, 180, 181
quantitative RT-PCR 183, 195
quantitative trait loci 178
quarantine 12
races 117
race-specific resistance 123
radicle growth 22
ranking 257
RAPD 76, 77, 287
recessive gene 119
recombinant inbred lines (RILs) 178
recurrent selection 102, 103
reduced stands 147
regional trials 111
regression analysis 276
relay cropping 229, 230, 233

resistance.............. 12, 88, 96, 136, 138
resistance alleles.............................. 101
resistance index.............................. 162
resistance QTL.............................. 161
resorcinol.. 23
response time.................................. 216
rhizosphere............. 284, 293–297, 302
rhizotrons... 180
ribosomal ITS DNA........................ 294
rice..................................... 4, 177, 286
RNAi............. 186, 187, 194, 195, 322
Rockefeller Foundation.................... 14
root cultures...................................... 67
root exudates................ 23, 35, 62, 244
rotation................................... 230, 235
rush skeletonweed........................... 303
sanitation...................................... 9, 13
saprophyte....................................... 309
SCAR..................................... 124, 287
secondary dormancy......................... 21
secondary haustoria.......................... 27
seed bank........ 220, 221, 297, 330, 333
seed coating...... 12, 148, 157, 291, 292, 296
seed density............................ 216, 270
seed dormancy.......................... 20, 221
seed dressings................................ 153
seed germination.............................. 63
seed industries................................ 336
seed production.... 9, 29, 202, 207, 210, 220
seed supply..................................... 263
seed treatments............................... 156
semilicoisoflavone B......................... 64
Senna... 319
sequence database............................ 38
Sesbania sesban...................... 229, 233
Shannon-Weaver indices................ 295
signal exchanges............................... 89
signal transduction......................... 181
slow release seed dressings............ 152
small interfering RNA (siRNA)...... 186
SMART technology.......................... 37
social scientists............................... 266
soft genes.............................. 315, 316
soil fertility............. 227, 230, 233, 274
soil moisture................................... 217
soilborne microbes.......................... 285
sorghum breeding............................. 87
sorghum midge............................... 256
sorghum......... 3, 8, 159, 199, 223, 231, 253, 286, 293, 328
sorghumol.. 50
sorgolactone............................... 50, 92
sorgoleone................................. 23, 24
soybean........ 62, 64, 80, 116, 222, 274, 290
species distribution............................ 7
SSH products.................................... 41
SSR markers.......................... 108, 167
stem biomass................................. 261
stemborer........ 250, 267, 268, 274, 335
storage pests........................... 267, 336
stover.. 226
stratification...................................... 49
Striga infestation................................ 5
Striga-poverty parallel......................... 6
Strigaway.. 149
strigol.. 49, 52
strigolactone.......................... 23, 49, 92
subsistence agriculture................... 328
suicidal germination.......... 54, 107, 218
sunflower................... 56, 94, 169, 330
suppressive soil...................... 217, 221
suppressive subtractive hybridization (SSH)................... 37
suppressiveness.............................. 218
synergists.............................. 313, 314
synteny... 178
synthetic stimulants.......................... 54
tactile signals.................................... 25
target enzymes[8].............................. 147
tef... 10, 210
termites.. 268

thigmotropic responses 24
tied ridges 201, 256
threonine 306
tobacco 304
Tolyposporium ehrenbergii 257
toxA promoter 316, 317
transcript profiling 183
transcriptomic studies 181
transgenic biocontrol 313
transgenic crops 331
transgenic maize 192
transgressive segregation 161
transparent cup system 190
trap crops 55, 265, 297
Triphysaria 34–42
Tripsacum dactyloides 94
trpC promoter 316
tryptophan 306, 315–317
tubercles 27, 118, 315, 317, 321
turcicum leaf blight 148
tyrosine 306
ugali 260
uncinanone C 244
underground attachments 108
Universal Cassette 316
Urea 223, 224
Uromycladium tepperianum 303
USAID 14
vacuole formation 188
Vacuoleless1 187
valine 303, 306, 308
vascular connection 25, 177, 189
vegetative compatibility
grouping (VCG) 287
vetch 93
virulence genes 309
virulence 301, 305, 308, 314, 315
vitexin 65
wealth ranking 267, 278
weevils 268, 335
worldwide expansion 80
WRKY genes 181, 182
Xanthomonas 304, 306
xenobiotic detoxification 40
xenognosin 23
xylem 26, 28, 119, 176
year x genotypes interaction 134
yield 206
yield increase 149, 277
yield losses 10
Zea diploperennis 109